THE EXPERIMENT MUST CONTINUE

PERSPECTIVES ON GLOBAL HEALTH

Series editor: James L. A. Webb, Jr.

The History of Blood Transfusion in Sub-Saharan Africa, by William H. Schneider

Global Health in Africa: Historical Perspectives on Disease Control, edited by Tamara Giles-Vernick and James L. A. Webb, Jr.

Preaching Prevention: Born-Again Christianity and the Moral Politics of AIDS in Uganda, by Lydia Boyd

The Experiment Must Continue: Medical Research and Ethics in East Africa, 1940–2014, by Melissa Graboyes

THE EXPERIMENT MUST CONTINUE

Medical Research and Ethics in East Africa, 1940–2014

Melissa Graboyes

Ohio University Press
Athens

Ohio University Press, Athens, Ohio 45701
ohioswallow.com

Cover image: Uncataloged archival materials at Amani Medical Research Station, Amani, Tanzania, 2008. Photo by author.
Cover design by Beth Pratt.

Printed in the United States of America
Ohio University Press books are printed on acid-free paper ∞ ™

225 24 23 22 21 20 19 18 17 16 15 5 4 3 2 1

Library of Congress Cataloging-in-Publication Data

Graboyes, Melissa, author.
The experiment must continue : medical research and ethics in East Africa, 1940/2014 / Melissa Graboyes.
p. ; cm. — (Perspectives on global health)
Includes bibliographical references and index.
ISBN 978-0-8214-2172-7 (hc : alk. paper) — ISBN 978-0-8214-2173-4 (pb : alk. paper) — ISBN 978-0-8214-4534-1 (pdf)
I. Title. II. Series: Perspectives on global health.
[DNLM: 1. Biomedical Research—history—Africa, Eastern. 2. Anthropology, Cultural—history—Africa, Eastern. 3. Biomedical Research—ethics—Africa, Eastern. 4. History, 20th Century—Africa, Eastern. 5. History, 21st Century—Africa, Eastern. 6. Human Experimentation—history—Africa, Eastern. W 20.5]
R850
610.72'4—dc23

2015034120

Dedicated to each of my families.
My parents and sister, who model and respect hard work
My large Italian family, acquired by luck of marriage
My own *piccola famiglia*, Alfredo, Silvia, and Giovanna.

CONTENTS

ILLUSTRATIONS

Figures

Maps

Tables

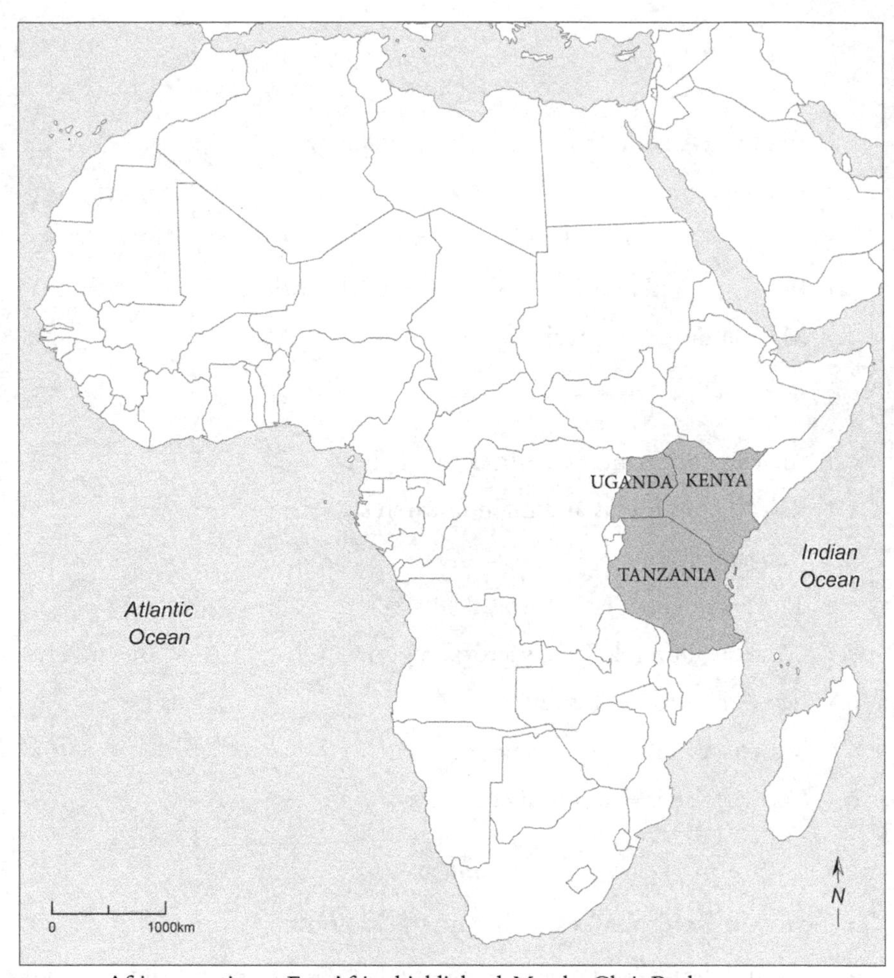

MAP 0.1. African continent; East Africa highlighted. Map by Chris Becker.

PREFACE

Gonja, Take One

It was sometime in the 1970s. Or the 1950s. Or maybe even the 1960s. In any case, it happened years ago. "It" happened in the town of Gonja, in the Pare area of northern Tanzania, and started with angry residents who were unhappy with researchers working in their village. The rabble-rousers were either a group of wholesome, yet angry, residents, or a group of unruly, pot-smoking youth who had recently returned from the war with Uganda. Their anger was directed at a set of researchers who were working in the village at night, either collecting mosquitoes or blood samples. Maybe the researchers were *mumiani* (bloodsuckers) and murderers, maybe not. In any case, they had made the bad decision to drive home that night rather than sleeping in the village. Their car was forced to a stop on a blocked road. Villagers appeared and began hurling stones. The car was damaged; the researchers sat inside, afraid. Flames appeared: the car was on fire. The researchers fled, and it was only due to the appearance of the police that no one was killed.

The story of Gonja was the first account I heard when I asked people about the history of medical research in East Africa. Researchers remembered some version of the story, local people in the region knew about it, and, depending on the teller, the story was used to emphasize any number of points. When told by current medical researchers, it was a morality tale of what happened when well-intentioned scientists encountered uneducated villagers. Among a certain group of researchers, the only lesson to be learned from Gonja was that African rural residents were uneducated, unpredictable, and had yet to learn the benefits of biomedicine and scientific investigations. In some of the narratives, the teller continued into the present, explaining how the bad behavior of the Gonja residents resulted in the building of one of the largest police stations in the district. It also meant that there was almost no research done in Gonja for decades, and it wasn't until 1993 that a research team was sent in to "check people's feelings." Only at that point—twenty, thirty, or forty years after the initial conflict—had villagers finally learned their lesson and begged for the researchers to return, to help reduce malaria and bring drugs, like in years past.

Although I was happy that people shared their recollections of Gonja with me, I was perplexed by how many competing versions of the story I heard in just a matter of days. It was a story that continued to be told, but it seemed none of the tellers were preoccupied with fine details. The dates changed, the people involved changed, and the rationale and moral changed. I didn't know what to make of Gonja, but I kept asking about it, kept thinking about it, and kept squirreling away references. I knew Gonja was important, but I wasn't sure why. So I set the story aside, continued with my work, and hoped that by the end of my research I'd have untangled Gonja's significance.

Surveying the Pathological Museum

Lieutenant Colonel William Laurie, the Director of the East African Medical Survey, exclaimed in 1952, "The African is a walking pathological museum."[1] Laurie was not the only one to consider the African as such, or to be excited by the myriad tropical diseases found in East Africa. The medical missionary Stanley George Browne used the same phrase when describing his work in the Belgian Congo in the 1940s, remarking, "[The native] is a walking pathological museum."[2] During a survey project in Kenya in 1937, the researcher in charge declared that each of the Africans was an "ambulant pathological museum."[3] In 1944, a colonial worker in West Africa stated, "There is no doubt that the African native is often a pathological museum."[4] It's unclear how common a refrain it was, but it would appear to be a phrase that circulated among medical researchers—a common remark expressing amazement at the collection of germs, pathogens, viruses, parasites, and other abnormal and unusual diseases likely to be found in a single African body.

The phrase captures much of what was wrong with the East African Medical Survey and belies an expectation of how researchers expected to interact with Africans. It was not unusual or unexpected that medical workers would look at sick Africans as objects. The phrase oriented researchers to focus on pathologies rather than bodies, and on sick body parts rather than sick patients. Research practices and the material culture of medical activities in the 1950s reinforced this tendency to think in terms of objects rather than people: medical tubes and vials suck and store bodily fluids, scissors snip samples of skin, needles drain blood, tightly lidded jars contain stool samples. These pieces of bodies, floating in formaldehyde, stored in glass, packed in ice, were transformed from being parts of people into data. As a modern manager of an international contract research organization stated dryly, "We don't see patients, we see data."[5] In fact, the objectification of sick bodies has been a central part of the medical profession. Medicalization of the body (defined as seeing something in medical terms, often unwarrantedly) leads easily into seeing the body as a set of objects, and to a general practice of objectification. The goal isn't just to separate the idiosyncrasies of individuals from the disease, but to seek objectivity and objective truths. Sick people are not necessarily helpful to science, but they are when they can be turned into data.

It is also worth remembering that the pathological museum was a real place. Many medical schools in Europe had these museums, and they were places where aspiring doctors and researchers—especially those planning on working in the tropics—could see examples of many diseases that would be impossible to otherwise see in their home countries. As with any museum, the pathological museum was a place to view, to gaze at the exhibits. When walking through a pathological museum, there was a one-way viewership: the objects were dead, cut to pieces, and preserved indefinitely; the medical doctor could view the pathology without shame or self-consciousness, could stare as long as he wanted. It's also worth remembering that specimens were often collected with the goal of sending them back to a pathological museum. In one sense, the sick African really *was* a walking pathological museum. From the researcher-*cum*-collector's point of view, the sick person could easily be reduced to a set of sick parts, each deserving of its own exhibit in a far-off gallery.

The samples for the museum, or the pathologies to be recreated as pieces of data, could not be collected without contact, a human interaction. The scholar of photography Christopher Pinney explains the concept of a "dialogic" period, as the space of time when the subject and photographer come together to create an image.[6] While Pinney references the moment in the creation of a photograph, the same concept applies to medical research. It's useful to think about research, and even a medical survey, as a discrete moment in time, a dialogic period characterized by exchange and interaction. The encounter relies on the participation of both parties; there must be a productive give and take. It is a moment that I refer to as a *medical encounter*, and which this book works to reconstruct.

Methods, Sources, and the Challenges of Fieldwork

Prior to beginning graduate school, I spent a year working in Tanzania with the public health organization Population Services International. My current interests in this topic were piqued during that time, especially as I traveled through the region and saw the ubiquitous advertisements soliciting volunteers for HIV/AIDS drug trials. These fliers inevitably advertised the study as the "cutting edge" or something similar, and I viewed them with a combination of frustration, disdain, and sadness. A closer reading of the fliers and background knowledge of the process of human subjects research quickly revealed that few of these trials were beyond the very early phases of testing.

Drugs to be sold in the United States must past through three "phases" of human testing in order to be approved by the Food and Drug Administration (FDA). Roughly, the first phase tests the drug—often on healthy volunteers—for serious side effects that could preclude its widespread use. In phase one testing, subjects who are taking no other medicines are especially valued because there is less chance of the experimental drug interacting with other drugs in the body and producing unusual side effects. These research subjects are referred to as "drug naive" and it's much more likely to find drug naive people in the developing world. Phase two tests whether the new drug is better than nothing, and is conducted on sick subjects. Drugs that have "passed" these first two phases by being mostly nontoxic and an improvement on doing nothing are allowed to progress to the final stage. Phase three involves testing the new drug against the best available treatment for the same condition. When there is reference to people participating in a "therapeutic" drug trial, or talk of someone in an experimental drug trial where they are miraculously cured, it is typically in reference to a phase three trial.[7] This is the only phase in which a sick person gets access to a new drug that has a decent chance of being effective, or at least is likely to be better than nothing. (There is also an informal phase four, when the drugs are already on the market but continue to be monitored.)

The drug trial advertisements in East Africa offended my sense of ethical behavior. While I understood the need to recruit people to these studies and the obvious benefits if effective drugs or a vaccine were found, I wondered if these ads were not falsely raising people's hopes. Most East Africans

I spoke with believed these projects were giving out *dawa*—medicine. Yet I knew that only people participating in phase three trials had a real chance at receiving new, effective medicines. People participating in phase one and two trials were volunteering to test drugs for potentially serious side effects and to see if the new interventions were better than nothing. It was a dubious use of the word *dawa*.

When I left Tanzania to begin graduate school at Boston University, I knew I was interested in studying the history of human experimentation in East Africa, but I wanted to combine historical training with a better understanding of global public health. After a few years, I had finished my history coursework and exams, earned my Masters in Public Health, and become conversant in Swahili, and I returned to East Africa for a year of research. During those twelve months in the field, and in subsequent summer trips, I conducted forty-three formal interviews, worked in more than a dozen different locations, gathered historical materials from formal and informal archives, and observed medical researchers in a variety of settings. I aimed to be as thorough as possible in researching my topic, occasionally adopting some of the ethnographic and direct-observation techniques of anthropologists. What became most obvious during fieldwork was that it is a difficult activity, full of unexpected challenges and detours.

As I discovered repeatedly, success in the field relied upon plenty of preparation; the work also benefited from a dash of serendipity. My first piece of luck came when I was allowed to participate in the Mosquito Ecology and Control Course in Tanga, Tanzania (run jointly by the Danish Bilharziasis Laboratory and Tanzania's National Institute for Medical Research, Amani Research Centre). The two-week course gave me newfound appreciation for the work of entomologists, and firsthand experience doing the research that I often read about in historical documents. Our entomological research work involved the physical labor of trekking through thick mud to find mosquito breeding sites and stomping around cesspits, the challenges of convincing homeowners to allow mosquito traps in their homes at night, and the tedious laboratory work of mosquito identification and dissection to establish whether the mosquitoes were malarial. Perhaps just as important, the course introduced me to a set of well-educated East Africans who worked in science and alerted me to the existence of valuable historical materials not in the national archives. They also provided invaluable introductions to colleagues throughout the region.

My time living and researching in East Africa made me much more aware of all the ways doing "good" (or at least accurate) history could be threatened. While in the port city of Mwanza, in the western part of Tanzania, I began reading about the work of the Filariasis Research Unit and its attempt to eliminate filariasis from Ukara Island in Lake Victoria. The documents were plentiful, and detailed a very obvious break in 1959. From 1956 until 1959, residents on the island had willingly participated in drug trials and other research activities. After that date, participation rates dropped off staggeringly. In a matter of a few years, Ukara went from being an ideal testing place to one where researchers loathed working. Through careful reading of the documents, I had figured out the main reason why: the Wakara had been accepting experimental drugs that the researchers had been advertising as "medicine" for over four years, but very few people had been cured. People were tired of receiving ineffective drugs and being lied to, and refused to participate. Since Ukara Island was only about forty miles north of Mwanza, and was reachable by boat, I decided to take a trip to flesh out my understanding.

A few weeks later I was on Ukara Island, speaking with two older men who remembered the filariasis project. When I asked about 1959, and why people suddenly stopped participating, they gave a simple answer: there was a new *mtemi* (local leader) who was not as excited about the research project as the old leader, and he had not instructed residents to cooperate. Although I asked the two men directly about whether the "medicines" given out by the researchers were effective, or whether the Wakara people were angry about being lied to, they looked at me quizzically.

The experience on Ukara Island reemphasized the importance of actually visiting a place and talking with the people who had lived through these events, and of searching out materials in more unusual places. I didn't entirely ignore traditional archives. I spent weeks and months in the Kenyan National Archives, the Zanzibar National Archives, the Public Records Office in London, and the Wellcome Library in London, in addition to accessing digital materials of the World Health Organization. But I quickly realized that materials in these places were unlikely to answer the questions about human experimentation I was most interested in. I didn't want to rely on official reports housed in the national archives, and be left "listening for the silences" of African voices. In this spirit, I prioritized visiting places where research stations were located, or where large-scale projects had taken place. After fourteen months of research in East Africa, I had gathered archival and oral data in a dozen different locations, ranging from mission hospitals to remote islands in Lake Victoria.

Two of the colonial-era research stations were in the northern Tanzanian town of Amani and in western Tanzania in Mwanza. (Amani has a fascinating, long history, having originally been built by the Germans as an agricultural and forestry research station.[8]) These two places ended up providing thousands of pages of uncatalogued documents that few—if any—other scholars have used and written about. The materials included the private papers of medical researchers who worked for the Pare-Taveta Malaria Scheme, the East African Medical Survey and the Filariasis Research Unit. There were confidential research reports, letters written by frustrated field workers to their bosses in the cities, notes in Swahili from angry residents who objected to research being done in their villages, and newspaper clippings reporting both the organizations' press releases and residents' reactions.[9] Very few of these documents exist in duplicate in other archives; many documents only reside in Mwanza and Amani.

I stayed for weeks in each location, and it was invigorating work, since each day led to new discoveries. But, for all the excitement of historical discovery, working in Mwanza and Amani was not without challenges. The documents were entirely disorganized, and after receiving approval to work with materials in Mwanza, I was furious when a mid-level bureaucrat denied me entry. When I found a sympathetic co-worker to unlock the room where the documents were, I spent another day working through the materials. I did, however, wonder if I had actually become a "thief"—which was what the angry bureaucrat yelled at me when he returned the next day to find me inside the office.

I wasn't so blind as to be ignorant of the irony, or the myriad ways my own challenges mimicked those of earlier medical researchers. There I was, accused of being a "thief" and "stealing" documents, writing about the challenges of medical researchers who were also called thieves and accused of stealing blood. The challenges of field research in East Africa often created morally ambiguous situations: situations where I had to figure out how to translate formal ethics into field ethics. Should I bribe someone to get access to materials? (No, but a heavy dose of persuasion and pestering was acceptable.) Was I a "thief" for figuring out a way to use documents one man had prohibited me from seeing? (No, because he had no formal authority to make that decision, and I had been granted access by those in charge.) Did people actually understand and value my project, or were they

FIGURE 1.1. Uncataloged archival materials at Amani Medical Research Station, Amani, Tanzania, 2008. Photo by author.

just letting me do what I wanted because I was a white foreigner handing out gifts? (Hard to say.) There were no easy answers, but this discomfort and self-questioning bred a deeper appreciation of the challenging situations any type of researcher encounters.

In addition to the archival materials, oral sources derived through semi-structured interviews were my other source of information. I conducted a total of forty-three formal interviews with people who participated in medical research (as subjects or members of the community who assisted in the research), professional medical researchers during the colonial or post-colonial eras, missionaries who helped researchers gather participants, and with East Africans who lived in communities where medical research had been conducted. Interviews typically lasted about an hour, although a few of the livelier ones went on for two to three hours. I asked questions about what "research" was; past experiences with medical researchers, or working as medical researchers; and opinions about difficult medical scenarios I described. Asking about research was complicated since the topic was not well understood by people. That usually led me to ask if the person had ever given blood, taken pills or received shots outside of the hospital, or met a roving "doctor" or "expert" who was doing "research" or an "investigation." Although I spent a lot of time conducting formal interviews, many of my best insights came from conversations with a mix of health professionals, young people, amateur historians, and the best chicken fryer in Zanzibar. These informal exchanges gave me a chance to talk about my research and have lively discussions without falling into the rigidity of a formal interview.

I analyzed the oral and written sources in dialogue with each other and paid close attention to places of discord—when the oral and archival sources were in clear conflict. In some cases, I was able to "right" these disagreements; in other cases, a level of ambiguity remains. I did not begin by assuming that my oral sources were any less accurate or "factual" than the written sources, nor that the value of my interviews was only in preserving people's opinions, impressions, or understandings of past encounters. In this way, I depart from the approach taken by Luise White in her groundbreaking and creative work on blood rumors in East Africa—a topic I discuss more fully in the conclusion of chapter 2.

Everyone formally interviewed consented orally after receiving a written description of my research and listening to me read the document aloud. When I audio-recorded interviews, I asked permission at the start and again at the end of the interview, offering to delete the recording if the person felt we had discussed overly sensitive topics. I took it as evidence that my consent process was working when some people refused to be interviewed. At the conclusion of the interview, I presented a gift that was typically worth about five US dollars—often sugar, soap, or tea, although it was sometimes cash. I typically conducted the interview in Swahili, although there was often another person present (usually an older male) who had facilitated the introduction and helped clarify any questions or confusions that came up. The interviews were transcribed with the help of Tanzanian research assistants in Mwanza and Zanzibar. I was responsible for all translations from Swahili into English, although I have double-checked difficult passages with native speakers. As for interviews not done in Swahili, a few were conducted in KiKara or KiKerewe and required an intermediary translator, and a few others were in English.

I was surprised to rediscover, even while speaking Swahili and coming with contacts, how hard it was to show up in a new place, establish yourself, explain your project, and hope people would at least tolerate—if not accept—you. As my interviewees reminded me, *I* was a researcher and struggled with many of the same issues researchers over the past half-century have struggled with—consent, benefit, and clarity of explanation—even if I was only asking questions and not collecting blood. And, just as with researchers from decades past, my methods in practice were quite different from what I had theorized. My questions (lovingly crafted in Boston with the oversight of many experienced professionals) were designed to be nonbiased, culturally sensitive, and nonthreatening. Yet those questions were tossed to the side as I saw their inefficacy firsthand. My *haute* methodology met its match in rural Tanzania through a series of challenging interviews full of evasive answers and misunderstood questions.

This research occurred under the watchful eye of Boston University's Institutional Review Board (IRB). The university's interpretation of federal guidelines meant that I initially collected signatures from nonliterate people and kept interview transcripts under "lock and key," even though I couldn't stop people from walking into my hotel room and out with my laptop. My methods produced viable results and a long list of things to do differently in the future. Most notably, I will keep in mind the conclusions I reached for this book. There is often a profound gap between formal ethics and field ethics; one must be nimble, adapt to local conditions, and take cues from the subjects one is working with, who must always be considered active and vital participants in the research enterprise.

ACKNOWLEDGMENTS

This book's subject matter speaks to the potential misuse of people, historically and in the present, and asks hard questions about why we do medical research, at what cost, who benefits, and whether those benefits are worth the risks we ask some people to bear. I felt a deep duty to do justice to this topic, to the stories people told me, and to not become cynical or immune to the worrisome things I found and heard. The information I collected over the years has not been easy to sit with. The constant rattling around of stories heard in interviews, and the heaviness of information gathered from the archives, reminded me that until I published this book, my debt to the many people who invested in me had not been met. This work is far from perfect, and the remaining shortcomings and errors are my own responsibility. However, I have done my best to fulfill my obligation to the many East Africans who spent time with me, the individuals who helped shape my thinking about this topic, the institutions that provided financial support, and the many friends and family members who supported this project by supporting me.

My time in graduate school at Boston University was formative and I thank my advisors James McCann and Diana Wylie in the Department of History and Michael Grodin at the School of Public Health. All were generous and helpful, and I consider their scholarship to be models for my own work in so many ways. Courses taken with George Annas and Leonard Glantz at the BU School of Public Health deepened my knowledge of human rights law and ethics and the history of medical research in general. I have only fond memories of BU's African Studies Center: Michael DiBlasi, Barbara Brown, Ed Bustin, Joanne Hart, Jean Hay, Sandi McCann, Judith Mmari, James Pritchett, and Parker Shipton helped train me and became good friends. I have a great admiration for this group's collegiality and generosity. I am also thankful to those organizations that provided funding: two years of a US Department of Education Foreign Language and Areas Studies grant and three years of funding from the National Science Foundation. Additional funding came from the Boston University Graduate Writing Fellowship Program,

Department of History, African Studies Center, and the Boston University School of Public Health. Research clearance and oversight was provided by the Zanzibar National Archives, the Tanzania Commission for Science and Technology (COSTECH), and Boston University's IRB.

Many people helped me during fieldwork in East Africa, but probably none more than my friend, Hamza Zakaria. He and his family made us feel at home in Dar es Salaam a decade ago, and then extended the welcome to Zanzibar. In Zanzibar in 2008, Juli McGruder, Charlotte Miller and Mattar Ali, and Erin Mahaffey and Adam Grauer all made life more fun and shared important information about history, medicine, and life in East Africa. Mwalimu Jecha at the State University of Zanzibar's language institute spent many hours helping me improve my Swahili. Mwanza and Bukumbi were so enjoyable because of the hospitality of Dr. Mugema and his family, and Mzee Kitaringo and his family. At the NIMR offices in Mwanza, Dr. Changalucha granted me access to the library and pointed me toward other helpful individuals. On Ukerewe and Ukara Islands, Mzee Majula and his son Dickson were excellent hosts. In Kenya, many thanks to Wenzel Geissler and his family for welcoming me in Kisumu. In Nairobi, Reuben Lugalia and Humphrey Mazigo provided great company and answered plenty of questions related to current medical research. I'm grateful for the help of the Tanzania National Institute of Medical Research workers: Dr. Leonard Mboera, Dr. Stephen Magesa and Dr. Yahya Athman, who all spoke with me at the early stages of my work. I am also obviously grateful to the many people who agreed to be interviewed.

Many generous colleagues have discussed ideas, read chapters, and provided sources. A very special thank you to my friend Daphne Gallagher for meticulously commenting on a bulk of the chapters; she is a formidable scholar and her thoughtful criticisms greatly improved this book. A real benefit of living in the Pacific Northwest is having Jennifer Tappan as a colleague, and my ideas are much more nuanced because of ongoing conversations with her. Mari Webel, on the other hand, is a long-distance colleague, but our Skype conversations are no less helpful. During the writing of my dissertation and beyond, James Webb was ready to discuss all things malaria and to provide encouragement and advice about tackling such a large project. At the University of Oregon, Vera Keller and the History of Science reading group provided feedback on an early chapter. Mokaya Bosire helped parse Swahili terms and meanings with me, and was always ready to talk about East Africa. Kristin Yarris has provided moral support in addition to being an excellent sounding board on issues of global health and medical anthropology. I was able to present parts of this work at the Health in Africa Workshop at the African Studies Association in 2012; the Institute of African Studies at Columbia University in 2013; and the University of Oregon's African Studies Lecture Series in 2014. I'd also like to thank my students in the University of Oregon courses "Health and Development in Africa" and "History of East Africa," who read draft book chapters, asked thoughtful questions, and served as constructive readers. Pieces of chapters 1 and 6 were printed in a special issue of the *International Journal of African Historical Studies*, and a section of chapter 4 was published in *Developing World Bioethics*.[1] My thanks to the editors for permission to reprint.

I had four excellent research assistants during the course of writing this book. In Zanzibar, Mohammed Idrisa did much of the interview transcription work; Zachary Gersten assisted in Boston; Hannah Carr worked with me over many months in Eugene; and Lindsay Murphy stepped in at the final, crucial moment. I am appreciative for all their assistance, and for their combined abilities to locate obscure sources, manage buggy databases, and handle inordinate amounts of email. Chris Becker carefully produced all the maps. The team at Ohio University Press was stellar: professional, punctual and meticulous. I enjoyed working with them, and they greatly improved the final product.

And, finally, I can thank in print my friends and family. Research and writing happened over nearly a decade and across three continents. Personal friends in Eugene helped make finishing this book easier to bear; many good times have been had with Katie and Grant Schoonover both on running trails and around the dinner table. Erica and Tom Collins, Daphne Gallagher and Stephen Dueppen, Lindsay Braun and Larissa Ennis, Heather McClure, Kristin Yarris, and Sharon Kaplan have all been great company and superb supporters when my motivation waned. There has also been a solid cohort of friends from graduate school who have tolerated many emails and phone calls full of questions; Arianna Fogelman, Lynsey Farrell, and Andrea Mosterman deserve particular thanks.

Finally, as my dedication referenced, each of my families has been instrumental in helping me to finish this project. My parents, Sue and Tony Graboyes, have been an inspiration in their own ethos of hard work, and my father was a constant (if sometimes nagging) reminder that I was not finished. The extended Famiglia Burlando—particularly Liliana Molano and Franco Burlando—showed great tolerance for working summer vacations. Our time in Italy is full of meals we don't have to cook, clothes that we don't have to launder, iron, or fold, and impromptu gatherings that we don't have to plan, but which involve my favorite Zii (Nino, Vittoria, Paolo, Daniela). I am nearly certain that it's only with this type of assistance that a large task—like the writing of a book—can occur while having a small child underfoot.

The person most deserving of thanks is Alfredo Burlando. I am very lucky to have a partner in life who is a constant source of intellectual stimulation and unquestioning support. Much of my time in East Africa happened with him by my side, and we have learned about the region together. Alfredo listened to many of these arguments take shape and was a willing reader even though my book includes no mathematical equations and few charts. Depending on the day and the need, he has cooked, cleaned, changed diapers, edited, and even helped with the odd footnote. He and Silvia have borne the brunt of long, odd, work hours without too many complaints; I love them both for that, and much, much more.

ABBREVIATIONS

AIDS	Acquired Immunodeficiency Syndrome
ART	Antiretroviral Therapy
CDC	Centers for Disease Control and Prevention (USA)
CIOMS	Council for International Organizations of Medical Sciences
CRO	Contract Research Organization
DC	District Commissioner (colonial East Africa)
DDT	Dichloro-Diphenyl-Trichloroethane
DEC	Diethylcarbamazine
DIBD	Division of Insect-Borne Diseases (Kenya)
DSMB	Data Safety and Monitoring Board
DO	District Officer (colonial East Africa)
EAC	East African Community
EAHC	East Africa High Commission
EAMS	East African Medical Survey (Mwanza, Tanganyika)
EMA	European Medicines Agency
EPI	Expanded Program on Immunization
FDA	Food and Drug Administration (USA)
FPA	Filariasis Prevention Assistant (Zanzibar)
FRU	Filariasis Research Unit (Mwanza, Tanganyika)
GAELF	Global Alliance to Eliminate Lymphatic Filariasis
GAVI	Global Alliance for Vaccines and Immunization

GCP	Good Clinical Practice
GMEP	Global Malaria Eradication Program (WHO)
GPELF	Global Programme for the Elimination of Lymphatic Filariasis
GSK	GlaxoSmithKline (UK)
HIV	Human Immunodeficiency Virus
IRB	Institutional Review Board
IRS	Indoor Residual Spraying
KAR	King's African Rifles (colonial East Africa)
KEMRI	Kenya Medical Research Institute
KNA	Kenya National Archives
LF	Lymphatic Filariasis
MDA	Mass Drug Administration
MDGs	Millennium Development Goals (United Nations)
MDR TB	Multidrug-Resistant Tuberculosis
MRC	Medical Research Council (UK)
NIMR	National Institute of Medical Research (Tanzania)
NTD	Neglected Tropical Disease
OHRP	Office for Human Research Protections (USA)
PC	Provincial Commissioner (colonial East Africa)
PEPFAR	President's Emergency Plan for AIDS Relief (USA)
RGA	Regional Government Authority
RHSP	Rakai Health Sciences Program (Uganda)
SCD	Sickle Cell Disease
STI	Sexually Transmitted Infection

TB	Tuberculosis
TNA	Tanzania National Archives
TPRI	Tropical Pesticide Research Institute (Arusha, Tanzania)
UNAIDS	Joint United Nations Programme on HIV and AIDS
UNICEF	United Nations Children's Fund
VSS	Vital Statistics Survey (Pare-Taveta Malaria Scheme)
WHO	World Health Organization
XDR TB	Extensively Drug-Resistant Tuberculosis
ZNA	Zanzibar National Archives

THE EXPERIMENT BEGINS

1 MEDICAL RESEARCH PAST AND PRESENT

The East African Medical Survey

> As an introduction it is essential to emphasize the difficulties met with in the carrying out of adequate medical surveys in East Africa. Many of the tribes are primitive and intensely suspicious: there is fear of witchcraft and it is not unusual for a request from us for even a specimen of faeces to be met by a firm refusal on the grounds that this specimen of stool is required for the performing of magic rites aimed at bewitching the donor of the specimen. The taking of a specimen of blood is especially resented: in one survey this resentment was so active as to lead to an abandoning of the survey. Medical officers face on one side the criticism that too many refusals will give a biased and incomplete picture, while on the other side they may be criticized for stirring up trouble in the areas in which they work. Ideally it is essential to carry out repeated exams of the excreta and of the blood . . . from each of the 4,000 natives on whom medical exams have been carried out. Any attempt to enforce this impossible standard would quickly arouse such deep resentment among the people that there would be no alternative but to abandon work. The most that could be expected is one specimen each of stools, urines and bloods from a large proportion of the natives examined and even this calls for much diplomacy and knowledge of the African way of life. Much credit is due to the medical officers for their perseverance even in the face of personal danger, as has twice been the case.[1]

Lieutenant Colonel William Laurie, the first director of the East African Medical Survey (EAMS), paints a fairly bleak—if accurate—picture of medical research in the region in the 1940s and 1950s. He characterizes it through the difficulties, fear, suspicions, refusals, and resentment that surrounded the work; even a simple request for a specimen of feces could be met by a "firm refusal." Conflict and misunderstandings were commonplace and, from his perspective, there was little to celebrate other than his brave workers. Laurie's honesty and frustration also indicate that, although this medical research happened in a colonial context and under unequal power conditions, Africans were no mere subjects of medical research. They were active participants in these encounters, forcing projects to change and adapt based on what they deemed acceptable. In doing so, Africans shaped the practical and ethical norms in the literal and figurative space of the "field."

The post–World War II project was part of what has been called the "second colonial occupation" of Africa.[2] The survey was the brainchild of Professor MacSweeny at London University and Professor George Macdonald from the Ross Institute at the London School of Hygiene and Tropical Medicine. The EAMS commenced in 1949, and, as was true of so many of the research institutes and individual projects in the region, the goals of the EAMS were never entirely clear, although they grew narrower during the organization's six years of work.[3] In 1949, the EAMS was described as having two distinct phases: first, that of "mapping of disease," and, second, an attempt at "selective elimination of disease."[4] By 1951, plans to eliminate disease had disappeared, and the organization shifted entirely to describing local conditions: to get "a complete picture of what actually is medically wrong with the African."[5]

Six locations across Tanganyika and Kenya were selected for in-depth surveying of thousands of East Africans. In each place, a team of researchers would descend to collect samples of stool, urine, blood, and skin to test for diseases such as anemia, worms, river blindness, malaria, and bilharzia. The medical researcher Hope Trant, who worked for the EAMS, was being both cynical and accurate when she called herself a "collector of specimens."[6] The number of Africans involved in each place varied between 2,000 and 6,000, and the science to support the "right" number of samples was not at all clear. In addition to the residents who were medically examined, all women were required to give maternity histories, and thousands of other people participated in community-wide agricultural, veterinary, dietary, or tuberculosis surveys.[7] In Tanganyika, research was conducted on Ukara Island in Lake Victoria, Bukoba along the shores of Lake Victoria, Kasulu and Kibondo districts in western Tanganyika, and Kwimba in Sukumaland. In Kenya, surveys were completed in the western region in Kisii and along the coast in Msambweni. There were hopes to conduct a survey in Uganda, but because of problems securing help from the Uganda Medical Department, no research was done there.[8] While those in charge claimed the sites were "representative" of East Africa, it seems many places were selected with an eye toward practical matters such as ease of access, existing infrastructure, and the presence of helpful local leaders. Nothing indicates these locations were representative in any meaningful way, or that conclusions relevant to these places could be convincingly extrapolated to the wider region.

MAP 1.1. East African towns and main research stations. Map by Chris Becker.

When the EAMS began, it was lauded as something novel. An annual report boasted "that never before had such investigations been planned on so broad and adequate a base." After five years of work, the fate of this never before attempted scientific inquiry was plain. The administrators of the project conceded, "experience has since shown that the base was neither broad nor adequate."[9] By 1955, seven years after officially starting work, the EAMS had failed to accomplish either of its two publicly stated goals. Researchers were not able to create a broad and adequate base of scientific data, nor had the massive collection of data led to scientifically informed conclusions that shaped policymaking in East Africa. Despite tens of thousands of samples analyzed, reams of medical forms filled out, thousands of hours of interviews, and work in six different locations, the survey could not say anything new about disease or health in East Africa. Or, as one critic put it, the information gathered by the survey "*should* be of the greatest practical value to the East African governments" and "*could* lead to the development of measures to solve East African health, social and economic problems."[10] But the truth was that the shoulds and the coulds had not been fully realized.

The EAMS was the first large-scale example of organized medical research across the region, but it was not the first or last case of human experimentation. Since Europeans arrived in East Africa in the mid-1800s, Africans have been exposed to Western medicine and biomedical research practices. East Africans were the human material necessary for research projects focused on malaria, trypanosomiasis (sleeping sickness), leprosy, onchocerciasis (river blindness), schistosomiasis (bilharzia), and lymphatic filariasis (elephantiasis)—just to name a few of the tropical diseases that captured colonial imaginations. This sometimes meant taking pills or being injected with experimental drugs, but more frequently it meant providing blood, urine, stool, or skin samples, or being examined, measured, poked, and probed. Sometimes research practices were as invasive as a lumbar puncture, and other times as seemingly innocuous as having the interior of your home sprayed with insecticide.

Africa has long served as a source of scientific knowledge—what Helen Tilley has referred to as a "living laboratory" and "natural laboratory," echoing the sentiments of colonial-era researchers in fields as diverse as ecology, forestry, and tropical diseases.[11] The impression of Africa as source of data and fertile testing ground is accurate in many ways. Over the years, discoveries have been made in the East African region with global repercussions, such as those pertaining to Kaposi's sarcoma and the nature of drug resistance with antiretroviral drugs.[12] A particular focus of medical research over the past century was malaria, and Kenya, Tanzania, and Uganda have been the home of multiple elimination attempts, indoor residual spraying experiments, and countless drug trials, including currently ongoing malaria vaccine trials. The postwar global attempt at malaria eradication also led to more unusual approaches, such as a 1961 experiment in the northern Tanganyikan town of Mto wa Mbu, where all of the town's salt was treated with the malaria prophylaxis, chloroquine. The project was "extremely popular" locally, and effective at reducing malaria rates. When the World Health Organization (WHO) abruptly stopped funding it in 1966, local residents were dismayed, but raised money to purchase the chemically treated salt themselves. The project ultimately faltered in 1972, and malaria rates returned to pre-experiment levels.[13]

Not all attention was focused solely on diseases endemic to East Africa. Tuberculosis (TB) was of global importance throughout the colonial era, and East Africa was one of many sites of TB research. In 1952, the UK Medical Research Council (MRC) ran a project inside the infectious disease hospital in Mombasa, where Kenyans sick with TB were given either the established treatment or an experimental one. In a published article the researchers wrote that the sixty-six patients involved were "unaware that they were receiving different treatments."[14] The MRC continued its TB research into the 1960s, basing experiments at hospitals and blurring the lines between treatment and research. One Kikuyu man, infected with TB, who was treated at the Infectious Disease Hospital in Nairobi in 1961, wrote to the Director of Medical Services in Kenya to complain that the hospital nurse never specified "whether it was trial treatment or ordinary" that he was receiving. Having found out that he had been enrolled in an experiment, he asked "whether a patient is to be forced to accept a trial treatment or to be requested to do so?"[15]

What do we make of these examples? They show the broad range of activities that qualify as human experimentation, but they also hint at the diversity and complexity of ethical dilemmas that mark medical research. Ethics refers to a whole branch of philosophy that addresses notions of morality and right and wrong, and medical ethics may bear on any number of diverse topics, including abortion, euthanasia, the distribution of scarce resources, and human experimentation. Research ethics have been defined as being "about ways to ensure that vulnerable people are protected from exploitation and other forms of harm."[16] Another, slightly broader definition is "how research scientists ought to behave towards their research subjects. Ethical rules govern the proper, moral, and desirable conduct of an individual or a profession; they have prescriptive, explicative, protective, and creative functions."[17] Sometimes the ethical questions arising from these situations are obvious—as with the enrollment of a subject in a TB drug trial without his knowledge or consent. Other times, they are less clear—such as what happens when an international agency like the WHO decides to withdraw funding, or when community members believe an activity is done as a permanent public health intervention rather than as a short-term experiment.

There are many examples of researchers and participants engaging with these types of questions and coming to nuanced ethical stances, while drawing upon language that sounds strikingly modern in the framing of individual rights, autonomy, and respect. As just one example, in 1907, a doctor in the Belgian Congo inquired as to whether Africans infected with sleeping sickness could be forcibly treated with the drug atoxyl, which contained high levels of arsenic and often led to blindness. The response stated that the use of such a medication "would be inhumane and its administration to unwilling victims would be contrary to elementary principles of 'natural law.'" If blindness was a possible result, "it could be used *only* with the consent of the victim, who would have to be forewarned of its danger."[18] The response takes account of many of the issues modern medical ethics would ask about such a high-risk treatment: accurate information must be shared about the potential dangers, and informed consent must be gained. Individual researchers could show a great degree of sensitivity toward the ethical dilemmas they faced while trying to carry out their work.

These cases of sensitivity are not as rare as one might think. Mixed in the archival record are bits and pieces of surprising, even jarring information regarding African subjects and European researchers not behaving as expected, unlikely alliances being created, power relations being inverted, practices being contested, and new norms and everyday ethics being created and remade in a sometimes collaborative, sometimes conflict-ridden process of give and take. Why did African chiefs use coercive methods to enroll their subjects in colonial medical research schemes? Why did some colonial researchers risk their careers to make arguments to their superiors about minimizing risk and increasing benefit to African participants? Who would guess that European medical doctors in the repressive Belgian Congo would engage in a thoughtful ethical analysis of whether a dangerous therapy could be forcibly given to Africans? How did patients at the Itesio Leprosarium in Kenya make a very modern-sounding appeal for better, more effective treatment, writing, "we are also has the rights . . . the right of human being are only one as the others [*sic*]"—invoking the now ubiquitous language of human rights?[19] Why did a European researcher try to criminalize the dissent of an African community nearly fifteen years after the creation of the Nuremberg Code? How was it possible that East Africans exerted so much control in these medical encounters, especially given the very real power inequities between colonial researchers and subject? From the earliest accounts, researchers were often alarmed by how the behaviors of supposedly passive subjects forced them to modify or cancel projects. There can be no doubt that East African "subjects" are a big part of this history of medical research, and not merely as pathological museums to be observed, or as the human bodies where exotic diseases are to be found.

Arguments

This book provides detailed, localized information about how medical research actually took place: how researchers behaved when arriving in communities, recruiting participants, managing risk and offering benefits, and, ultimately, concluding their experiments and leaving. It also asks how East African communities and participants made sense of these encounters. It lends historical depth to modern questions of medical ethics and brings to light a host of ethical questions that continue to resonate today. Questions such as what makes a subject a "volunteer," what types of conditions are "coercive," how much individuals and communities must know about the short- and long-term risks of experimental interventions to be truly informed and consenting, the types of benefits that are meaningful and appropriate, obligations when ending research projects, and questions of overuse of populations. The significance of these questions might be better appreciated by first clarifying what medical research is, and then understanding its scale in East Africa.

Medical research is a sustained inquiry into a particular health-related question that is answered with the systematic collection of data or through experimentation, and where the goal is to create new, generalizable knowledge.[20] In addition to this standard definition, the present study also considers projects that were often labeled as "schemes" or "interventions" to improve public health, but where bodily samples such as blood, urine, stool, skin, or spinal fluid were taken with a primary goal of gathering data. I also consider activities that were labeled as public health interventions when the methods used were experimental or actually changed the disease environment in unpredictable ways. My last departure from standard definitions of research is that I adopt a different vocabulary, referring to East Africans as "participants," since they were rarely passive recipients of medical interventions; they were not "subjects" (the commonly used term) but active contributors in the medical encounter.

In terms of scale, medical research was a major part of many East Africans' initial exposure to biomedicine. By my conservative estimate, in the fifteen years between 1945 and 1960, more than 200,000 East Africans participated in some form of medical research—which generally meant submitting to a bodily examination and/or giving a sample of blood, urine, or stool.[21] That number represents a bare minimum. In 1950 alone, more than 100,000 people in the southern province of Tanganyika and the West Nile District of Uganda were examined for diseases such as leprosy and sleeping sickness.[22] These exams often involved undressing and being palpated. Even if one disagrees with the idea of calling this research, these kinds of large-scale activities clearly extended the touch of biomedicine.

If we focus only on research that required bodily samples, the numbers are still significant and justify the estimate of 200,000 people involved. The EAMS collected blood specimens from 25,000 people in 1951–52.[23] The Filariasis Research Unit collected 50,000 blood slides from across the region in 1954–55.[24] For another sense of scale, we could focus on the work of a single agency during a single decade. During the 1940s, the Tanganyikan Medical Department collected blood and urine samples from 3,000 schoolboys and army recruits while testing for hookworm and bilharzia. More than 3,800 people underwent blood testing for research related to sexually transmitted diseases. Over 7,000 blood slides and nearly 1,000 stool samples were collected while investigating sleeping sickness and hookworm. In all, blood samples were collected from more than 30,000 people. Even this partial account puts the number of Tanganyikans who gave blood, urine, or stool samples to this single agency in less than a decade past 43,000.[25]

The size of these projects, and researchers' tendencies to want to work in the same place over a period of time, meant that some populations moved from having been in very superficial contact with biomedicine to being heavily used in just a few years. Researchers involved in these large projects wrote of being "afraid of milking the same cow too often" and cautioned each other against conducting too many projects in the same place simultaneously.[26] In Kagunga, on the lakeshore just south of the Burundian border, 6,000 blood slides were taken in 1952; three months later, a thousand people were examined again. Three thousand more were examined in 1954, 1955, and 1956.[27] In 1950, in Kibondo District in western Tanganyika, blood slides were taken from 9,000 people, more than 25 percent of the total population.[28]

The sheer number of people who participated is one indication of the importance of medical research, but the number also leads to more interesting questions: How did these thousands of encounters shape East Africans' opinions of biomedicine and the colonial enterprise? How was, and is, research understood by those East Africans who were participants? And what does this book—which is a history, and begins to reconstruct an emic perspective of East Africans' understanding of medical research—have to say to current debates related to human experimentation in the global south, cross-cultural medical ethics, and ongoing miscommunications between researchers and subjects?

There are four main arguments running throughout the book. First, historically and in the present, East Africans perceived research very differently than researchers did—to the point that it is questionable whether people knew they were participating in research. Second, despite the fact that there was no shared sense of what constituted research or why it was done, researchers and participants both tended to talk about these encounters in a transactional way, as a form of exchange. The third theme is the conflict resulting from putting points one and two together: because there was not a shared sense of what research was, but both sides were judging the encounter as a type of exchange and had expectations about what was fair and appropriate, there were frequent disagreements and occasional cases of spectacular conflict. Finally, since East Africans were active participants in these encounters, it follows that the ethical norms that came to characterize field ethics in East Africa were not just dependent on the desires of European researchers, or on the result of theoretical ethics being placed in a field-research environment. Rather, the dialogic period and the medical encounters, which necessitated the participation of both researcher and subject, led to the creation of a hybrid ethical form, which may be referred to as "everyday ethics" or "field ethics."

Research Is *Jambo Geni Sana* (a Very Foreign Thing)

There was, and is, very little shared understanding of the key components of what medical research is, who does the research and why, and what constitutes the risks and benefits. As is explained in chapter 2, many East Africans continue to discuss medical research in terms of blood, and, more specifically, state that research consists of taking blood, and that the "risk" of research is losing blood from the body or having it circulate outside the body. There is also a widespread misunderstanding that the benefits of participating in research are the medicine (*dawa*) that is given out by researchers. Medical researchers are often labeled as a generic type of expert, and are on occasion likened to traditional healers who have the potential to both harm and heal. The reasons why research is done are particularly hazy, with many people claiming the goal is to discover disease and treat individuals. Therapeutic misconception—the belief that research is being done to benefit the individual—is rife, and creates serious questions about the quality of participants' consent.

Anthropological work from Kenya indicates that these types of confusion exist into the modern period. In general, "the concept of research and of different studies remains difficult to get across."[29] A group of scientists and anthropologists working on the Kenyan Coast have admitted that despite lengthy and thorough explanations of their projects to residents, they consider "incomplete levels of understanding, or 'half knowing,'" as almost inevitable.[30] This group has gone so far as to question whether it's a gulf that can be bridged, noting that "it may be very difficult, arguably impossible" to help some participants understand the nature, goals, and activities of specific medical research projects.[31] These are some of the researchers most committed to trying innovative new strategies to ensure research is understood, which makes these findings particularly troubling, and even more important to acknowledge.

Research as a Transaction: Gifts and Commodities

One of the surprising areas of agreement both historically and in the present is how both subjects and researchers discuss medical research as a type of transaction or exchange. Colonial researchers enmeshed themselves in a series of economic exchanges that often started the moment they arrived in a field site and needed transportation, food, and lodging, and began to either pay generously or haggle while complaining about unfair prices. Researchers linked themselves to communities via favors, salaries, and what they offered residents to participate in the projects. Economies sprang up for items desired by researchers which would not otherwise have been traded openly: it became common to exchange blood for pills, urine for car rides, stool samples for medical examinations.[32] This web of debt and indebtedness helped tie a foreign researcher to a community and contribute to trust or distrust.

Even with a shared conceptualization of medical research as a transaction, there is still gray area in defining the type of transaction. An object moving from one person to another can fall anywhere within the extremes of giving voluntarily, with no expectation of anything in return, to forcibly taking something without payment. Some might frame this as a choice between Mauss's gift or Marx's commoditization.[33] On the one hand, medical research could be characterized as being dependent upon commoditization of pieces of the body such as blood, skin, urine, and stool.[34] These items were physically separated from an individual, had a value, and were exchanged between the person who "produced" the substance and the researcher who valued it. On the other hand, it's doubtful that these exchanges were wholly about commoditization. The anthropologist Parker Shipton has argued persuasively that in Luo country in western Kenya, "there is no systemic 'gift economy' or 'commodity economy' but rather exchanges that weave in and out of these principles."[35]

Moral economy may perhaps be a more appropriate framework for considering these transactions. The anthropologist Ruth Prince describes how the term captures "the shared mores and values with which people evaluate their relations with others, from economic transactions to the obligations informing social hierarchies and patterns of accumulation, including relations between political elites, states, and citizens."[36] It should come as no surprise that as blood and other valuable substances moved from a body to outside a body, there were often disagreements about what type of exchange was taking place, whether it was equitable, whether the person was to trusted.[37] These transactions allowed for judgments to be made about researchers, for trust to be established and relationships to be created—all out of a transfer of bodily products from one person to another. However, if the exchanges of medical research were sometimes considered gifts and other times considered commodities, it meant that there were countless opportunities for miscommunication. Modern findings about medical research in the West African nation of The Gambia found that scientists and communities often came to "radically different framings" of the exchanges they were participating in, which could lead medical researchers to believe "that what they take from subjects is a gift rather than part of a transaction, and thus act in ways that from the other side appear to be stealing."[38] If a participant believes she is participating in an inequitable exchange and may have been the victim of theft, it is no surprise that conflict often results.

Research as a Space of Conflict

"Conflict," broadly defined, runs through each of the chapters; nearly every case study has moments of tension or places where research is put in jeopardy. The reasons were not always clear, since many of the projects at first glance look benign. These moments of conflict illuminate two important points. The first is that disagreements between researchers and East African participants were frequent, but that the arrangement of power was unpredictable. The fact that there was conflict indicated that people were not the docile subjects some past accounts have implied, and power was rarely effectively exerted *upon* subjects for very long. More frequently, participants demonstrate an ability to resist effectively, or at least mitigate, the control of individual researchers or larger projects. (I am wary of labeling these cases of researcher and community conflict as "resistance" since the term is too often used as shorthand for resistance against the colonial state. I am far more comfortable explaining it in terms of the very real dissatisfaction and discontent bubbling up because of specific projects, particular researchers and discrete interactions.[39]) The second point about conflict is less obvious, and often overlooked in the literature about colonial science. That is, that colonial researchers were frequently in conflict with each other and that their disagreements illustrate the diversity of opinions existing within a supposedly unified and homogenous organization. The information presented throughout the book on the divisions in colonial opinion forces us to develop a more nuanced perspective on how colonial science functioned in East Africa.

One of the limitations of my argument is that I cannot speak to divisions within a community, and whether the resistance researchers perceived as all-encompassing was actually so unified. In many of these case studies, if there were significant fissures in villagers' opinions (breaking on the lines of gender, class, religion, ethnicity, age, etc.), the data I have do not reveal it. This does not mean such fissures did not exist, but the dissent in these communities was widespread enough to present what looked like a unified front to bewildered and frustrated researchers. If the divisions in public opinion had been big enough, it seems someone—the researchers, the chief, a sympathetic villager—would have tipped the medical workers off. Researchers had no qualms about trying to appeal to whatever sympathetic factions they could find, but the data rarely reveals such divisions. It is also worth noting that I do not formally address how gender shaped (and shapes) the medical research encounter in East Africa. This is not an inadvertent oversight, but a result of the data I was able to gather: a majority of the researchers and participants available to be interviewed were male. Among the women I did speak with and who emerged in the archival documents, I was not able to discern themes or arguments on the basis of gender that felt adequately evidenced. It is a topic that I hope to pursue in the future, especially given the rich materials provided by Dr. Hope Trant, who is discussed in chapter 5.

Research and the Production of Everyday Ethics

This book's final argument is that what came to be the norms of medical research—the accepted practices, how work actually got done, and the myriad compromises and modifications that required—were the product of negotiations. Clearly stated, researchers and participants were both responsible for establishing the norms of day-to-day medical research practice. East Africans regularly surprised and frustrated researchers by being such active research participants, and their preferences and demands forced researchers to modify scientific plans. Africans absconded from projects that didn't fulfill their needs, shut down projects that didn't align with their interests, or modified them in ways that better accommodated their own expectations of fair benefit, acceptable risk, or norms of the body and healer-patient relationships. As one example, plans to take blood in the middle of the night to test for lymphatic filariasis were abandoned after people refused to participate. Instead, blood was taken in the early evening, even though this was a less accurate method for testing and led to biased and problematic results.

In recent years, the distinguished medical anthropologist Arthur Kleinman has called for a shift in medical ethics so that they are "the outcome of reciprocal, participatory engagement across different worlds of experience."[40] I agree with the sentiment, but I disagree with one of the assumptions: that current medical ethics are not the result of participatory engagement. Many, if not all, of the case studies in this book show that East Africans have made their preferences known, shaped practices in the field, and worked to create new forms of medical ethics that accorded with their own wishes and expectations. The "everyday ethics" governing medical research interactions in East African communities are the result of what the anthropologist Wenzel Geissler terms an "ethics of collaboration."[41] While formal ethics as stipulated in various human rights documents are well intentioned, they tend to discount or ignore "what happens when one person responds to the other in open-ended, face-to-face relations that occur *within* the field."[42] In response to this shortcoming, anthropologists have introduced the terms "everyday ethics" and "field ethics" that focus on "the ethics guiding the everyday life of research."[43] A few words of caution here: by arguing that African subjects were active in establishing norms, I am not claiming that these interactions were fair, just, or ethical. Research was often deceptive, coercive, and exploitative, and there were power differentials between government-sanctioned researchers and colonial subjects/East African citizens that we must remain attentive to. Although I do use the terms "everyday ethics" and "field ethics," they are problematic precisely because they sound benign, and may allow us to be too complacent when discussing exchanges that were deeply unequal.

Book Scope and Organization

The temporal and geographic scope of this book is unusual in that historians do not typically analyze the present, and anthropologists and bioethicists rarely look systematically to the past. The seven-decade span beginning in 1940 allows us to take account of change over a broad swath of time, to construct both historical and modern analyses, and to locate much continuity. Although there were cases of medical research in the region prior to 1940, it grew significantly in the following decades; it was only post-1940 that there came an infusion of British colonial funds and the creation of a more coordinated regional bureaucracy. (Discussions of medical research earlier in the century, as well as of the changes occurring in 1940, are found in appendix B and a related journal article.)

It is also unusual for Africanist historians to take a regional approach—one that I feel is more than appropriate for this topic. Post–World War I, all of East Africa was administered by the British (Tanganyika and Uganda as protectorates and Kenya as a colony), which led to the creation of regional research institutes, shared medical experiments and personnel, and a set of ethical questions around human experimentation that were more similar than different. Ecologically, the countries share many disease environments, and conditions such as trypanosomiasis (sleeping sickness), malaria, schistosomiasis (bilharzia), onchocerciasis (river blindness), and lymphatic filariasis were all widespread, as were conditions such as dysentery, diarrheal diseases, leprosy, ulcers, and yaws.[44] The fact that diseases did not, and do not, respect national boundaries created an incentive for shared research projects; it also sometimes incentivized heavy-handed research practices that ignored the needs and norms of East African communities. There is also something deeply practical about studying this region as a region, as projects continue to be shared and shaped across national boundaries, and I maintain that despite East African diversity, these are places more similar than different. A regional approach may also make the modern relevance of my arguments and conclusions more readily apparently.

I feel obliged to point out that, despite taking a regional approach, I have benefited greatly from other scholars' micro-histories and medical ethnographies focused on specific communities. Literature on the topics of health, healing, and disease in East Africa, such as by Steven Feierman, John Janzen, and Meghan Vaughan, and works covering other areas of sub-Saharan Africa, have influenced my thinking and framing of this book.[45] I've also been inspired by a very dynamic set of new medical ethnographies focused on eastern and southern Africa. Recent books by Johanna Crane, Stacey Langwick, Julie Livingston, and Claire Wendland, in addition to the bountiful writings of Susan Reynolds Whyte, have all shown how biomedicine is constantly engaged with, or perhaps entangled with, the African people and places in which it operates. These works have persuasively shown that African communities have clear ideas about what fair relationships should look like, carefully choose which biomedical ideas and interventions to adopt or ignore, and are often cognizant of the global inequities and geopolitics that shape the interventions they are offered.[46] I draw upon these detailed works throughout this book to help me place regional trends in a local context.

❖

The book is organized to mimic the progression of the medical research encounter from researchers' arrival in a community to the final dismantling. Each of the major sections of the book addresses a moment in the research encounter: from arrivals ("Researchers Arrive"), to recruitment and consent practices ("Consent or Coercion?"), to the balancing of risks and benefits ("Balancing Risks and Benefits"), and, finally, to how experiments are ended and whether there are longer-term obligations ("Exits and Longer-Term Obligations"). A "section" of the book should be read as a single unit, since each begins with two case studies that raise themes more fully discussed in the analytic chapter that follows.

The historical and modern case studies that begin each section are meant to illustrate the continuity of ethical questions inherent to medical research in the region over the past century. These case studies are not examples of "good" or "bad" research. These narrative vignettes should, though, lead us to question assumptions about moral progress, since many of the same challenges facing medical researchers of the 1940s persist in 2010, and many of the most challenging ethical questions remain inadequately answered. The hope is that by placing historical examples in close proximity to modern ones, there will be fruitful reflections on the unresolved ethical problems of contemporary global health research in low-income settings, while also generating a deeper appreciation of the long history of problematic medical encounters in the region.

Chapter 2 presents an internal (emic) view of medical research from the perspective of East Africans. Although East African perceptions of and reactions to medical research projects are integrated throughout the book, this is the space where I introduce and explain some of the central misunderstandings. The following section, "Researchers Arrive," highlights the process of arriving and initial interactions between researchers and a community. It focuses on two different disease elimination attempts, one in the Lamu Archipelago in the 1950s and another on Zanzibar Island in 2001. "Consent or Coercion?" reviews the practices of consent and delves into debates about therapeutic misconception in the region, as well as the broadening of consent beyond just an individual researcher and subject. These case studies present a UK-sponsored TB drug trial in Kenya in 1961, and a discussion of the important role over the past twenty-five years of African fieldworkers in medical research projects on the Kenyan coast. The next section, "Balancing Risks and Benefits," discusses the disagreements about what constitute appropriate amounts of benefit and risk during medical research projects. The historical narrative charts the work of Hope Trant and the East African Medical Survey in 1954, and the modern case study focuses on the circumcision trial conducted in Rakai, Uganda, in 2005. "Exits and Longer-Term Obligations" discusses the difficulties of ending large-scale and long-term projects. The modern example focuses on the testing of the new malaria vaccine occurring since 2009, while the historical example dissects a failure to eliminate malaria in 1955. In conclusion, chapter 7 discusses the modern global medical research industry and moves more firmly into the realm of normative ethics, offering judgments about the ethics of some of the practices I've reconstructed and described in earlier chapters.

The book's appendixes include a glossary of Swahili terms and an essay discussing further readings on human experimentation globally. Additional materials, developed to be used in undergraduate settings—including a set of teaching activities and digitized primary source materials—are available on my website (http://pages.uoregon.edu/graboyes/).

PERCEPTIONS

2 EAST AFRICAN PERCEPTIONS OF MEDICAL RESEARCH

Mama Nzito, Dead Kids, and Bilharzia Research

I did not find Mama Nzito—she found me. I was in a small village outside the port city of Mwanza, searching for the oldest man in the village. A group of elders assured me that this particular man would be able to answer all of my questions, and they sent me off with vague directions. When I passed Mama Nzito working in her garden and explained who I was looking for, she offered to lead me to the tiny house. Inside, I discovered that the oldest man in the village really was quite old: he could barely hear, or speak above a whisper. Even so, I went ahead and tried to interview this Mzee (*Mama* and *Mzee* are both used to address elders). I pulled out a wrinkled consent form, an audio recorder, my notebook, and started speaking.

Since he couldn't hear my questions, the Mzee's first answers were entirely off topic and nearly unintelligible because they were delivered in a raspy whisper. Finally, in frustration, Mama Nzito began answering the questions I was directing at the old man. When I asked whether he had ever participated in medical research, Mama Nzito provided the response. In the 1960s, white "experts" (*wataalum*) came to her primary school, located in a village just down the road from where we were talking. After meeting with the headmaster and gaining his approval, these men gave the students "medicine" (*dawa*) in the form of injections and pills to "treat" (*kutibu*) their bilharzia.[1] As Mama Nzito understood it, these medicines were just a cover; the researchers weren't really there to treat any child's disease. Their real intention was to enter the school, steal the students' blood, and sell it. Worst of all, some children died in the process of blood extraction.[2]

When Mama Nzito finished talking, many questions remained: Why would the headmaster agree to let these experts steal the students' blood? When and how did the children die? How was the blood stolen? Where was it sold, and to whom? I asked her these questions, and Mama Nzito had no answers. But, while she couldn't answer my questions, she also didn't back down in the face of my questioning. She sensed my growing skepticism and told me—with impeccable logic—that just because she didn't know the specifics of how blood was stolen and what was done with it didn't mean that it wasn't true.

During the course of the hour-long interview, Mama Nzito spoke repeatedly of *damu* (blood), its value, and the government's role in the stealing of it. As she discussed the work of medical researchers in her area, she also posed many questions without clear answers: "You will ask yourself: Why does he want my blood? Where will he send it? What will it be used for? . . . They will take your blood, but they won't return with answers."[3] Linked with her discussion of damu were frequent references to dawa. As she saw it, dawa was a way to lure people to give blood, a common payment for blood, and an excuse to get into a place, like a school, where blood could be obtained. Mama Nzito raised the topics of damu and dawa together frequently enough to lodge both words firmly in my mind.

The story of children murdered by blood-stealing researchers should have been easy to dismiss—it followed the model of baseless and hard-to-believe rumor—but I found it difficult to dismiss as such. In this case, I discovered that a version of the story was well documented in local archives. Medical and police reports showed that in 1965, six children died at a school just up the road from Mama Nzito's village after receiving an injection of a bilharzia drug.[4] The deaths occurred at Busirasonga Primary School in Sima, in Geita District in western Tanzania. Six out of 123 children died after receiving an injection with a drug intended to treat bilharzia. The Ministry of Health called the administration of drugs at Busirasonga Primary School "mass treatment"; publicly, the deaths were attributed to poor-quality or inappropriately administered medicine. While it is impossible to know the actual cause of death, it is plausible that the mass treatment was part of a research project testing small variations in dose or treatment schedule—lending credence to the local idea that the children died at the hands of researchers.[5]

It's worth noting that just one year before the deaths, the drug given as mass treatment in Busirasonga was still being tested by the East African Institute for Medical Research. The 1963–1964 drug trials of TWSb (sodium antimony dimercaptosuccinate) were conducted on school children in the Mwanza region to determine appropriate doses. A group of children were given the drug at school, while others were admitted as patients in the hospital and received much higher test dosages. Being part of the inpatient trial meant receiving up to five injections per day, and many children experienced side effects of anorexia, nausea, and vomiting. As the combination of the hospital stay, the frequent injections, and the obvious side effects made people increasingly nervous and angry, mothers pulled some of the children out of the project.[6] The East African Institute for Medical Research was based in Mwanza and had been very active in testing bilharzia drugs in the region in the years prior to 1965. It's quite likely that even if a particular family did not have a child who had received a drug either at school or in a hospital, they knew someone who had. The idea of "researchers" or "experts" arriving at a school with drugs in hand, with the sickness or death of children as the result, seems to have been well accepted and almost expected. In my own interview with an older couple, they recounted how "We'd hear that today they coming to the schools to test blood [*kupimwa damu*]. The parents would not send the kids to school because they didn't want them to be killed . . . but maybe this is wrong."[7]

Bwana Matende, Blood Stealing, and Filariasis Research

When I asked about medical research, many people told stories of Bwana Matende.[8] *Bwana* is the Swahili word for "Mister" and can be used as a sign of respect; *matende* is elephantiasis, which is a common symptom of lymphatic filariasis. Thus, one translation for Bwana Matende is Mr. Elephantiasis.[9] More important than the name, though, was the perceived true work of Bwana Matende: creeping around in the middle of the night, stealing African blood, and selling it internationally. He was a white doctor or researcher who worked in a lab in Mwanza near the government hospital. He stole and dealt in blood, and also gave out dawa. Mr. Elephantiasis sucked (*kunyonya*) blood—never to drink, but to sell. Bwana Matende was not a vampire but an unrepentant businessman.[10] It was while pursuing his main goal of collecting blood to be sold that he inadvertently killed Africans. His unlucky subjects would be "finished" (*kumaliziwa*) and the body disposed of. The African blood was sold abroad for white people to make extremely potent dawa used in Europe and the United States. Bwana Matende was most active in the 1950s and 1960s, and was no longer in the Mwanza Region today.

As his name made clear, Bwana Matende focused on the disease of filariasis. One of the peculiarities of lymphatic filariasis is that, for accurate testing, blood samples must be taken between eleven at night and two in the morning, when the microfilariae are active in the peripheral blood.[11] This medical necessity created a set of conditions that brought researchers during both the colonial and immediate postindependence eras into villages in the dead of night, where they would round people up in the center of town or go door to door, and take blood that they then stored in small vials. Those vials were carefully placed inside coolers, put into vehicles, and driven away to some unknown place, for an unknown use.

During interviews, stories of Bwana Matende frequently broadened into discussions of the connections between blood, medicine, money, and the government. In meandering accounts, people explained how African blood was stolen, that blood was turned into medicine, and that medicine was sold to Europeans or rich locals. The stories also had to explain why the government would allow citizens to be killed and their blood stolen. When asked if the government approved of Bwana Matende's work, Tanzanians responded with a version of "Eh—the Government? He *is* the government!"[12] As one man told me, if a person was unlucky enough to be taken by Bwana Matende, the police wouldn't help. Since Bwana Matende was part of the government, the case would be closed, and the police officer would write that the death was due to "bad luck."[13]

This linking of Bwana Matende with the government is important. He was active before, during, and after the Tanzanian independence and consolidation process of the early 1960s, and when I pressed people to be more specific about which government Bwana Matende was working for, they responded by saying "government is government."[14] Any government could be bloodthirsty or act as a profiteer on the back of its citizens—that was not a characteristic singular to the colonial state. By giving the government a role (even that of tacit bystander) in blood stealing, people implied that blood stealing, murder, and profiteering were open secrets.[15] In fact, the Swahili word *siri* (secret) was often invoked during discussion of medical research, the government's complicity in blood stealing, and the larger nature of government and its relationship to its citizens.

Aspects of the stories told about Bwana Matende resonate with the history of medical research in western Tanzania from the late 1940s through the 1960s. In the late 1940s, the Filariasis Research Unit and East African Medical Survey were established in the port city of Mwanza on Lake Victoria, and they continued to operate in the area through the 1960s; the original building still houses medical researchers today. The Filariasis Research

Unit was particularly active, and thousands of East Africans came into contact with its members as they conducted large-scale surveys (to establish prevalence rates), tested new drugs, determined appropriate doses of effective therapies, and then attempted to provide mass treatment.[16]

In the decade between 1950 and 1960, well over 50,000 Tanganyikans had their blood taken by researchers during the darkest hours of the night.[17] On Ukara Island alone, drug trials conducted in 1950 involved more than 35 percent of the population giving blood and taking pills. Thousands of other residents in villages around Mwanza also had blood samples taken at night and received experimental therapies in the form of pills and injections.[18] Bwana Matende stories circulated and gained currency as European and African researchers entered villages under cover of darkness, took blood, and then quickly departed.

The Bwana Matende stories influenced how East Africans interacted with, and understood, medical researchers as a group. I asked a few different people how they could know whether a researcher was Bwana Matende or just a typical researcher—or how they might know when it was safe to participate, and when agreement could lead to blood theft and death. Mzee Thomas answered by explaining that, when you mix clean and dirty water, the water may still appear clean, but you know it is actually dirty. As he saw it, Bwana Matende was like a drop of dirty water infecting all researchers: as soon as Bwana Matende was present and people knew about him, all medical researchers were infected.[19] Or, as Mama Nzito explained to me about the perils of keeping bad company, "If you sleep in a place for five minutes, you will start to stink like that place."[20] What this meant in practice was that if Bwana Matende existed (which many people believed, or at least couldn't disprove), then medical researchers as a group were to be suspected. These suspicions were heightened by perceptions of government involvement.

The obvious overlaps between two different narratives—official accounts taken from government reports and research documents, and the "unofficial" understandings of East Africans participating in medical encounters—force us to recognize very different explanations of the same event. They also raise a set of challenging questions: Why do East Africans choose to talk about medical research in terms of damu and dawa and money and the government? What are the implications of the continued circulation of these stories, even though Bwana Matende supposedly finished his work fifty years ago? How has the residue of past projects, misunderstandings, mistreatment, and deaths stemming from medical research and public health projects shaded the present? How has this history of medical research shaped people's understanding of, and participation in, current medical research projects?

❖

In this chapter I show that stories about blood theft are firmly associated with biomedicine and biomedical research, and have been shaped by medical and public health encounters. I use the stories as a starting point to illustrate some of the ways modern East Africans choose to talk about medical research: through stories of blood theft, the invocation of researchers' ability to kill or cure, the firm characterization of researchers as a generic group of experts with questionable ties to the government, and understanding medical research almost entirely through a lens of blood. The continued circulation of stories like that of Bwana Matende, and more general understandings of medical research that do not match up with Western definitions of it (either in terms of who is conducting such research, or what it consists of), profoundly influences the behavior of modern East Africans. Yet it is difficult to sort out precisely how and when East African understandings of blood, dawa, medical research, and its risks and benefits have changed. Modern East African understandings, and my portrayal of these notions, are clearly accumulated reflections based on decades of past experiences.

While this chapter builds upon a rich literature focused on East Africa, the themes presented speak to the work of social scientists and medical researchers beyond the region. Other large-scale medical campaigns carried out over the past century—often labeled as public health interventions, but where the activities were still experimental—suffered from similar instances of misunderstandings. The best-known and best-documented cases created unexpected, unintended, and devastating consequences that changed the human disease risk environment and set the stage for new epidemics. One obvious example is the Hepatitis C epidemic in Egypt that stemmed from schistosomiasis (bilharzia) control efforts in the 1950s.[21] Another episode, with medical doctors being given permission to administer a province of French Cameroon in the early 1940s, led to an increase in sleeping sickness cases and generally poor health outcomes.[22] Likewise, it has been suggested that French and Belgian policies in West and Central Africa of mass treatment for sleeping sickness not only contributed to Hepatitis C epidemics in the first half of the twentieth century, but created ripe conditions for the amplification of HIV from a local disease to a pandemic.[23] While we continue to focus on East Africa, we must keep in mind that unfortunate outcomes coupled with deep misunderstandings were not unusual far beyond the region.

In response to the general question, "What is medical research?" most elders over sixty remarked on how strange and unusual it was. In Swahili, it was *jambo geni sana* (a very foreign thing), a thing of *ajabu* (wonder), and involved learning about *vitu mbalimbali* (faraway things).[24] Interview responses indicated that people's general understanding of research was still mostly descriptive and relied on associating research with specific people and their jobs (doctors, witch doctors, government employees), places (laboratories), and tools (microscopes, needles). Many people also described research in terms of the transactions researchers engaged in: they desired blood and bodily samples, and gave pills and injections in return. Residents rarely, if ever, identified research as the testing of new drugs or procedures. During only a handful of interviews was it mentioned that research could result in the development of new medicines. Only one man questioned how new therapies were produced. In the midst of discussing the smallpox vaccine, he told me residents in his village were happy to be vaccinated, but then went on to ask himself, "But where was it tested first? Perhaps in another place?"[25] In fact, most of the individuals I formally interviewed believed research *was* being conducted *in order* to help or treat them.[26] As one woman told me when opining that a healer was the same as a researcher, "even this *mganga* [healer] is a researcher because he investigates your disease and then he cures you."[27] What is important is that she specifies that a mganga researches *your* disease (*ugonjwa wako*) and then he *cures* you (*anakutibu*). By her definition, research identifies and cures diseases in individuals.

The rest of this chapter presents information from interviews and archival accounts that focus on how medical research and researchers were, and are, understood. I begin by discussing who researchers are associated with and what research is compared to, and provide some concluding thoughts about modern misunderstandings, including therapeutic misconception.

Who Are Researchers?

Many people explained the act of researching by trying to figure out who a researcher was or to whom he might be similar. Comparisons were made among healers (mganga, pl. *waganga*), biomedical doctors (*daktari*, pl. *madaktari*), other "experts" (*mtaalam*, pl. *wataalam*), government officials, and witches (*mchawi*, pl. *wachawi*). Occasionally people recognized the differences between the work of researchers and doctors or healers. As one man explained, "a researcher looks at things to discover if they are there, a doctor cures the things that are there."[28] Another man, when asked whether doctors and researchers did the same work, responded, "Without the researcher, a mganga can't cure you. . . . Doctors can't start to cure you without the researcher."[29] In the case of these responses, researchers were discoverers, and there was a symbiotic relationship between the work of researcher as discoverer and that of mganga/healer as curer of disease.

One way to piece together local perceptions of medical research is to look closely at the words being used. At the turn of the century, there was no specific word in Swahili capturing either the substantive act of research or the verb "to research." Currently, most people use one of three words to describe the work of researchers, translatable as: to search or seek; to examine; and to spy. Table 2.1 shows the three major root verbs that are and have been used to discuss medical research.[30]

Table 2.1. Swahili words for "research" and translations

Simple definition		*Related words*	
kutafuta	to look for, seek, search	*kutafiti*	to pry into, examine, criticize
		utafiti	curiosity, inquisitiveness
		mtafiti	a prying, inquisitive person
kuchungua	to peep, spy, examine, pry	*kuchunguza*	look carefully into
		uchunguzi	prying, criticism, curiosity
		mchunguzi	a critical person; seeker of knowledge
kupeleleza	to spy, examine secretly, pry into, be curious about	*upelelezi*	a person who investigates, examines; a spy

In a simple sense, the sets of words describe increasingly intensive forms of searching as you read down, in addition to having a more negative connotation. *Kutafuta* is the least value-laden, and is used widely to describe looking for objects, people, or more esoteric ideas, such as a better life.[31]

Kuchungua implies that the person has used an element of prying in order to gather information. Both *kutafuta* and *kuchungua* connote that the search for information is overt and even public, even if it is not appreciated. *Kupeleleza*, however, invokes a sense of criticism, almost always implying that this form of investigation involves a degree of furtiveness and covertness. For example, a Tanzanian assistant involved in medical work reported how "It was this spying on houses [*kupeleleza vyumba*] that upset" the public.[32]

Moving back up the table, *mtafiti* continues to be a word of choice when talking about researchers. Still, although relatively benign, it too hints that an acceptable activity has been carried out on the wrong scale: searching is fine, but searching too insistently is frowned upon. One man explained, "A researcher [mtafiti] is an important person because he indeed is the one who discovers everything [*anayegundua kila kitu*]."[33] This supports the dictionary definitions of who a mtafiti is, yet there is a hint of excess since a researcher knows not just about disease, or bugs, but about "everything." Even when one of the most neutral words, *mtafiti,* is used, the subtext of prying and invasiveness remains. There is one other word I heard used to describe researchers, and it is the only term that was positive: *mtaalam,* which can be translated as "expert" or "specialist." Mtaalam (or *mtaalamu*) is connected to the term *elimu*, which is "knowledge or learning." Someone who is referred to as mtaalam is educated, learned, and well-informed; he is a scholar or sage.[34] When researchers were referred to in this way, it was less of a direct description of their work and more of a general commentary on their education and expert knowledge.

Healers and Harmers

There were two competing narratives about researchers that emerged, each focusing on very different aspects of their work. In the stories told by Mama Nzito and others, the focus was on blood theft and the murder of innocent Africans. Researchers were powerful experts who were sanctioned by the government and feared by locals. They were unsavory characters bent on making money, even if it meant sacrificing human life. In contrast, another set of stories—which appeared in nearly every interview, frequently alongside or intertwined with the first, malignant narrative—described benevolent researchers.[35] In their positive descriptions of researchers, East Africans noted the similarity between researchers and doctors: they both gave out medicine and helped the sick recover.[36] These behaviors meant research was described as *kitu kizuri*—a good thing.[37] These starkly different characterizations of medical researchers were put forth by most of my interviewees, who saw no inherent contradiction between researchers as potential murderers and researchers as benevolent distributors of drugs. Researchers could be involved with both curing and killing. This willingness to maintain two contrasting impressions of researchers is likely tied to the fact that one of the figures a researcher might be compared with—a mganga—is commonly thought of as a powerful individual who can both harm and heal. As a variety of ethnographies from across East Africa have shown: "In the eyes of ordinary people both good and bad aspects of the doctor can be found in the same person. The mganga has power to protect and harm. . . . The common epitaph is 'How can s/he cure witchcraft, if s/he is not an expert in it!'"[38]

The Zaramo of coastal Tanzania recognized that, "in practice, a *mchawi* [witch] may also be a mganga," and, on the coast, both *uchawi* (black magic) and *uganga* (white magic) "may be practiced by the same individual."[39] Thus, the very ability to heal requires knowledge of how to harm. In some cases, a mganga is considered an actual witch, since "although he seems to be using his powers to help others, one cannot be sure about all his activities."[40] Only by understanding what causes illness (or, how to cause illness) can cures be discovered. And while a good mganga should always use his power in a positive way, there is always the potential for that ability to be abused. The mganga is treated, therefore, with a mixture of respect, caution, and fear—not so different from the way researchers were viewed.

For most of East Africa, the Swahili word *mganga* (or its analogue in other Bantu languages) is a broad term that can reference the healing performed through magic, the use of herbs or other medicines, bone setting, or divination. The Rhodesian medical doctor Michael Gelfand was particularly well positioned to comment on the role of a mganga, having been trained in biomedicine and then focusing on African medicine. He wrote:

> European society has no one quite like the nganga [mganga], an individual to whom people turn in every kind of difficulty. He is a doctor in sickness, a priest in religious matters, a lawyer in legal issues, a policeman in the detection and prevention of crime, a possessor of magical preparations which can increase crops and instill special skills and talents into his clients. He fills a great need in society, his presence gives assurance to the whole community.[41]

Ludwig Krapf's 1882 Swahili dictionary defines a mganga as "a medicine man who uses magic."[42] Maureen Malowany finds the definition significant because Krapf "was able to recognize both aspects of African medical practice: the diagnosing and treating of environment-caused diseases and the equally potent treatment of spiritually caused illness."[43] Among the Zaramo, the mganga's role "is to heal, and they combine all methods of therapy in fulfilling that role, be they herbal, communal, religious or magical."[44] People going to a traditional mganga know their "case will be considered more completely than would be possible at a government hospital" and that a healer would typically treat the patient as part of a larger social and cultural whole.[45]

As healers and witches are understood to rely on the same skills, once researchers were compared with healers, it was not such a stretch to compare them to witches. There are many overlaps with the perceived behavior of witches and the observed behavior of researchers. Witch doctors often work at night and want blood; their uses for blood are socially unacceptable and they profit from working with highly personalized substances. Researchers also worked at night, collecting blood samples by going door to door or collecting night-biting mosquitoes by walking around in the bush. For both witches and researchers, blood was valued above all other substances and its use was shrouded in secrecy.[46] People hesitated to make this comparison directly to me. However, in explaining the actions, beliefs, and fears of *others*, words related to witchcraft often arose. Residents would hide from giving blood since they were afraid it would be used to *kuroga* or *kurogwa* (bewitch) them or would be used for uchawi (witchcraft).[47] There were also veiled references to how "those beliefs" (*imani hizohizo*) caused people to suspect the researchers.[48]

Researchers Are Researchers

While comparisons were drawn between researchers and witches and healers, researchers were also viewed as a homogenous group of people: they all arrived in cars, wore uniforms, were assisted by Africans, asked questions, ventured into the bush, used specialized and foreign tools, were educated, often spoke English, and could be overly curious in ways that offended local sensibilities.[49] As a group, these researchers were busy collecting blood samples, either gathering villagers in the middle of the night or going house to house. Other researchers gathered everyone in a public space and asked people to partially disrobe so their skin could be evaluated for signs of leprosy. Still others had little interest in people and chose to focus on insects or cattle, venturing into swamps to collect mosquitoes or smearing livestock with liquid medicine. Although there were clearly many differences, those differences were not seen as especially consequential. The work was considered to be essentially the same.

This collapsing of all different types of researchers into a generic group was partially a result of the large amounts of research being done in certain parts of the region. Many colonial-era medical research projects employed dozens of specialists: entomologists, parasitologists, medical doctors, nutrition specialists, and nurses, in addition to a bevy of assistants and translators. The large number of researchers per project was compounded by the sheer quantity of projects starting, stopping, merging, and overlapping.[50] In early 1955, the district officer in Taveta, Kenya, had to explain to the leaders of the ill-fated Aptitude Testing Project why local reception was so chilly.[51] He recounted the history of their participation in different government-sponsored research, public health, and agriculture projects over the prior ten years. As he could personally attest, the WaTaveta people had already labored to implement irrigation schemes, given thousands of blood samples for parasitological examinations, and participated in multiple agricultural surveys. The multitude of projects going on in the region was not unusual. In the Pare region of Tanzania, just across the border, a government official praised the WaPare in 1952 as being very cooperative when it came to research and government campaigns. He noted that "They all fought for [plague] inoculation; the Banana Wilt campaign has caused no trouble and the filariasis sampling was done with no lack of volunteers."[52] On Pate Island in 1956, human blood sampling was occurring just a week before a livestock survey, and the government official noted, "it would be bad psychology to deal with both surveys simultaneously."[53]

In each of these cases, the officials made distinctions between the different departments conducting research in the areas. But, for rural East Africans, there was likely no such distinction, as more and more researchers entered their villages and homes with questionnaires, needles, and tubes to collect all sorts of samples and information. Just as researchers were conflated into a generic set of experts, so were their tools. Rubber tubes often appear as a "formulaic element" in many of the previously documented stories of blood theft—long, flexible tubes used to suck the blood of a person.[54] Older women who lived in Nairobi reported that, in the 1920s, men would enter their houses as they slept, "carrying 'a sort of sucking rubber tube.'"[55] But rubber tubes were rarely—if ever—used for blood taking. For most medical research projects, only a tiny bit of blood for a slide is needed, and that drop of blood is taken by pricking the finger. (See figure 2.1 of a child being finger pricked and tubes sitting in the foreground.) For a more substantial blood draw, a needle was inserted directly into the vein of the arm, and blood was collected in a vial.

Although medical researchers didn't use tubes, entomologists did. More than likely, the rubber tubes appearing in many of these stories are the aspiration tubes used by entomologists for mosquito collection. During human landing catches (when a person waits for a mosquito to land on him so it

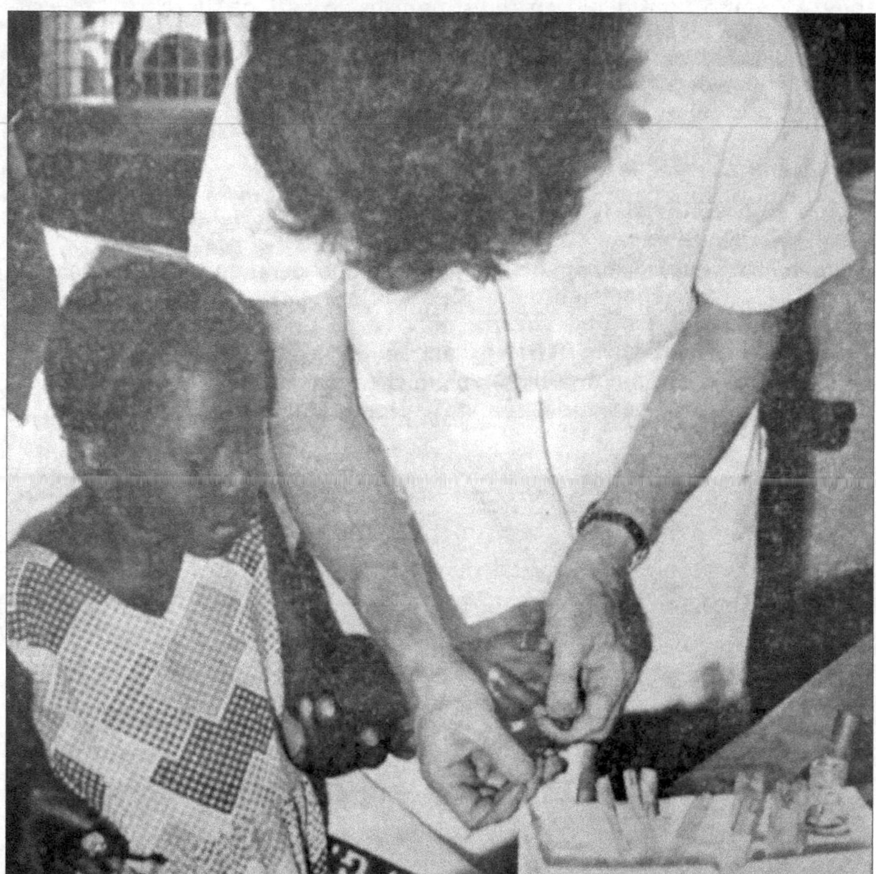

FIGURE 2.1. "Collecting blood in capillary tube for the C.M.R." *Source:* East African Institute for Medical Research Report, 1958–59. Crown Copyright material is reproduced with the permission of the Controller of HMSO and the Queen's Printer for Scotland.

can be collected and analyzed—ideally, before it bites), scientists use these long rubber tubes to suck the mosquito into a holding chamber before blowing it back into a netted trap to carry back to the laboratory for analysis. Thus, it is often entomologists who are seen creeping around at night in the "bush" (*porini*) to observe or destroy mosquito breeding sites, moving in and out of houses to set up and collect mosquito traps, and carrying around tubes that they suck on—all behaviors often attributed to blood-stealing medical researchers.

Researchers Are Government

In the same way that researchers were indistinguishable from each other, they were also considered an indistinguishable part of the larger government. Researchers appeared the same as other government officials in many important ways: they arrived in places bearing stamped and sealed letters of permission from government agencies, they received the chief's assistance, and behaved like other state employees. The perceived link between researchers and the government was captured with the phrase *ajili ya serikali*—because of, or on account of, the government. When people used that phrase, it implied that research happened because the government made it happen, allowed it to happen, or forced it to happen—which is a relatively accurate statement. A majority of research projects were funded with government money. Any big project in East Africa had to have government permission, and researchers relied heavily on local chiefs in order to complete their research. Starting in the late 1940s, when the East African Council for Medical Research was established, even projects undertaken by private individuals came to be entwined with government.

These government-affiliated medical researchers were burdened with the negative views many people associated with the larger government. Citizens' impression of the government was explained using a play on words in Swahili. *Serikali* means "government," but it sounds and looks remarkably similar to the phrase *siri kali*, which means a potent/dangerous secret. The very word to denote government implied to people that it was made up of dangerous secrets. The work of researchers was often referred to as just one of the many "secrets of the government" (*siri ya serikali*). The theft and sale of African blood by researchers was assumed to be part of these secrets. When I asked Mzee Thomas about whether the modern Tanzanian government was aware of the blood stealing, he responded, "This now, is a secret of the government [siri ya serikali]." When I asked whether this was "only the government of Tanzania, or all governments," he laughed and explained, "All! Every government in the world, even yours."[56]

When it came to describing the government and its relation to residents, *kali* (fierce) was closely related to another word—*nguvu* (strength or authority). Typically, anyone who was kali also had nguvu to back them up, or it was the nguvu that allowed them to be kali. Nguvu bred fear, or at least suspicion, and references to nguvu typically involved feeling the negative weight of government.[57] Because of the nguvu of the government, orders didn't always have to be made explicit. The perceived close connections between researchers and the government often led to a sense that villagers felt compelled to participate in medical research projects. While no one claimed government control was so complete as for every act of defiance to be punished, interviewees did emphasize the strength of the government (*nguvu ya serikali*). One woman claimed that during the colonial period even babies sucking at the breast would stop to pay attention when the government spoke.[58]

Another dimension to the public's understanding of researchers comes through the use of the phrase *amri sio ombi*—"orders, not requests." The district commissioner, the chief, and the researchers did not *request* people to participate in research trials, they *ordered* it. Communities' responses to those *amri* (orders) varied from cheerful participation to outright refusal. Smallpox vaccination campaigns in the late 1960s elicited both types of reactions. When I asked one woman (a mother of thirteen, of whom seven had died) if anyone had refused to be vaccinated, she stared at me with disbelief and clearly stated, "In the face of illness, a mother can't refuse."[59] Yet sometimes such willingness came in the wake of amri. One man was emphatic that no one had any doubts about getting the vaccine, since smallpox was so dangerous, but the government gave the amri just in case.[60]

When I pressed interviewees about whether anyone could openly disagree with the government or chief's orders, I was repeatedly told that no one would.[61] Still, although they claimed no one could disagree, there was plenty of discussion of the tactics used to discipline dissenters. Those who would not follow amri would be grabbed and gathered together, forced to participate. Punishment included fines and confiscation of property such as cattle.[62] Two men finally admitted there was one person who would disagree with orders: "a madman."[63] Because of the nguvu of the government, orders didn't always have to be made explicit and only madmen would question them. While this was certainly a characteristic of East Africans' relationship with the British colonial government, it was one that continued into the independence era. When I would ask for clarification about which "government" was being referenced, interviewees would shake off the question: *serikali ni serikali*—government is government—no matter who is in charge.

One of the major areas of disagreement between medical researchers and East Africans was about whether the researchers were actually doing government work, and whether science ought to be considered in relation to everything else around it. The researchers may have acknowledged that they were technically employees of the state, but they maintained that their work was apolitical and ahistorical—not shaped by or a result of anything other than the practice of science and objective scientific data. The researchers maintained that key questions such as the form science took, which projects were funded, what were considered viable research questions, or where projects were sited were answered entirely on the basis of objective, unquestionable data. Science was not affected or shaped by the larger social and political world that affected everyone and everything else. As the previous discussion indicates, however, East Africans saw medical research as firmly entwined within and impossible to separate from larger government activities or motivations.[64] In the 1990s, as Tanzania began to adopt the new malaria drug sulfadoxine/pyrimethamine, some people were very angry and reported that they "were part of an experiment staged by the Tanzanian government to see how the drug kills or how many people would be killed by using SP."[65]

While East Africans often sought to understand medical research by better understanding who medical researchers were—and often came to the conclusion that the researchers were government, if not government-affiliated—that was not the most common link people made. More than associations with government, with force, and with secrets, there was one substance most people thought of when they talked about research: blood.

What Is Research?

Most East Africans describe "research" in terms of blood. Mzee Mwendadi, who was a driver for Mwanza-based medical researchers in the 1970s and 1980s, explained research as "the taking and checking of blood."[66] People frequently spoke of how blood was "looked at," "examined," or "checked." "Disease" or "bugs" were searched for or "discovered" in the blood, by unclear methods. This strong linking of medical research with blood is in keeping with findings from other parts of Africa. The West African nation of The Gambia has been a site for British medical research dating to the 1940s.[67] There, when villagers were asked if they would prefer medical research without the blood taking, they responded, "But blood is necessary in all medical research!"[68]

Most respondents agreed that research comprised multiple steps: first, take blood; then, discover disease; finally, distribute medicine.[69] One woman explained, "When they finish taking blood, they search for disease and they cure you, and you go home."[70] These stages are in keeping with what has been reported in western Kenya. When secondary school students were asked to write a composition about "research," they described how "first 'blood is taken'; second 'it is studied with the microscope'; finally 'we are told who is ill and given medicine.' (Some well-informed children added a fourth one: 'you get a Ph.D.')"[71] Yet, are we sure that the damu that is mentioned by so many East Africans is really equivalent to the blood of biomedical accounts? Since damu is the substance most East Africans associate with medical research, it's worth asking whether we really understand what it is, and how a definition of blood that's broader than Western biomedicine might affect people's views of medical research.

Research Is Blood

Modern Swahili/English, Swahili/French, and Swahili/Italian dictionaries define *damu* first and foremost as "blood," but go on to mention menstruation, blood relationships (children), blood relatives, and the proverb "Blood is thicker than water."[72] In each of these areas, the definitions and connotations of blood do not stray too far from a biomedical understanding; meanwhile, in most American and European language communities, blood can be used as a reference point to family, in addition to being used idiomatically. Swahili dictionaries going back to the 1880s reflect roughly the same usages. It is clear that, as in most parts of the world, a reference to "blood" can refer to both the physical substance and to kin and regeneration. That blood is also linked with menstruation makes implicit reference to the fluid's role in female fertility and an individual family's regeneration.[73] In some parts of East Africa, moreover, menstrual blood and its regular flow also signify the overall health of the society.[74] However, Swahili dictionary definitions do not mention probably the most salient components of East African understandings of blood: that blood is often used as a proxy for general physical health, and that the distribution of blood outside the body is dangerous and should be avoided because it can jeopardize an individual's physical, spiritual, or mental health.

Many East Africans make assessments of their physical health based on the general quality of their blood. Blood can be weak, strong, run quickly or slowly, and be sick or healthy.[75] People are born with either weak or strong blood, and those with weak blood are more likely to become sick and less likely to make a full recovery after being ill. For example, on the Tanzanian coast, many mothers explained their children's frequent illnesses by saying that the children "did not have enough blood in their body [*damu hana* or *damu imepungua*]." The mothers explained this lack of blood by citing numbers they were told at the local clinic, that a child had "only 40 percent" or ".5." These figures referred to the hemoglobin levels that nurses and doctors had mentioned during consultation with the mothers.[76] In explaining their children's poor health, the mothers drew upon existing ideas of the relationship between the quality of blood and health, but also integrated biomedical information into their explanation, since the 40 percent and 0.5 indicated, to them, an actual lack of blood in the body.

As the above example implies, current conceptions of blood are neither static nor uninfluenced by biomedical ideas. Prior to the introduction of biomedicine in the late 1800s, it was common practice for East Africans to utilize a variety of healers including diviners, herbalists, and those specializing in Islamic medicine. As Europeans arrived, the search for effective treatment and persuasive explanations for diseases broadened again to include missionaries, who were considered just another set of "immigrants and traders."[77] Even outside the cosmopolitan Swahili Coast, there was a strong tradition of accepting foreign specialists, treatments, and explanatory systems. Among the Iraqw of Northern Tanzania, "the incorporation of an alien way of looking at and acting on illness" was not at all new.[78] In Uganda, as Susan Reynolds Whyte argues, there is "reason to believe that the exotic has always played a part in Nyole and other East African medical systems."[79] In interviews with modern residents of Dar es Salaam, elders said that the "traditional," or widely accepted, therapies for some diseases had significantly changed over the years—a change they attributed to "the greater presence of biomedicine in their lives."[80]

Although we might be tempted to perceive here a replacement of one (traditional) system of thought with a biomedical (modern) one, that would not be accurate. It would be more precise to see ideas around health, disease, and healing that now exist in East Africa as syncretic.[81] To discuss anything as "syncretic" is to imagine two or more distinct systems coming together to form something new—a third system—that borrows bit and pieces from each. This new, syncretic set of ideas concerning health is a patched-together mosaic of medical technologies, systems, and concepts. It is a product of a long history of medical pluralism that has involved centuries of contact with Ayurvedic and Islamic medicine in addition to biomedicine. It incorporates local ideas of witchcraft and sorcery, political and societal health being manifest though individual bodies, and a broad conception of "normal" health and appearance linked to day-to-day functionality.

There are clear examples of East African understandings of blood assuming syncretic forms. "Anemia" is translated in Swahili as *upungufu wa damu*, which is, literally, "deficiency of blood."[82] Traditional remedies dating back to the 1890s focus on "building" or "strengthening" the blood by eating "hot" foods such as beans, leafy greens, and raisins, which biomedicine identifies as being iron-rich.[83] Thus, well before the first vitamin was identified in 1910, and before chemical analysis showed that these particular foods were iron-rich, it was enough to have a category of hot foods that would "strengthen the blood" through the production of the blood humor.[84] Yet, biomedical understandings of blood are still constantly rubbing up against preexisting ideas, and that friction creates new explanatory models. When a boy died on the Kenyan Coast in the 1980s, people explained that he had "no blood in his body" and opined that the boy should have eaten "hot" foods. Still, in discussing his death with a foreign anthropologist, they were adamant that either hot foods *or* vitamins would have cured the boy, since both strengthened the blood.[85]

This system of medical syncretism forces biomedicine to exist with other, potentially contradictory beliefs. A clear example is many people's understanding of malaria. From a biomedical perspective, malaria is spread by female anopheles mosquitoes, which carry a parasite from infected person to uninfected person. Every malaria infection can be classified into one of four types, produces a particular (and predictable) set of symptoms, and can be successfully treated with a number of different drugs. Yet, although there is widespread awareness of these biomedical expectations about malaria, they are not the only—or the dominant—local understandings of the disease. Multiple public health projects have found that most Tanzanians know that mosquitoes spread malaria.[86] But the knowledge is not exclusive. Medical anthropologists have shown that, while people agree that mosquitoes cause malaria, there are also other well-known and widely accepted causes. As Susan Beckerleg reports, "the view that mosquitoes cause malaria by introducing *wadudu* [bugs/parasites] into the blood stream is not well accepted. . . . And even where accepted, the theory has to coexist with apparently contradictory causes such as changes in the wind."[87] In central Tanzania, people recognized mosquitoes as causing malaria, but also felt that exposure to "hot sun" and "hard work" could lead to malaria.[88] In Ifakara, Tanzania (in the southeast), anthropologists found that people's explanations for malaria often wove together notions of witchcraft with "knowledge of the biomedical cause of malaria." They found that in addition to noting mosquito bites as a cause of malaria, other modes of transmission included drinking or wading through dirty water and/or being exposed to hot sun.[89]

One disagreement between biomedicine and indigenous health concepts centers on blood regeneration. From a biomedical perspective, blood regeneration occurs naturally: the body produces new blood in the same way that the heart beats or the lungs take in oxygen—without conscious thought on the part of the person. Thus, in this framework, the loss of small amounts of blood for donations or medical tests is not considered dangerous, and in most medical research projects blood taking is labeled as "no" or "low" risk. However, for many East African groups, blood regeneration is considered difficult if not impossible, and occurs only through conscious changes in diet or avoidance of certain behaviors. As it has been described among the Haya of western Tanzania, certain foods increase the blood (meat, green leafy vegetables, and fish) while others decrease it (coffee and citrus). Additionally, "hot" activities such as working in the sun or excessive sex can cause "illness such as feverish chills, which are characterized by a lack of blood."[90] In general, women and children are thought to have weaker blood and to be more at risk during procedures like blood donation or surgery because of the difficulty of blood regeneration.[91]

Disagreement about the ability or inability to regenerate blood means that East Africans and biomedical researchers are likely to come to radically different assessments of the risk of giving blood. This is partly linked to physical health: whether one will have enough blood in the body to be healthy and strong. A second concern regards the risk of having blood move through unknown hands in unknown places, opening oneself to the risk of witchcraft.[92] The general feeling is that there are only a few acceptable occasions when blood can circulate outside the body, such as during marriage ceremonies or rituals to mark blood brotherhood or blood friendship. Moreover, even while blood is shared at such times, these are fraught exchanges. It is the very risk involved in sharing blood that emphasizes the depth of relationship with the person the blood is being shared with. By contrast, when the prospect of giving blood to an unknown foreign medical researcher is raised, an East African is likely to come to a very different assessment of the risk involved in such an encounter.

Stories about blood theft and medicines made from human blood have a long history. For at least the last 130 years, one frequently mentioned nefarious use for blood is as an ingredient for *mumiani*.[93] The word *mumiani* is often translated today as a person, someone who is a "bloodsucker."[94] But, when the term was first recorded in the 1880s, it referred to a potent medicine made from human blood. Krapf's 1882 Swahili dictionary defines mumiani as "a fabulous medicine which the Europeans prepare, in the opinion of the natives, from the blood of man."[95] In 1923, someone writing to the Swahili paper *Mambo Leo* noted that the word *mumiani* was a foreign one, but that people knew of it:

> as we have heard, Mumiani is a medicine. Should a person fall and break a bone, any bone, if he is administered with this medicine the bone will heal. Whether this is true of false, those who say will know. Certainly there are those who say the medicine truly exists, especially around Lamu. I have no real need to contest this medicine, except for the way it is [said to be] obtained.

The meaning was largely unaltered in 1939, when Johnson's dictionary described mumiani as:

> a dark-coloured gum-like substance used by some Arabs, Indians and Swahili as a medicine for cramp, ague, broken bones. . . . It is used as an outward application, also when melted in ghee for drinking as a medicine. It is said to be brought from Persia, but many natives firmly believe that it is dried or coagulated human blood taken from victims murdered for the purpose, and when a rumor is started that mumiani is being sought for, the natives in a town are filled with terror and seldom go out of their houses after sunset.[96]

It is widely accepted that the term *mumiani* is an import, although from where is unclear.[97] The fact that the medicine was made from human blood was repugnant but not surprising. As Simeon Mesaki notes, "Since human life is the most precious commodity . . . the most powerful *dawa* (medicine) may be sought from human flesh and blood."[98] Modern accounts from the Swahili Coast note similar beliefs about sorcerers using "human beings preserved half-alive as medicines."[99] In western Kenya, when discussing "research" and "blood," one child "suggested that research and blood collection serve to produce new medicines," while another girl wrote of how "Whites used to make medicines from blood and bodies."[100]

These concerns about the loss of blood have been captured most clearly in relation to East Africans' reluctance to participate in blood donation or blood bank programs since the technology was available in the mid-1940s. In colonial Nairobi, when Kenyan soldiers were encouraged to give blood, many refused, with half mentioning "fear of losing blood that could not be replaced."[101] The same fears were expressed around blood donation in western Tanzania, where 35 percent of the people polled at a variety of public and private hospitals in Mwanza Region viewed donation as harmful and believed that it could damage health.[102] Clearly, damu is a concept that is broader than just "blood," and one that carries with it a different set of perceived risks that would affect a person's willingness to give blood. But, as the discussion of mumiani alluded to, there are also connections between blood and medicine: blood as an ingredient for medicine, and medicine given in return for blood in the medical research exchange. Furthermore, the possession and distribution of dawa (a powerful substance) makes researchers powerful and also dangerous people.

Research Is Medicine

As with damu, the Swahili word *dawa* benefits from a translation more nuanced than just "medicine." Dawa is best thought of as something powerful, something that can have a good or bad effect, as an agent that causes a change. Among the Pogoro of Tanzania, medicine/dawa is "a generic category which refers to substances with transformative potential."[103] The concept of a "medicine" or the use of the specific term dawa is broad enough to refer to insecticides to kill mosquitoes, pills to treat a case of malaria, or an amulet meant to protect against witchcraft.[104] An important characteristic of dawa is that it can be either curative or harmful—its most important quality was not the type of change it affected, but merely that it had the power to change a person, thing, or situation.[105] "Medicines change the state of the person, either by curing, protecting and empowering or, for victims of witchcraft, by weakening, draining and poisoning."[106] The decision about whether dawa would harm or heal depended on the person using it—a healer who wanted to cure, or a witch who wanted to harm—and their knowledge and ability to use the medicine. "The special transformative powers of particular medicines are not intrinsic to the plants comprising them, but depend on the powers of the person who made them. Any plant, it is said, can become medicine in the right hands."[107] That belief was further reaffirmed through direct observation and experience. There is record of at least two herbal remedies used in East Africa prior to European colonization (*Abrus precatorius* to cure eye ailments, and *Myrsinaceae* root to treat worms) that are highly effective at appropriate dosage, but poisonous in higher quantities. It was only with specialized knowledge that a dawa could be guaranteed to be curative rather than dangerous.[108]

Is It Dawa or Not?

"Is it dawa or not?" was what one man living in the northern Tanzanian town of Mto wa Mbu asked when all of the salt sold in town was treated with chloroquine as a way to try to reduce malaria transmission. It was a fair question without a simple answer, since the salt was supposed to be dawa but didn't end up being a very good one. The man's question resonates on a larger level, however, and could be asked of all researchers showing up in villages with pills and syringes in hand. Was what they were handing out dawa, or not?

In the context of medical research, East Africans identify dawa as compensation for giving blood, and that dawa is believed to be powerful, effective, and curative. The biggest disconnect between the concept of dawa and that of medicine is the ability of the word *medicine* to be modified in a way that explicitly states or connotes *experimental medicine.* Despite the myriad modifications of the word *dawa* shown in table 2.2, the one constant is an assumed efficacy or potency. Thus, most problematically, there is no such thing as an *experimental* dawa. This difficulty of translation, and the assumptions about the potency of dawa, create a challenging situation where it is often unclear whether researchers are handing out dawa (i.e., effective medicine) or not.

Table 2.2. Swahili words for types of *dawa* (medicine) and translations

Swahili	*English*
dawa ya kichocho *dawa ya malaria*	medicine for bilharzia medicine for malaria
dawa ya kinyeji *dawa ya kienyeji*	traditional medicine herbal medicine
dawa ya daktari	doctor's medicine Western medicine
dawa ya hospitali	hospital medicine Western medicine
dawa ya ki-China	Chinese medicine (i.e., pills made in China)
dawa ya mtafiti *dawa ya mzungu*	researchers' medicine white people's medicine
dawa ya kutapika *dawa ya kutapisha*	emetic medicine for vomiting
dawa ya kuhara *dawa ya kuharisha*	purgative medicine for diarrhea
dawa ya kunywa	medicine to drink medicine for internal use
dawa ya kutia *dawa ya kupaka* *dawa ya kubandika* *dawa kujisugua*	medicine to apply medicine to lay on medicine to stick on medicine to rub on/medicine for external use
dawa ya miti *dawa ya miti shamba*	plant medicine herbal medicine
dawa ya zamani *dawa ya kisasa* *dawa ya siku hizi*	medicine of the past medicine of today (Western) medicine of these days traditional medicine/modern medicine
kidonge/vidonge *tembe*	pills
dawa ya kuchanja	medicine of cutting vaccination
sindano	injection
msumari	nails injections
dawa ya mmbu *dawa ya nyumba*	medicine of mosquitoes medicine of houses
dawa ya Nivaquine	medicine of Nivaquine

When I asked current medical researchers about how you might say "experimental medicine," I was told emphatically that uneducated villagers would not understand such a concept. Medicine implied efficacy, otherwise it would not be called medicine. At my prodding to consider a hypothetical situation when one might need to convey the concept of experimental medicine, they suggested creating phrases using the verbs *kujaribu* (to try, to test) and *kufanya kazi* (to do work, to function), in addition to using the conditional form of verb conjugation to emphasize that something *might* happen and the uncertainness of a particular outcome.[109]

Multiple factors play into the perpetuation of this misunderstanding of medical research and confusion about what an experimental medicine is. It starts with people identifying the act of research as the act of taking and analyzing blood. The research encounter is framed as an exchange where a researcher takes blood and the subject is given medicine. The medicine is assumed to be effective since it is the payment for having given blood. This means that there may be a complete inversion of how the researcher and subject identify and weigh the risks and benefits of research. An East African may consider the drug as the benefit of research, while the research team, Institutional Review Board, or national ethical review boards will see the taking of an experimental drug as a *risk* of research. Furthermore, while the East African may see giving blood to a foreigner as a risky endeavor because of threats to personal health and the potential for witchcraft, the research team may consider blood taking (and the possible identification of disease) as a benefit of participating in the project. This conflicting understanding of what actually constitutes the risk and benefit of research almost guarantees that the East African participant and medical professional will come to different conclusions about such risks and benefits. It also means that by labeling both experimental and proven drugs as dawa there is guaranteed to be therapeutic misconception.[110] Therapeutic misconception is the name given to the erroneous belief that participation in medical research will benefit the subject personally.[111] Such misconception chips away at the *informed* and *understanding* components of modern consent practices, and calls into question the overall ethical nature of research. This is discussed in greater depth in chapter 4.[112]

By focusing on how East Africans talk about medical research, the language used, and the stories told, it becomes clear that there are real differences over basic questions like who a researcher is, what kind of work he does, whether he is helpful or harmful, and the role of blood within the research encounter. Stories like Mama Nzito's account of dead school children and the narratives of blood-stealing Bwana Matende are typically discussed by medical workers as nothing more than rumor. Labeling these stories as rumor allows for the narratives to be dismissed as false and fantastical. Yet, a careful focus on the very words used to discuss research, and the language used historically, indicates a general sense that research involves prying and snooping and sometimes even spying. This understanding of research is in keeping with tendencies in other countries, but is also a product of East Africa's medical history and researchers' own explanations over the past half-century.[113] It also means that stories of blood theft are given fertile conditions to grow.

Throughout East Africa, and Africa more widely, rumors have circulated for decades about medical researchers who steal and sell African blood. Among historians and anthropologists, the stories have often been explained as representing vague fears about the postcolonial condition and global inequalities, a rise of the occult, or resistance against the colonial state.[114] The stories are assumed to be untrue; as one scholar wrote, mumiani "of course do not exist."[115] Luise White, who has provided the first thorough accounting of blood-stealing stories in the region, maintains that the falseness of the stories is what makes them meaningful.[116] As the anthropologists Fairhead and Leach point out, "rumour has become shorthand for an idea that can be dismissed; that needs to be replaced with proper 'facts.'"[117]

Most international medical and public health organizations would agree with White's assessment that the stories are not true and with Fairhead and Leach's assessment that rumors must be replaced with information. In the public health realm, the stories are signs of African inexperience with Western medicine and signal the continued presence of traditional beliefs. Public health workers want to minimize, ignore, manage, spin, or step around stories of blood theft—not engage with them. A report put out by the UNICEF office in Kenya addressing anti-vaccination rumors clearly stated that their goal was to persuade the critics to stop spreading stories.[118] The working assumption of most biomedical practitioners—doctors, researchers, public health acolytes—seems to be that disagreement with their position is the result of ignorance. This has been referred to as the "public ignorance model," where disagreements between scientific experts and the public are always assumed to be the result of "inadequate public understandings."[119] Assumptions about rural populations' ignorance and the need to educate in order to encourage their participation in projects meant to benefit them are not limited to the public health realm. The same narrative also circulates in development circles. World Bank projects in Tanzania often draw upon government officials who "regard themselves as an educated elite, responsible for telling 'peasants' how to develop because they 'don't know anything,' attitudes replicated throughout the civil service."[120]

Current discussions of those who choose not to participate in medical research or other biomedical interventions sound suspiciously like discussions of "unenlightened" Africans from the colonial era. For many in the medical community, the assumption remains that rumors will "abate with 'proper' biomedically oriented scientific education."[121] From a biomedical perspective, there is no rational reason for people to be against modern public health or research interventions, be they vaccination campaigns, blood banks, or low-risk medical research projects. Thus, a crop of education campaigns have begun to enlighten Africans regarding all sorts of safe and appropriate biomedical procedures.

A study of blood donation refusals in Nigeria declared, "Most of the reasons given were based on misconception, misinformation and ignorance" and concluded that "massive public health and literacy campaigns . . . to inform and educate the rural populace" were needed.[122] Doctors at the Bugando Medical Center in Tanzania came to the same conclusion, recommending a blood donor recruitment campaign that would focus "on clearing wrong conceptions about blood donation through providing information on all aspects related to blood donation."[123] This zealous approach of combatting ignorance with education isn't confined to history or foreigners. A Tanzanian medical researcher declared that people living on the islands in Lake Victoria "just didn't understand [medical research]—they didn't have any education."[124]

But is better or more education about biomedicine really the solution? Evidence presented in this chapter would indicate the answer is no. East Africans are opting out of medical research and public health interventions because they assess risks differently, or have an entirely different understanding of what constitutes the medical research encounter. These differences are significant—they highlight huge chasms that have profound implications for whether modern East Africans can be considered freely consenting volunteers in medical research projects, how risks and benefits are judged, and how therapeutic misconception has become a common part of the research experience.

Many early medical workers presumed that, as people were exposed to biomedicine, traditional systems would be replaced. In 1937, the Kenya Annual Medical Report presumptuously announced that European medicine was "thoroughly established throughout the colony."[125] Yet more recent research has shown that "education and extensive use of biomedical services appears in many cases to have had limited impact, if any at all, on the popularity of traditional medicine."[126] Expecting East Africans to suddenly adopt a biomedical model wholesale—rather than continuing to create a syncretic system based on centuries of contact with foreign healing systems—is naive and contrary to basic historical and anthropological evidence.

RESEARCHERS ARRIVE

Historical Narrative

❖

"INSPEAKABLE ENTOMOLOGISTS"

H. H. Goiny and a Failed Attempt to Eliminate Lymphatic Filariasis, Pate Island, Kenya, 1956

On February 5, 1956, entomological field officer H. H. Goiny of the Kenya Medical Department arrived at Pate Island in the Lamu Archipelago.[1] He came with a clear mission: to do his small part to help eliminate the disease of lymphatic filariasis from the island. He also came with three African assistants to help him, and after a thirty-six-hour boat ride from the Lamu harbor, where he had dragged his own equipment through the mud and slime, they disembarked, exhausted. Since no one met them at the port they wrestled their supplies to shore and set up tents. The following day they would start their work: mapping mosquito breeding sites, trapping mosquitoes indoors and out, and preparing for the future phases of the elimination attempt that would include spraying DDT inside homes, testing local residents' blood to estimate the prevalence of the disease, and administering drugs to all islanders.

Just because no one met Goiny at the port didn't mean the islanders didn't know of his arrival. In fact, as Goiny was setting up his tent a meeting was in progress with the *mudir* (one of the traditional leaders on the island). The meeting was about the villagers' strong belief that Goiny was there as a government emissary intending to do them harm. When the meeting concluded, a group of local Arab men arrived at Goiny's tent to pointedly ask who had given him permission to camp on "private property."[2] Goiny protested and the men grudgingly allowed him to stay, but he had been put on notice: the village was not happy to see him.

The following morning, February 6, a delegation of villagers invited Goiny to a meeting to "discuss his plans." He arrived to find hundreds of people (with "sullen, sulking countenances") gathered for a "deliberately planned and punctiliously organized defiance meeting." Residents "packed

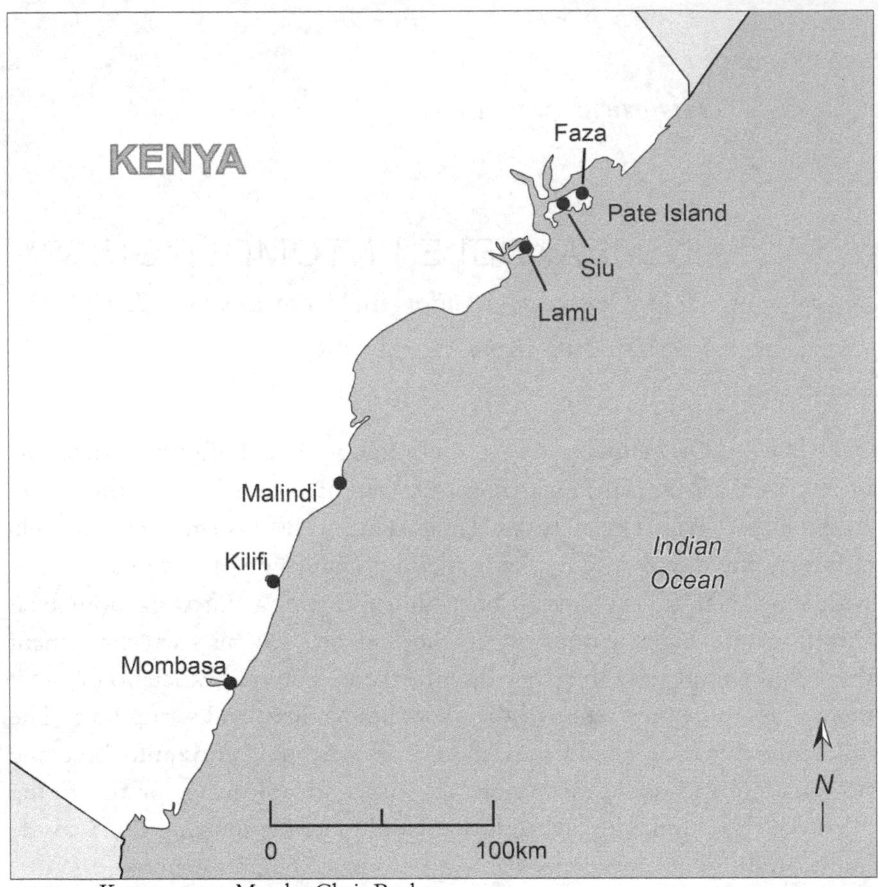

MAP 3.1. Kenyan coast. Map by Chris Becker.

the verandah of the communal store" and filled "every inch of available ground." Community leaders emphasized that the whole village was present and they had "unanimously" decided that "every man, child and woman" would oppose all forms of "domestic control." They would resist his work "by all and any means"—including violence. Goiny left with "the formal assurance" that his work "would be resisted and prevented by the entire population."[3]

Goiny had arrived on Sunday and on Monday morning he had attended the protest meeting with hundreds of residents. On Monday afternoon, another delegation of men arrived threatening violence if he tried to enter any homes, alerting him to additional meetings that had been held at mosques where a "resolution" had been endorsed to obstruct Goiny's work "by all and any means, not excluding . . . violence to the European." On Tuesday morning, when trying to do outdoor mosquito catches and identify breeding sites, a group of villagers followed Goiny and his assistants, heckling them. An African health officer was almost pushed down a well by a group of angry youth with adult villagers looking on. Goiny was outraged and on Wednesday morning he went directly to complain to the mudir. On his way to the mudir, Goiny stumbled on yet another protest meeting happening at the mosque in a nearby village. This led him to conclude that there was a "concerted movement to defy and defeat the planned anti-filariasis investigations on the island." As Goiny parodied the meeting in a letter to his boss:

> Prayers and sacrificial slaughterings were to be offered up for the deliverance of the island from the wiles and schemes of those in-speakable [*sic*] entomologists, harbingers of Allah alone knows what pernicious forms of government interference in our domestic affairs in the disguise of benefactors.[4]

After three days on the island, Goiny was "stripped of the last rags of any illusions" he had about the real attitude of the villagers toward the scheme.

By Thursday, letters in Arabic began arriving at the district commissioner's office, explaining that the Pate Islanders would not carry out the orders of the "Sanitory Doctor [*sic*]."[5] By the time the letter had been translated and read by the district commissioner on Friday, Goiny and his assistants had already left the island. The arrival had been quickly followed by a premature exit, and the project—long before actually beginning—was on the road to ending.[6]

Virtually unmentioned prior to Goiny's arrival on Pate was the fact that his was not the first medical research project on the island, or even the first lymphatic filariasis elimination attempt. Ten years prior, in 1946, optimistic researchers from the London School of Tropical Medicine arrived at Pate intent on eliminating the disease. Their work ground to a halt as villagers refused to participate, petitioned the *liwali* (the British-recognized traditional leader of the Kenyan coastal area), and forced the researchers into an early retreat. The government euphemistically declared the investigation "postponed" due to the islanders' truculence, and that project was also over before it began.[7] And while Goiny and others in the medical department had either never known or forgotten about this prior attempt, the islanders had not. As the researchers were surprised to discover upon speaking with residents, there was "violent opposition . . . hostile attitudes and hot tempers still simmering from ten years prior."[8]

Residents made clear in the following weeks in a myriad of ways that they loathed government interaction, and it didn't matter what was being promised. Based on past public health campaigns and the failed 1946 attempt, islanders had decided they were tired of the rough treatment, heavy fines, and destruction of personal property that had often accompanied the government's goodwill gestures and public health programs. Past public health measures on the Kenyan coast included anti-mosquito campaigns shortly after the turn of the century, which involved house-to-house visits by the liwali's representatives. In 1913, anti-plague campaigns included the threat of forced quarantine of railway workers, who were believed to be particularly susceptible. In 1933, the Ministry of Health and local board of health in Mombasa discussed a policy of "slum clearance" as a form of malaria and mosquito control. In the 1930s and 1940s, prosecutions for violations of the mosquito bylaws were "numerous and complaints about their harsh application equally so."[9] From the perspective of local residents, it was far better to continue living with a disease they had grown accustomed to than to risk additional contact with an untrustworthy and heavy-handed government. This local logic was alternately perplexing and galling to the British officials.

Less than two months after Goiny's failure the Medical Department had given up entirely on the elimination attempt. After much cajoling, only one village on the island had consented to indoor residual spraying and treatment with drugs. Within two years, though, the whole project would peter out quietly without elimination being accomplished or even neared. As in 1946, residents of Pate Island succeeded at changing the Kenyan colonial government plans, modifying international research agendas devised by tropical disease experts, and bringing another researcher's grand arrival and hopes for disease elimination to an abrupt end.

Modern Narrative

❖

A "REMARKABLE ACHIEVEMENT"?

A Lymphatic Filariasis Elimination, Zanzibar, 2001

World Health Organization workers arrived on the island of Unguja (often called Zanzibar) in the Zanzibar Archipelago off the coast of Tanzania in 2000 to begin work on an ambitious yet straightforward plan: to eliminate the disease lymphatic filariasis (LF). The levels of LF on the island were among the highest on the globe: roughly 15 percent of the entire population was infected; in some villages, nearly 30 percent of the population had microfilariae, indicative of the disease in their blood. The disease would be attacked using precise, modern techniques targeting the parasite inside the human body. If everyone on the island—roughly a million people—took a single pill once a year for five years, a disease that had long plagued the island's residents could be defeated. Getting one million people to all take a pill on the same day would constitute one of the largest "mass drug administrations" ever attempted anywhere on the globe—and present a daunting set of logistical hurdles. Yet according to the WHO, if more than 75 percent of Zanzibaris took the drugs at the same time and then continued to do so for another four years, LF could be defeated. This would be a clear victory for local and international public health organizations, would save many Zanzibaris from disfigurement through grossly enlarged scrotums, legs, arms, or breasts, and would reap untold economic benefits.

The Zanzibar campaign in 2001 was one part of a larger WHO effort to eradicate LF globally through the newly created Global Programme for the Elimination of Lymphatic Filariasis (GPELF). One early success of the GPELF was to negotiate a donation from the pharmaceutical companies GlaxoSmithKline and Merck & Co., which promised to provide free drugs for as long as was needed.[1] This agreement paved the way for a global strategy that relied on distribution of free yearly treatment in poor countries.

The WHO worked closely with officials in the Zanzibar Ministry of Health to help coordinate the logistically complex campaign. The WHO made clear that, while they were providing technical support and expertise, it was the Zanzibaris who would be responsible for the actual work—and there was a lot of work to be done. The plan described how on a single day—October 27, 2001—nearly every Zanzibari (excluding the pregnant, very young, and very ill) would be given a single pill, which would kill all of the microfilariae existing in their bodies.[2] Three thousand specially recruited and trained Filariasis Prevention Assistants (FPAs) would each be responsible for visiting fifty households to administer the drugs. Public service announcements, maps with lists of households assigned to each assistant, the distribution of pills to the regions where they would be used, and transportation all had to be coordinated.

The plan was based on the logic of treating every single resident living in an LF-infected region regardless of whether they actually had LF. By treating everyone, there would be no need for testing to identify who actually had microfilariae in their blood, but it would also mean subjecting healthy people to the treatment's side effects and using far more of the drugs than was strictly needed to treat those actually infected. Five years was determined to be the likely length of time required. The yearly drug treatment would kill all baby worms (microfilariae), which would mean stopping transmission to the mosquito vector, and adult filariae had a maximum life expectancy of five years. When there were neither microfilariae nor adult filariae in the body, the disease would be eliminated.

In addition to the daunting logistics of getting a pill to everyone on the island, there was also the small matter of convincing every Zanzibari that s/he should actually take the pill. A degree of local resistance was assumed, and the FPAs were expected to make two visits to each of their fifty households prior to their final visit, when they would watch everyone swallow the single pill. The first two visits to the households were meant to register everyone living in the house, explain side effects of the drug, and address any concerns, in addition to showing the pills and explaining that each person would need to swallow the drug in the presence of the FPA. The WHO explained how each of the FPAs had been specially trained to address local concerns. Yet, there were many questions that didn't seem to have satisfactory answers: "Why are the drugs free? Why not test our blood first? Why you, and not a medical doctor? Why lymphatic filariasis and not malaria?"

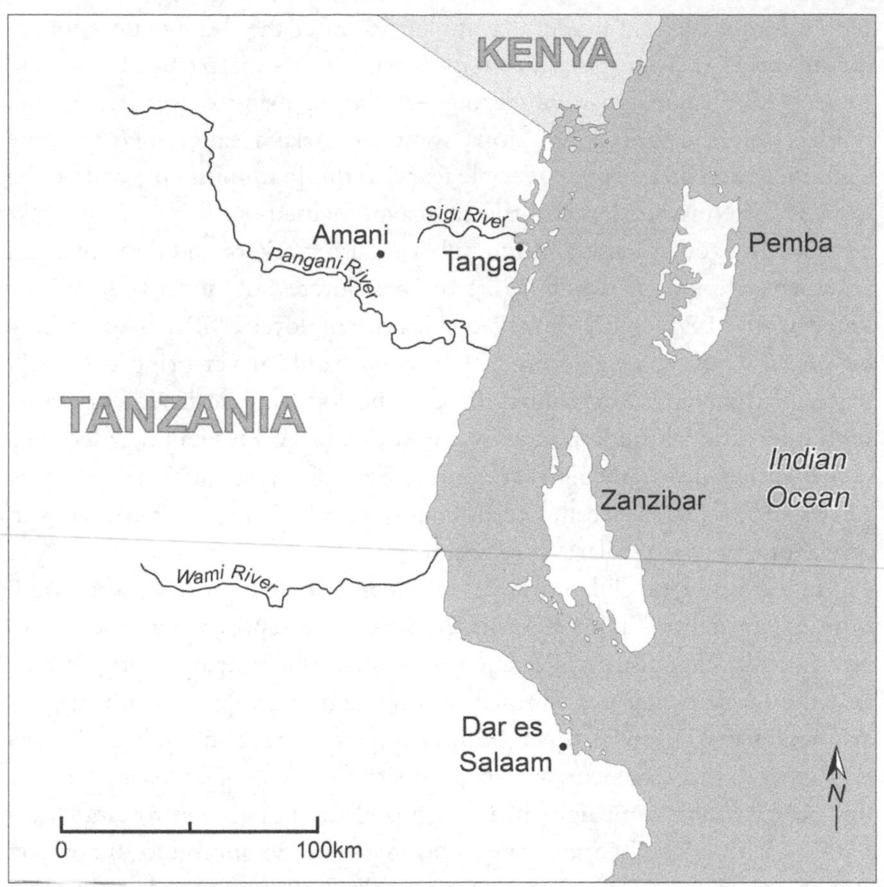

MAP 3.2. Tanzanian coast. Map by Chris Becker.

Meanwhile, rumors began to circulate that the free tablets to be handed out were a form of birth control or had "unknown side effects." Rather than greeting the WHO and Ministry of Health workers with appreciation, residents bluntly told the public health workers, "We don't need these drugs. Please don't come here."[3]

When raising doubts about the campaign, residents recounted failed public health schemes from prior decades, particularly the massive failures of the malaria elimination attempt in the 1960s. Beginning in 1954, a WHO-led campaign in Zanzibar had focused on reducing mosquito populations, and successfully cut malaria rates from 7 percent to 1.7 percent.[4] But, with the island going through dramatic political changes, including a violent coup, the WHO abandoned the project in 1968. Soon after, the disease returned with a vengeance and residents were left to suffer the deadly effects of rebound malaria.[5] The similarities between the 1960s malaria campaign and the 2000s LF campaign were glaring—at least to many Zanzibaris. As one resident told the WHO, "We don't want the malaria experience to come back. We are afraid of that. . . . People fear that this [commitment] will not be sustained."[6] Noisy criticism of the campaign focused on the perceived links between the government and the public health workers, and directly asked whether government would act in the best interest of citizens. Zanzibaris stated to the FPAs and Ministry of Health employees, "If it is something beneficial to us, you [government workers] would never bring it to us."[7] There also remained difficulties in aligning local public health priorities with international funding interests. Although the WHO had decided on a global elimination campaign targeting LF, people were quick to note that LF was not a real concern, and pointedly asked, "Why are you giving us drugs for this when malaria is killing us?"[8]

In retrospect, possibly more jarring than residents' worries is the health officials' apparent obliviousness. In the very same report where the WHO reported islanders' distrust and requests to cancel the campaign, they claimed that the modern filariasis elimination campaign reinforced "Faith in government. Faith in international health campaigns. Faith in medicine." They also asserted that Zanzibar was an ideal site for the project because previous public health campaigns had "sensitized the population to large-scale public health efforts."[9] Such statements evidence the profound disconnect that continues to exist between many public health and medical workers and the larger public. The report goes on to predict that, "if LF is pushed out of the islands, faith will be revived in health initiatives. . . . If it works, it will show the people that the government takes care of their health."[10] The grand predictions about restored faith in public health institutions and government beg the simple question: What would become of this presumed faith if LF was not eliminated from the islands?

What actually happened on that October day in 2001 when drugs were first distributed? As would be expected with such a large campaign, problems popped up in the final weeks and days leading up to the mass administration. The donated drugs arrived in Zanzibar only eleven days prior to the planned date. Rather than three thousand FPAs, more than four thousand were eventually needed and hastily trained. Even with the increased number of assistants, most FPAs had to visit sixty to seventy households rather than the planned fifty. Although some people refused to take the drugs, local authorities handled many of the incidents. As one *sheha* (local official) offered, "Everyone took the tablets except one man. He ran away twice, but I will go back for him."[11] Potentially more problematic, on the actual day for drug distribution there were widespread drug shortages, and administration continued into the following day. The good news was that 76 percent of the total population took the pill: the campaign had succeeded in its goal. As the WHO report summed up, "the people of Zanzibar had made a rational cost/value decision."[12] Even skeptics could agree that the goal for the first year had been accomplished. If mass drug administration could happen consistently in future years, LF would be eliminated from the island.

Between 2001 and 2007, Zanzibar continued yearly mass administration and distributed more than five million total doses of albendazole and ivermectin to the 1.1 million residents of the island. Over those six years, 70–80 percent of the total Zanzibari population received drugs.[13] After the fourth year of mass drug administration (MDA), two sentinel sites measured only 1 percent and 0 percent microfilariae prevalence rates; after the fifth round of MDA, both sites had 0 percent prevalence.[14] At that point the program was considered to be in the "terminal phase," which consisted of maintaining zero transmission.[15] The activities in Zanzibar indicate that it is possible to interrupt transmission and that MDA can be an effective strategy.[16] It remains to be seen whether the parasite densities are low enough in humans (and that introductions of new infections from the mainland are rare enough) for all LF transmission to stop. While Zanzibar has been declared a successful example, claims about successful elimination are notoriously slippery, since the term implies that the disease will be gone *permanently*. We must wait and hope that this will be the case in Zanzibar.

The WHO Progress Reports for the GPELF remind everyone of the clear path countries must take to attack the disease: begin by mapping the disease foci, undertake mass drug administration for five years, and, after this, a period of surveillance and eventual verification of disease elimination. Official WHO reports and plans make no mention of failures that might jeopardize the global campaign. In fact, quite the opposite sentiment is put forth. The plan remains to fully eradicate LF by 2020 and the claim is that the goal is half accomplished. International publications as early as 2006 were touting the program's "remarkable achievement."[17] From some angles the news does look promising: among the fifty-three countries globally that have begun mass drug administration, thirty-seven have already distributed the drugs for five or more years as recommended. In Africa, ten countries have administered at least five years of drugs over 100 percent of their geographic area.[18] Yet, of those thirty-seven countries, only five have moved into the surveillance phase that implies the disease has likely been eliminated.[19] Those five countries—Sri Lanka, the Cook Islands, Tonga, Vanuatu, Niue—account for an amazingly small proportion of the global burden of LF.[20] In Africa, the only country that is mentioned as having moved into the surveillance phase is Togo. (Zanzibar is not mentioned as a country that has moved into this phase because it is part of Tanzania, and the remainder of the country has not been nearly as successful as the island of Zanzibar.)[21] There is something comforting in the linearity implied in these steps—that diseases really can be eliminated by following a simple master plan—but such formulaic prescriptions ignore the many uncertainties that continue to characterize eradication attempts. It remains unknown if five rounds of MDA will actually lead to halted transmission and permanent elimination of LF in most countries, and it is unclear how to keep areas free of LF in the longer run. It also remains largely unacknowledged that the history of past attempts in each place—whether failures of malaria elimination, or successes of other public health programs—will be important in determining how receptive local people are to these internationally backed activities. Although these short-term successes in Zanzibar are important and praiseworthy, it remains to be seen whether the program in Zanzibar actually "represents an excellent model for other countries."[22]

3 FIRST ENCOUNTERS, FIRST IMPRESSIONS

This chapter focuses on the chronological start of the medical research encounter by describing researchers' initial arrival in a village. These narratives of elimination on Pate Island and Zanzibar point to a fundamental misunderstanding at the root of many "first encounters." I look at how researchers arrived and introduced themselves to the communities, and at what information researchers shared with communities and how they shared it. I present detailed information about two historical attempts to eliminate lymphatic filariasis (LF) on Pate Island in the Lamu Archipelago, which was introduced in the preceding narrative, and on Ukara Island in Lake Victoria. I also discuss the modern attempt to eliminate LF in Zanzibar and eradicate it globally. The two historical campaigns had very different outcomes that allow us to ask a series of difficult questions: Did scientists purposefully misrepresent their work? Was it folly on the part of both sets of islanders to refuse the help of researchers who were trying to eliminate disease? What was the logic of turning down a seemingly low-risk medical intervention? Were the miscommunications and conflicts that plagued the programs on Pate Island and Zanzibar inevitable?

These case studies of failed elimination attempts and aborted arrivals make a few important points illustrated in the snapshot of Goiny's five days on Pate Island. First, communities had far more ability to change projects—or end them entirely—than most colonial medical researchers were willing to recognize. Second, although the medical departments and research teams often pretended that their work went on in a historical and political vacuum, East Africans disagreed. Local residents saw the research project as fully enmeshed in recent history, local politics, and their ongoing interactions with the government. Science was neither special nor solitary. While researchers assumed they were making first impressions, community members often viewed their work as repeat performances of past public health and medical research incursions. The chapter begins with information about the disease lymphatic filariasis; details the differences between what was planned by the Department of Insect Borne Disease (DIBD) workers versus what they shared with Pate residents; looks at how science interacted with local perceptions of government and local history; and, finally, examines how the failed attempt in Pate compared with a very similar project on Ukara Island. I conclude by returning to the ongoing global campaign to eradicate lymphatic filariasis.

Filariasis

Lymphatic filariasis (*Wuchereria bancrofti* in East Africa) is a parasitic disease transmitted by *Anopheles* mosquitoes and which has long been found in the region.[1] Once a mosquito bites an infected person and ingests the parasite, the mosquito becomes a vector, able to transmit the disease to other humans through subsequent bites. The parasitic worms live and procreate inside the human body and produce millions of new microfilariae that lodge in the body's lymphatic system. The lymphatic system regulates the fluid balance between tissues and blood, and damage to the system can cause swelling of the scrotum (hydrocele), or a swelling of the legs that results in thickening of the skin (elephantiasis). The disease is not typically fatal, and it takes many years—often decades—for a person to develop obvious symptoms. For a full depiction of the transmission cycle, see figure 3.1. Testing for filariasis requires blood samples (from either a vein or a finger prick) to be taken at night (typically after 10 pm), since it is only in the evening that the microfilariae become active in the peripheral blood; daytime blood tests produce false negatives.[2] The nocturnal periodicity of the microfilariae in the peripheral blood corresponds to the night biting habits of mosquitoes, timing that encourages transmission of the parasite. By 1956, when the campaign on Pate Island began, effective treatment in the form of DEC (diethylcarbamazine) was available.

In the 1950s, the medical community disagreed about how important a disease filariasis actually was, and how many resources should be devoted to fighting it. While the Kenya Medical Department plowed ahead with elimination plans, the mission of the more circumspect Filariasis Research Unit (FRU) was to establish the true effects of LF: in their words, to see if filariasis was "so great a threat to the welfare and economy of the peoples that it would be justifiable to recommend that large-scale programmes of control be initiated."[3] The difference in approaches between the more hands-on Medical Department and the more research-oriented FRU was

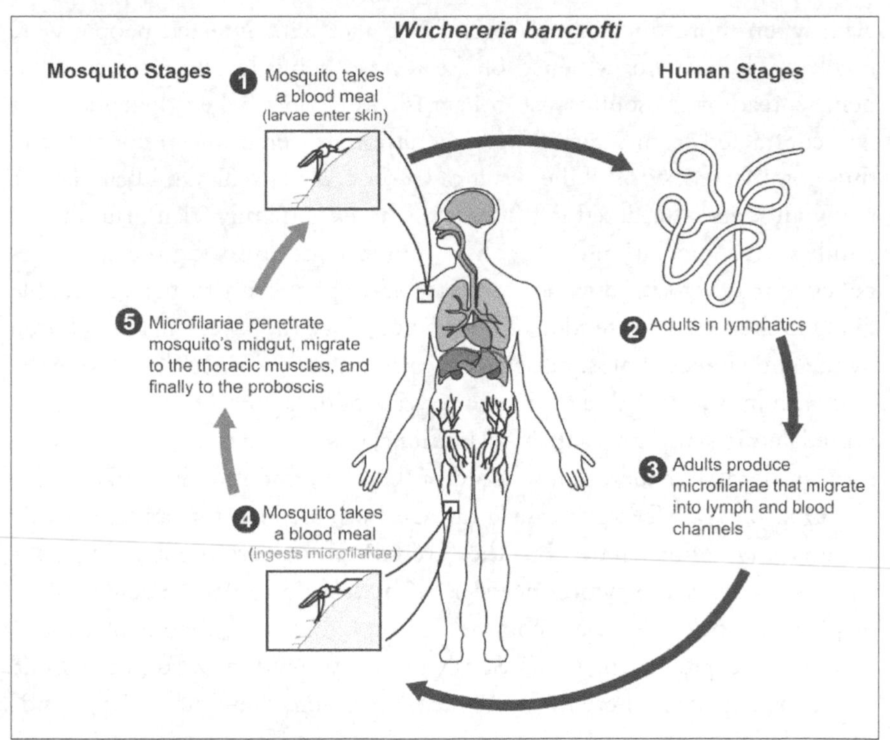

FIGURE 3.1. Lymphatic filariasis disease lifecycle. Produced by Chris Becker.

pronounced enough to elicit derision. A 1956 annual report boasted that the DIBD (part of the Kenya Medical Department) does "not confine our activities to 'counting the hairs in the hindquarters of mosquitoes.' We do not spend all the time in the laboratories but venture into the open air."[4] What was partially a disagreement on how important and burdensome a disease LF was also highlighted the very different missions of the two organizations. The researchers at the FRU saw themselves as working slowly and methodically to question and test every assumption; the DIBD was concerned with achieving measurable improvements in public health through a decrease in vector-borne diseases such as LF, malaria, and river blindness.

Although there was widespread agreement about the cause and treatment of LF among the European medical community, most East Africans did not share the researchers' understanding of the disease. Many communities considered filariasis to be two distinct diseases based on the two major symptoms—elephantiasis (*matende*) and hydrocele (*mabushe*).[5] The diseases were typically attributed to the intervention of God or witchcraft, even when more direct causes such as contact with infected people were invoked. There is no evidence of the disease being locally understood as being spread by mosquitoes.[6] On Pate Island, people believed elephantiasis was contracted from infected persons, although *who* came in contact, and thus got infected, was still the result of God or Allah's will. On Ukara Island, many villagers considered it a disease of the *ukoo* (family/clan), running in families over generations.[7] There is a certain logic to seeing the disease as genetic, or otherwise directly transferrable from person to person: people living in the same area tended to be affected, since they were all being bitten by the same infected mosquitoes. It was also unlikely for people to make the link with mosquitoes due to the many years of mosquito bites that were required before symptoms appeared. Local healers treated the conditions with roots or herbs, but there is no record of these treatments being effective.

On the East African disease landscape, both elephantiasis and hydrocele were clearly abnormalities, but they weren't particularly debilitating ones. In many ways these two conditions of hydrocele (enlarged scrotum) or elephantiasis of the leg were common enough to "demand no explanation" and to be accepted as "natural."[8] Surely no one wanted these conditions, but, in the many places where LF was endemic, sufferers were often not considered to be diseased because they continued to function in society. A person's ability to work, socialize, marry, and raise children kept the condition from being highly stigmatized and made it seem "normal" as judged by levels of functionality. Abnormality was often defined by being nonfunctional within a society. This acceptance of these conditions was so strong that among some western Tanzanian groups, a man with hydrocele was "respected for the size of his testicles," and, along the Kenyan coast, some people believed hydrocele actually increased virility.[9] This created a clear mismatch in the perceived importance of the disease, and whether it was considered worthy of public health attention. The European researchers were likely to refer to LF as a "curse" even though local communities seemed to consider it more like a common abnormality.[10]

What is also striking is how differently the disease was experienced in different parts of the country. In the late 1940s, the researcher Hope Trant was shocked to discover two very distinct responses to living with the disease. In Kyela in Southwestern Tanganyika, those with elephantiasis limped noticeably and complained loudly. On Ukara Island in Western Tanganyika, the Wakara who were infected continued hoeing fields and carrying compost, "only feeling sorry for themselves when their condition is being commented upon in a medical examination."[11] It remains an open question how debilitating the disease actually was and how much it limited functionality in daily life. Local perceptions of the disease stress individuals' physiological and social functionality: physically, people continued to participate in the labors of day-to-day life; socially, people were accepted by society. This acceptance of LF was true in the past and continues into the present. In the previous section's modern case study from Zanzibar, people continued to function with the disease and still tended to prioritize other public health and medical concerns above LF.

Local understandings of filariasis are important for figuring out why the elimination attempt was so forcefully refused. In order for an elimination campaign (or any public health/medical research project) to have local support, people have to feel it is important and relevant. Filariasis was neither a deadly disease nor extremely debilitating or stigmatized; it was unlikely to be identified as a first choice for elimination by residents. Instead, people likely would have preferred researchers to focus on other—deadly—diseases such as smallpox or malaria, or provide a true public health service like reliable access to clean water. As the LF campaign was in progress, a note written in Arabic from Pate Island residents stated that "there is scarcely [*sic*] of water in our villages and due to our poor conditions we beg your assistance," making a clear plea for water rather than activities targeting mosquito-borne diseases like lymphatic filariasis or malaria.[12]

Elimination: The Publicized and the (Secretly) Planned

Goiny arrived on Pate Island as a mid-level employee of the DIBD (a unit of the Kenya Medical Department), the plan of which was to eliminate filariasis on the island using a combination of entomological and medical methods. An entomological approach to elimination involves targeting and killing the mosquitoes that transmit the disease. If the mosquito population is reduced enough, transmission will cease even if humans remain infected. A medical approach targets parasites inside the human body, which is considered a reservoir of disease. When a majority of human infections have been treated and there are few microfilariae circulating in anyone's blood, there is less worry about the number of mosquitoes. The department combined both strategies, targeting both the mosquito and its breeding sites, as well as the parasite inside the human body.

The DIBD spoke confidently to villagers about how they would achieve elimination, but they claimed that the attempt would *only* target mosquitoes. Islanders were only asked to agree to "the search for and collection of mosquitoes in all stages of development."[13] Yet this claim was at best half true. From the very beginning, the DIBD had planned to take human blood samples between 10:00 pm and midnight to determine levels of infection and then to "dose" nearly everyone on the island with a worm-killing drug for a period of years to eliminate the parasite's human reservoir. Additional nocturnal blood testing would also be needed to make sure the drugs were working and overall levels of infection were decreasing. The medical work that targeted the parasite in the human body would be concurrent with the environmental modifications targeting mosquito habitats. Houses would be cleared of water storage containers that served as mosquito breeding sites, outdoor breeding sites would be destroyed, and the interiors of houses would be sprayed with DDT. The researchers openly admitted to each other that "searches in houses are bound to require the removal of some of the conglomeration of junk and dirt to be found in most homes . . . [and] the presently used water containers, tanks and drums . . . would all be ordered to be destroyed."[14] The level of "human" involvement in the project was expected to be large, but this was never discussed with the island's residents.

In 1956, when Goiny arrived at Pate Island, LF had never been eliminated anywhere in the world.[15] The research team knew that each of the proposed interventions—targeting humans, mosquitoes, and the environment—was crucial if elimination were to be even a possibility. They also knew that, even with an integrated approach, there was no guarantee of success. Despite knowing the experimental nature of their work, they concealed this information from residents, calculating that it would ease their way into the community. Our modern understanding of what is needed to eliminate a vector-borne disease like filariasis validates many of the techniques planned by the DIBD. Today, scientists are well aware of the threats of reinfection from mosquitoes or people entering from outside the treated area, declines in drug efficacy, and DDT resistance. But, although the DIBD had the correct scientific techniques and effective drugs, the project failed. The next section will explain the reasons why islanders refused to participate. Their rejection of the project and its ultimate failure had little to do with the efficacy of specific interventions, and much to do with how science, the researchers, and government were perceived.

Explaining Failures Comparatively

How do we explain the islanders' staunch refusal to participate in the elimination attempt and the Kenyan colonial government's quiet retreat from the island? Both defy easy explanation. It is worth comparing this to another historical attempt to eliminate LF, on Ukara Island in Western Tanzania. There, the Filariasis Research Unit engaged in field surveys and drug trials sporadically between 1951 and 1961.[16] Their activities also failed to eliminate LF, but they did "succeed" at not being kicked off the island right after arriving. Ukara Islanders flocked to participate in the mass administration of anti-helminthic drugs and tolerated the regular presence of entomologists, parasitologists, and other scientists coming and going and asking for their participation in a variety of projects. There were two major differences between the campaigns that go a long way toward explaining the disparate outcomes on Ukara Island and Pate Island: first, the behaviors of researchers when they arrived and how well they integrated with the community; second, the history of past government interactions in the area. Although public opinion could always change, in many cases the initial relationships and interactions foretold either smooth or rocky relations in the future.[17]

The attempt to control LF on Ukara Island grew out of a surveying effort organized by the East African Medical Survey. In 1951, nearly six thousand residents (out of sixteen thousand living on the island) were examined by researchers and gave night blood samples in order to determine the level of LF infection. The surveying and small interventions related to LF and bilharzia went on through 1956, when a more concerted effort began. The Filariasis Research Unit took over the work and began giving islanders a monthly dose of the anti-helminthic drug DEC (diethylcarbamazine) and doing entomological surveying to identify vector breeding sites.[18] Residents were split into three different groups, each of which received a slightly different dose. The goal was to determine the shortest and thus cheapest effective course of drugs that would eliminate the microfilariae in the human body—information that could be shared with other LF-infected regions. It is unclear from the researchers' notes and papers whether any of these small drug trials actually generated useful results. It is clear that the findings—if there ever were conclusive ones—were not used to inform policy in other parts of East Africa. Around 1958, the project stopped treating Ukara Islanders for LF and began observing how long those people would remain LF-free. This required regular (at least yearly) blood testing of people who had previously received drugs. Researchers quickly ran into problems, though, as the previously cooperative population began to refuse to give blood samples. The research team complained about this change in attitude in the 1958–59 report, and stated it was still a problem when doing reexaminations in 1960–61.[19]

Alec Smith was the entomologist stationed to work on the elimination attempt in Ukara; he was the "parallel" of Goiny, although the approaches they took to their work were very different. Smith settled in to live on the island for months at a time, and, upon arrival, worked on becoming visible and delivering tangible benefits to islanders.[20] He did this by building a laboratory, which brought him into contact with Wakara workers and allowed him to employ workers and distribute wages. During the process of building—during which he labored alongside the Wakara men he employed—he took the opportunity to explain his entomological research to the workers long before he actually started doing it. This paid dividends and Smith recalled how, "even long after completion of the laboratory, I was known and greeted all over the island, and permitted to engage in mosquito collection in people's houses, not only in daytime but also at night."[21] When writing his memoirs, Smith identified both the construction project and his ability to pay wages as key factors in the general acceptance of his work.

Although the need to build the laboratory was a stroke of luck, a prior medical researcher on the island had already done much to give people a positive impression of biomedicine. Dr. Hope Trant of the Filariasis Research Unit and the East African Medical Survey preceded Smith on Ukara Island by a few months, and she had worked hard to gain the community's acceptance and cooperation. She was a trained doctor with decades of experience in southern and eastern Africa, and after her arrival in October 1950 she spent so much time treating sick islanders that her boss frequently complained.[22] Although Trant had the potentially difficult assignment of collecting blood, urine, and stool samples, she won the islanders over. She lived in modest accommodations, interacted with the Wakara, and made herself useful in tangible ways. In interviews on Ukara Island in 2008, people still talked about the "Mama" Doctor with great pleasure.[23]

Trant knew that one of the simplest ways to smooth an otherwise bumpy arrival was by making tangible contributions, being serious about delivering benefits to the community prior to asking for anything. In many rural parts of East Africa in the 1950s, there was limited access to biomedicine and many diseases were not effectively treated with local remedies.

On Ukara Island, Trant discovered a latent demand for effective treatments. On one of Trant's first days on the island, she saved a woman in childbirth by providing antibiotics to stave off infection. The news spread quickly and Trant's commitment impressed the islanders—especially the fact that, every day for the next week, she walked fourteen miles round trip to check on the woman, who ultimately survived.[24] No one could deny that she had helped save someone, and residents who were initially suspicious had to reassess her presence and work on the island.

It was harder for researchers who were not medical doctors to be tangible about the benefits they were providing by studying mosquitoes. Smith described the unequal nature of research as entomologists practice it, noting that they "required, and usually received, a great deal of help from villagers but had little to offer in return except a much repeated official statement about the expected benefits to the people that would flow in due course from one's work."[25] Early into Smith's work in Tanzania in the 1950s, he decided that promises about future programs were too abstract and opted to stock a simple pharmacy of chloroquine (to treat malaria), aspirin, and an antiseptic solution as a way to provide more immediate help to those who helped him.[26] Both Trant and Smith recognized that to cultivate good will they needed to provide something in return.

The emphasis on Trant and Smith and their individual behaviors does not discount the importance of past encounters with government and other researchers. But personal interactions certainly had an influence on their reception; tangible benefits in the form of jobs, wages, antibiotics, and free medical care were part of a mixture of factors that determined a researchers' initial reception. It is also worth remembering that, in 1950, Ukara was a relatively unconnected place. It took a full day or more just to reach the island from the nearest major city of Mwanza. There was no British official located there, no official buildings or services, and only a handful of missionaries. When Trant wrote the Medical Department to beg for basic supplies, even they couldn't deny that Ukara represented "a gap" in the colony's medical services.[27] That gap meant there were fewer bad memories and plenty of space for the government to step in, provide services, and develop a good reputation.

Why Refuse? Explaining the Failure on Pate Island

Having covered basic information about Ukara and Pate Islands, it is now worth considering specific reasons why the project was so forcefully rejected on Pate Island. Goiny and the rest of the DIBD considered the islanders' refusal to participate in their project a sign of irrationality. Their most frequent explanation turned on the tropes of Africans as uneducated and primitive, lacking the knowledge needed to understand the complexities and benefits of the project. Members of the DIBD long complained that if they had "a more enlightened community" there would be no major obstacles to elimination.[28] As they saw it, their job was not just to conduct research, but to "enlighten the people" and hope that "perhaps, by degrees, the light will dawn upon them."[29] As Brayne-Nicholls put it, their duty was to "break down the inborn prejudice of the people to any form or progress."[30] But while the researchers were comfortable assuming that resistance was due to islanders being in the figurative dark, there are three other, more persuasive, explanations for why Goiny's work was so coolly received. First, the local history of the Kenyan coast and the Lamu Archipelago was extremely relevant. To this day, the coast remains a cosmopolitan place with a diverse set of healing practices; while this *could* have led to a greater tolerance for biomedicine, it actually meant that, in the midst of so many healing traditions, biomedical interventions were met with great skepticism. Second, the researchers' reception was linked to the waning authority of "traditional" leaders, which was partially a result of these men becoming de facto colonial employees and losing the respect of residents. Finally, Goiny and his research team were seen as being part and parcel of the larger colonial government system, and islanders were adamant that they wanted no additional contact with the colonial state in any form. The sections below will discuss each of these factors in turn.

Refuse Because of Local History

The Lamu Archipelago has a unique history relative to the rest of East Africa. This meant that work on these coastal islands was likely to be difficult, due to its cosmopolitan past, its ongoing skepticism of biomedicine arising from the presence of alternative healing systems rooted in Islam, and a contentious colonial history.[31] Despite the distance from central Kenya, the islands were frequently reminded of their colonial condition in the form of visiting British officials, extensive public health programs, the presence of biomedical hospitals, and a physical connection to the mainland through telegraph lines, roads, boats, and airstrips.[32] The human diversity contributed to a degree of cosmopolitanism, with immigrant communities from across the Middle East, India, and mainland Africa settling along the coast. These groups each brought their own ideas about healing, and the coast became a melting pot of medical systems that included Ayurvedic, humoral, and Islamic beliefs. This wealth of diagnostic and treatment systems meant "locals were often skeptical of western medical treatment."[33] That skepticism extended to British-run hospitals, which people perceived as places where bodies were often desecrated against families' wishes (through the performance of autopsies).[34] At one point, islanders successfully prevented a hospital from opening—another response that the government found entirely irrational.[35]

The presence of Riyadha Mosque-College in Lamu also reinforced Islamic medical ideas, since this institution focused on the teaching of Islamic medical practices and humoral medicine.[36] Treatment practices on the Swahili coast often draw upon the expertise not only of a traditional healer (*mganga* or *tabibu*) but also of an Islamic scholar (*mwalimu*). Overt forms of Islamic medicine include the use of Qur'anic scripture as medicine, either by copying verses that are then put inside amulets and worn on the body, or by writing a verse in saffron, then soaking the paper in water and having a patient drink the liquid.[37] In addition to the presence of healing strategies rooted in Islam, there was also less adoption of biomedical interventions. Multiple scholars have found a "continued reliance on healing philosophies that have *not* integrated biomedical notions of how the body operates," and Beckerleg goes so far as to argue that Western medicine remains a "culturally alien import."[38] This doesn't rule out modern forms of syncretism, such as when "Ibuprofen tablets might be taken with a glass of saffron colored, holy text-infused water," or the fact that "forearms strapped to intravenous drips are often adorned with amulets known to ward off the evil eye." But these cases of syncretism are less common than in other parts of the region where biomedicine has been more widely embraced.[39]

The Lamu Archipelago was also recognized as a particularly difficult place to work and local histories often paint it as an area where residents were known to have an independent streak. Although there were many commonalities found among people on Pate Island, it would be wrong to consider it a unified or homogenous place. Each of the major towns was considered a distinct political entity, and they had been fighting among themselves for hundreds of years.[40] One history of the Archipelago notes the last few centuries could be characterized by "attacks and counter-attacks, intrigues and counter-intrigues," in addition to "insurrection," "conflict" and "dispute."[41] There was a general level of contentiousness that pervaded many aspects of life and which was not restricted to the realms of biomedicine, nor always directed at the colonial state. In 1956, police were sent to a Lamu mosque when residents nearly came to blows in a disagreement about whether a certain phrase in the Friday prayer should be said aloud or under the breath.[42] Even when compared to how the entire coast was perceived, Pate Islanders were thought to be particularly intractable in the face of colonization, retaining "an unmitigated sense of independent-mindedness."[43]

There is also the most obviously relevant local history: that of the prior (failed) attempt to eliminate LF in 1944–45. That effort had also been "abandoned due to the opposition of the islanders."[44] And although Brayne-Nicholls knew of the failure in March 1956, he admits that, the first time he went to the island, he knew nothing of that history. There was no reference to the "Affair" in past annual reports of the Medical Department or the DIBD, so there was little chance Goiny would have become familiar with this history by rereading his institution's own formal documentation. If Goiny did know about this history, he did not seem to take any special precautions, nor does he mention it in his notes about his arrival and initial work on the island.

Refuse Because of Middlemen

Colonial researchers tended to assume that their arrivals were just about them, and that the community's reactions were in response to them alone. But the arrival of researchers and the community's reception were tightly linked to the position of the local African authority, the man who was the "traditional" leader: the chief, the *mudir*, the *liwali*, the *mtemi*, or the local equivalent.[45] African men (and they were most typically men) who served in traditional leadership positions often acted as middlemen—functioning as intermediaries between the colonial state and the communities where they lived.[46] The British governed East Africa through a system of indirect rule, which relied heavily on the perceived traditional authority of local leaders. Rather than using British officers, African men became part of the colonial government, translating orders and enforcing the rule of law in distant places. These middlemen were called upon to help prepare communities for the arrival of researchers; thus, to fully understand how researchers arrived and were received, we have to understand who was preparing communities and in what ways.[47] As researchers' own accounts make clear, a new project would have to explain the objectives and methods of operation to the district officer, obtaining "the necessary introductions and authorization." Researchers would then begin visiting "the various localities in which it was proposed to operate and interview the local Wakili, the Dresser, the Schoolmaster, and any other African official who would be concerned, and whose help was needed."[48] These visits were often made with African assistants and local leaders in tow, to work as translators and intermediaries and demonstrate their support of the colonial state if it was not already obvious.

On Ukara Island, the Filariasis Research Unit realized much of its success was dependent upon the power and authority of the local leader, the mtemi. They admitted in an annual report, "the number of people turning out at night to have their fingers pricked when nothing is given to them for their trouble depends entirely on the power of the headman or chief."[49] Historically, the Wakara seemed likely to follow the mtemi's orders and to believe that those orders were in their best interest. In recent interviews, residents frequently mentioned how they "trusted" the mtemi, and explained that their trust was rooted in the idea that he would only agree to a project if there were benefits for residents. One man explained how islanders were often afraid of researchers, but with the protection of the chief (*himaya ya chifu*), and the chief's approval of a project, the islanders would agree to participate.[50] The Wakara's stated trust in the local leadership coincided with high levels of coercive power: people freely admitted that it was impossible for residents to go against the mtemi's orders and that, as mentioned in chapter 2, dissenters would be fined or punished.[51] A few elders did cannily point out how the mtemi's orders always seemed to "follow the advice" of the government—leading them to indirectly call into question whether the mtemi was acting primarily in the best interest of his subjects, and how independent his decision making really was.[52]

The image of a powerful local leader as effective middleman on Ukara Island contrasted with conditions on Pate Island. There, the mudirs had far more limited control over villagers. Although both of the two mudirs (Sheikh Mohamed Saad, who served through June 1956, and Sheikh Abdallah Khatib, who served from June 1956 onward) held *baraza* (public meetings) trying to convince villages to accept Goiny's presence, they were unable to sway villagers supposedly under their control and even unable to stop them from making threats against the research team. Goiny had hoped that help from the mudir at Faza could result "in winning over the recalcitrant elements of the island's population," but he ultimately had to recognize that "the results of his best efforts remained largely negative."[53] The two liwali of the coast (Sheikh Al'Amin for Lamu in 1956; Sheikh Hyder el Kindry for the coast in 1960) were also often unable to sway public opinion. One explanation for the compromised authority of the "traditional" leaders is that people lost their faith in these leaders once they had become de facto government employees by participating in the British colonial system of indirect rule.[54] As Goiny's presence became more contentious, the mudir took to talking like a state employee, where the obligation was to follow protocol but not necessarily to deliver results. He wrote to the British district commissioner, "as instructed, I held a baraza at Faza . . . and tried my level best to explain." He admitted his failure and enclosed a letter from angry residents, noting it "is self explained."[55] In that particular situation, the mudir was little more than a conduit, passing messages between angry islanders and surprised officials. He was unwilling to risk whatever authority he still had to force residents to participate.

Even when these authority figures were unable to change public opinion or force participation, colonial officials did on occasion recognize their critical role and difficult work. In July 1956, Goiny remarked on Sheikh Al-Amin's "rare knack" for " handling the islanders (of which I have witnessed some impressive displays)."[56] In August 1956, after activities had started again, the provincial commissioner noted the success and considered that it reflected "great credit" on the liwali of Lamu and asked to convey his "thanks and appreciation."[57] When mass treatment began in Faza in January 1957, "the cooperation of the population was remarkable," which was credited to mudir Sheikh Abdallah of Faza.[58]

There are no easy generalizations to be made about middlemen such as the mtemi and mudir. They were not always as malleable and willing to promote colonial projects as British officials would have liked. In places with strong local dissent, chiefs and mudirs were just as likely to become the spokespersons of villagers as they were of government. Middlemen were mercurial and pragmatic, weather vanes of public opinion, and intent on protecting their own position of power within the community. Goiny didn't complain when he saw the mudir attending a protest meeting on Pate Island against his project; he knew the mudir was in an "admittedly rather delicate position in the community" and was unable to stay away.[59]

These middlemen were canny and savvy characters who walked a fine line between enforcing the orders of colonial officials and becoming the mouthpiece of community members. It was frequently the case that government might take away their salary, but local people might run them out of town.

There was at least one other set of middlemen who were critical to the research endeavor: the research team's African assistants and technicians. Henry Gigiri, Michael Ikata, and Faros Enos were all posted on Pate Island; a number of unnamed Health Office laborers also assisted. All of these men endured difficult work conditions, and one health office worker was reportedly almost pushed down a well.[60] Goiny described these East African men as working with a "paralyzing sense of insecurity," and he worried the men might suffer "unjustifiable . . . bodily molestation." The men beseeched him "not to post them in separate villages in complete isolation from each other."[61] There was a real element of danger to these men's work, although it's unclear if the local dislike was more rooted in their being colonial employees or the actual work they were doing. Although Gigiri, Ikata, and Enos limited themselves to walking outdoors and doing daytime mosquito catches, islanders ominously warned them not to enter houses unless they were "prepared to take the consequences."[62] In Faza, the mudir advised Michael Ikata and Faros Enos to be "as discreet and cautious as possible."[63] They also faced the same types of criticism that were leveled at Goiny. Michael Ikata had made an earlier preparatory trip and, as soon as he began the tour, "groups of pickets had formed . . . ready to assault them at their first attempt to enter a house or hut."[64]

On Pate Island, these assistants needed to enter homes, set traps, gather mosquitoes, engage with local people, and serve as an initial buffer between islanders and European staff. In other projects that involved more overtly physical contact with patient-subjects, the assistants were responsible for physically interacting with the research participants: taking blood, palpating bodies, explaining procedures, ushering them into makeshift laboratory and examination spaces, assigning numbers, filling in paperwork, and likely fielding questions and providing answers that the European researchers were not in a position to hear or respond to. The East African Medical Survey team wrote in their 1956–57 annual report that the African assistants working on the drug trials were "tireless in their endeavors to get the drug to every person by house-to-house visits," and noted that if the project was to be reliant on "persons 'working to time'" (i.e., working just their set amount of compensated hours), "insufficient coverage would be obtained" and the project would not succeed.[65] These men often worked long hours and, although they derived a certain amount of prestige from being employed by the state, having specialized training and regular wages, they were often roughly received and were the objects of suspicion. The islanders made their disapproval clear: in the case of the assistants on Pate, once the men arrived from the mainland, all prior offers of accommodation were "politely and evasively, but nevertheless conclusively, withdrawn."[66] These African assistants were lumped in with the larger set of oppressive, untrustworthy researchers. While it is unlikely that these men were ever solely responsible for projects failing (for which there were many causes), they were likely one of the crucial factors for determining whether a project was to succeed.[67]

Refuse Because It's Government

Villagers' most intense, earliest-stated, and most frequently invoked claim was that Goiny and his team were government.[68] Villagers saw the elimination attempt as firmly enmeshed in the social, cultural, and political milieu, that is, firmly enmeshed in the everyday. The project was viewed as dangerous and unappealing not only because of what individuals had to do, but because of who was making the offer. The colonial government was untrustworthy and local communities typically avoided contact. From their perspective, Goiny was clearly part of the government: local authorities publicized his visit prior to arrival, he appealed to the government when islanders weren't participating the way he wanted, he wore a uniform, and he issued orders he expected to be followed. As others writing about science-society relations have pointed out, it is often not the "science" but the "trustworthiness and credibility" of the institutions that influence whether or not people participate in an activity.[69]

From a local perspective, there were multiple ways the project would lead to more (unwanted) government intervention. First, villagers believed that if so much money was going to be spent on eliminating elephantiasis, it would doubtless lead to increased government control through additional taxes and fines to pay for the project.[70] This scenario had already played out in the nearby town of Mombasa: when a mosquito control program began in 1928 and quickly ran out of money, the Health Committee began taxing residents to meet the program's expenses.[71] Second, if staff arrived from the Medical Department there were fears that "everyone would be prosecuted and fined" for violating the minutiae of the health code. This was yet another fear that had been proven true. The Public Health Ordinance of 1913 gave the state far-reaching powers that included "removal of nuisances . . . demolition of insanitary areas, closure and demolition of houses unfit for human habitation."[72] By invoking a need to protect or improve the public's health, the government had the ability to enter homes, destroy private property, and even demolish houses. Past experiences with the Health and Public Works Departments' campaigns targeting plague, malaria, and yellow fever had led to just such destruction. The anti-plague measures were considered "interventionist" and "offensive." When plague was suspected, all of a household's clothing was burnt on the spot; straw roofs were euphemistically "removed" (destroyed) to "allow sunlight in" and compensation for all of the lost property was neither timely nor adequate.[73] In terms of the LF campaign, if the Health Department began house-to-house inspections, it was not hard to imagine water containers that served as breeding sites for mosquitoes being destroyed, and the owners being fined.[74] And, in fact, DIBD workers admitted amongst themselves that in the "previous campaign a fair number of receptacles were condemned."[75]

Although Goiny considered the project as a "government-sponsored venture of a non-political and strictly medico-scientific nature," the separation between science and politics was disingenuous. Goiny knew from the start that elimination would require cleaning up buildings, modifying water storage practices, and reorganizing sewage disposal—activities that were identical to those used by the colonial health department. In letters of protest sent to the mudir and forwarded to the British officials, Goiny was referred to as the "Sanitory Doctor [sic]," a government worker linked to the public health department, with a penchant for heavy-handed techniques. And there was good reason for residents to believe that the researchers who looked like government, yet claimed not to be, actually were government. For one thing, the workers from the Medical Department did look like government, or at least the part of the government that carried guns and forced people to follow orders. In East Africa, men employed by the King's African Rifles (KAR), the colonial medical departments, and the police all wore strikingly similar uniforms: khaki-colored button-down jackets with leather belts, shoulder straps, and short pants. Hats topped off the outfit.[76] In Kenya, even if the public health workers were not mistaken for the military, they were clearly viewed as "men of government." As Geissler reports from interviews with Kenyan public health workers, the uniforms they wore through the early 1980s consisted of brown shorts and shirts with a cross and "MD" for Medical Department.[77]

The assumption of the DIBD was that people would support a government-sponsored intervention if it placed few demands on them, but this was a profound miscalculation. Pate Islanders barely considered what they would have to contribute to the campaign before refusing to participate. The colonial researchers assumed science would be considered on its own merits, but islanders couldn't help but evaluate the project in relation to everything around it. The Pate Islanders were broad in their assessment strategies, seeing the interconnectedness between government science, public health, recent history, and present actions. This is similar to how residents in West Africa have assessed modern medical research projects. When deciding whether to participate in a trial run by the Medical Research Council (UK), anthropologists noted that "people's decisions are taken in a field of uncertainty and speculation in which wider confidence (e.g., in the Gambian government, or in Gambian field workers) or worries (e.g., about the duplicitousness of white people) are relevant."[78] With recent history in mind and taking a broader assessment of the risks of participants, it is less surprising that Pate Island residents had little interest in allowing government workers to establish themselves on the island.

Why Give Up? Because of Mau Mau

Having spent time discussing the reasons Pate Islanders acted in the way they did, it's also worth asking why the colonial government responded in the way that it did. Why did it allow Goiny to slink away rather than fortifying the research team and forcing residents to participate? Why not make an example of Pate Island with a show of government force? Ironically, the answers to those questions undermine Goiny's original declaration of his project being apolitical. In fact, the decision about how to handle Pate was shaped by colonial political conditions and concerns. In 1956, at the same time Goiny launched his boat for Pate Island, Kenya was in the midst of the Mau Mau Emergency, a Kikuyu-led civil war with an oppressive British response.[79] During this crisis of the colonial state, British authorities were on the lookout for potential threats to their rule. At a time when the British were involved in a brutal and repressive response to the Mau Mau revolt, why would they so easily yield to the "act of defiance" on Pate Island?[80]

Nothing in the documents indicates that the events on Pate Island were connected to Mau Mau. But, after a rough few days, Goiny couldn't help but see the specter of this larger civil conflict on his small patch of earth. He wrote to his boss about "the mass organized gatherings, ceremonials, sacrifices and invocations of a religious or pseudo-religious character." He claimed that there was a "striking similarity between the local methods for discrediting and undermining government influence" and that used by the Kikuyu, and that the overlaps were "unmistakable."[81] Goiny considered it a legitimate possibility that by pressing their work they could be "touching off . . . a seditious movement comparable to the Mau Mau rebellion in scope, if not extent."[82]

As the residents' truculence became known and filtered to the desks of other health and regional officials, two very different responses emerged. On the one hand, more cautious officials delicately pointed out that, because of Mau Mau, the colonial government shouldn't risk another insurrection, so the researchers should tread lightly in Pate. This implied changing the project or ending it, with the goal of keeping the islanders happy. On the other hand, precisely *because* of Mau Mau and the government's struggles to contain it, other officials argued for a firm response. The government shouldn't risk letting another part of the country get out of control, and a swift response to the villagers in Pate could send the right message. Brayne-Nicholls, the district commissioner of Lamu, who supported the second approach, ominously declared, "something can be done and will be done, to bring sense into the heads of the people of Pate, Siyu and Faza."[83]

At some point, the response became less a question of public health and more about government control. All of the discussion about eliminating disease and improving the islanders' health had disappeared—to be replaced with conversations focused on colonial control, expressions of authority, and whether residents were behaving as colonial subjects should. As notes flew between the male officials involved, the rhetoric escalated. Ronald Heisch, at the DIBD, wrote to DC Brayne-Nicholls for two weeks straight, snidely questioning his ability to control the islanders. The first letter pointed out his early impotence and offered, ironically, "if people of Patte are openly defiant, I am sure you will be able to deal with them."[84] With the islanders still refusing to participate in the DIBD project, Heisch wrote again a few days later, falsely commiserating: "the Patte islanders have always been difficult." He then noted, "their behavior seems an act of defiance to government" and concluded by cruelly conceding, "I suppose it is difficult to discipline them. A pity."[85] Brayne-Nicholls was unable to get the islanders to participate and Heisch continued with his barbs: "the insubordination of the Patte Islanders must be very annoying for you."[86] Heisch was relentless in his criticism, indirectly asking why the district commissioner of the coast was unable to control people under his jurisdiction, needling him in the hope of getting him to apply heavy-handed techniques to force the islanders to participate.

The way that Pate was ultimately dealt with—a slow, quiet retreat—was not unlike Goiny's original plan that was meant to minimize the chance of a rebellion. As the district commissioner, it was Brayne-Nicholls's job to maintain peace and order, not to push through research projects or public health programs that would lead to mass protests. While Heisch tried to bait him into tough dealings with the islanders, Brayne-Nicholls's original threat to "bring sense into the heads" of the islanders came to naught. Mau Mau was both distracting and worrying for officials and, while it had not factored into the planning of the project, it was considered once problems arose. There was too much at stake for the government to fuel another incipient rebellion, especially for the questionable goal of improving ungrateful islanders' health.

The successes of Ukara and the failures of Pate shouldn't be placed too heavily on the shoulders of the individual researchers involved. Entomologists like H. H. Goiny and Alec Smith didn't plan projects, and as mid-level employees were merely expected to follow orders and get their work done. Neither was failure entirely due to the effort or non-effort of local African leaders. Local history and specific experience and characteristics, including prior interactions with government, local cultures of healing, the training, skills, and interest of the field researchers, and whether projects were designed in a way to deliver early and tangible aid to residents, were all involved and influenced peoples' encounters. Only when assessed collectively can these factors help explain the very different outcomes of these two LF elimination projects.

One may also invoke luck in explaining Ukara Islanders' greater success. The researchers were lucky that the island had less contact with the Tanganyikan government, that public opinion of colonial interventions wasn't negative or hardened, that a local leader still retained the power to issue orders that would be followed, and that they had sent Dr. Hope Trant, who made a point of prioritizing community relations over the more efficient research program desired by her superiors. Peaceful community relations were possible in both places. The "uncontrollable" factors around past local history and views of the colonial government in Pate meant it was going to be harder to run a project there, but not hopeless. It is impossible to accurately predict the exact mixture of factors that would have won Pate Islanders over, but it likely would have included researchers spending money locally and behaving generously, offering treatment for conditions that people felt were worthy of attention rather than just targeting mosquitoes, and being more responsive in identifying and assuaging local anxieties when they arose. If some combination of these things had been done, there could have been a very different outcome.

On the other hand, we cannot place too much responsibility on Goiny, who spent a mere handful of days on Pate Island. The British colonial government's difficulties on Pate Island had not started in 1956, or even 1946. The conflict between the islanders and the DIBD had to do with the nature of the coast and coastal residents' interactions with centuries of overseers and would-be colonizers who included not only the British, but also the Zanzibaris, Omanis, and Portuguese. Although Goiny was caught off guard by his reception in Pate, the response wasn't unpredictable, and he was enmeshed in something far larger (and with deeper roots) than community dissent over a plan to collect mosquitoes. The residue left by previous encounters with public health and medical officials deeply shaped his reception. In some ways, Goiny may be painted as a victim of circumstances and history he knew little about. Without that knowledge it was nearly impossible for him to change the tone, quality, or character of the relationships between the DIBD and the Pate Island residents.

The arrivals of researchers onto East African islands in the 1950s and the filariasis elimination attempts recounted in this chapter—an unsuccessful one on Pate, and a slightly more successful one in Ukara—are typical of what many researchers' arrivals looked like in East Africa. There were plenty of hurried and premature exits, leaving communities no better off than before, but with annoyed and suspicious residents the next time a stranger arrived making grand promises. The example of Pate was not particularly exceptional: it was common to misrepresent work and to present research as a public health program. It was also not unusual for a research team to enter naively and optimistically, with a lack of knowledge about local history and past biomedical interventions.

The Pate example also makes clear exactly how divergent community members' and scientists' views were when it came to situating, understanding, and evaluating medical research. For the Pate Islanders, research teams were placed in their larger political, social, and economic context and were evaluated in relation to a much wider and longer web of past interactions with similar people or other government officials. This was nearly the opposite of how the scientists thought about their work, since they explicitly framed it as apolitical. However, even this sentiment doesn't hold up to close scrutiny, since it becomes clear that the handling of Pate was affected by the Mau Mau insurgency. Colonial officials in the medical department clearly considered the greater political context. There were also divergent perspectives on what it meant to be "government." For the researchers, identification as government was positive, signaling their commitment to disease elimination for the improvements it would make to public health. For Pate community members, the researchers' affiliation as government was threatening; past interactions with the government had not been forgotten, nor was there any desire to increase contact.

One of the notable things about Pate was the islanders' ability to alter the colonial research agenda. The researchers may have decided when and how to arrive, but Pate Islanders decided if the project would even begin. Community members, as indispensable subjects in these projects, had a profound ability to change research agendas and end them entirely. Officials who planned these projects were often based in offices in big cities—far from the messy realities of conducting field research—and they had a tendency to forget that, in many parts of the country, the best-laid plans were still subject to approval from residents. Reading from at least one vein of the literature in African history, what occurred in Pate would appear unusual or unexplainable. For those who consider colonial medicine a "tool of empire," the state's assumed heavy-handedness should have left Pate residents violated and forcibly turned into colonial and medical subjects.[87] In many ways, it is indisputable that medicine was a tool of empire—an implement that the colonial government consciously wielded to alternately cow and impress subjects. Yet these arguments overestimate the power of the state. The fine details of field research projects indicated that colonial medicine wasn't nearly as hegemonic or oppressive as some past work has implied. More accurate would be the idea of "hegemony on a shoestring," a hegemony that was carried out cheaply, incompletely, and with less than totalizing effects.[88] Evidence from Pate Island indicates that it was the islanders who held the upper hand when determining if projects would continue, not the colonial medical department. At the very least, these communities should be seen as formidable opponents in being able to resist and respond to colonial activities.

The opinions of the Pate Islanders in 1956 were neither static nor written in stone, and there was no reason to believe their views on government or science or biomedical interventions couldn't change. While Goiny's arrival and research activities were aborted, it did not portend the future. When Goiny's work stopped in 1956, it was unclear exactly how to proceed. As Brayne-Nicholls wrote a month after Goiny had been kicked off the island, "I have called off the investigation for I am sure that to pursue the matter against what appears to be the common will, however misguided, would not be in the best interests either of the Government or the islanders."[89] Heisch responded and agreed that "it would be bad tactics to try and force matters."[90] But local opinions shifted, elimination techniques changed, new researchers and local leaders emerged to shepherd future public health activities. Ultimately, through a variety of subtle shifts, Pate Islanders agreed to participate in a large-scale LF elimination program that included environmental control of mosquitoes, mass drug administration to islanders, and repeated night blood draws to establish prevalence levels. Such activities moved in fits and starts, but they did lead to a noticeable change in the disease environment. Data from Pate Island indicates LF prevalence of 40 percent in 1920, dropping to 16 percent in the late 1950s (after islanders agreed to mass drug administration with DEC), and falling to under 3 percent in the late 1980s.[91] A final lesson to be remembered, then: even while recognizing the import of past interactions and the residue left behind, these do not preclude minor or massive shifts of opinion and action in the future.

CONSENT OR COERCION?

Historical Narrative

"FORCED TO ACCEPT TRIAL TREATMENT"?

A Tuberculosis Drug Trial, Nairobi, Kenya, 1961

In early March, 1961, Julius Mwangi, a Kikuyu man, arrived at the King George VI hospital in Nairobi, Kenya, seeking treatment for an active case of tuberculosis.[1] Upon examination he was found to have extensive pulmonary tuberculosis and a cavitated right lung. This diagnosis meant Mwangi had a secondary case of tuberculosis, that the lung had already been damaged (cavities were apparent), and that there was a high risk that he would spread the disease to others.[2] Mwangi was admitted for "ordinary treatment" and a week later was transferred to the Infectious Disease Hospital to the care of Sister Margaret Millar.[3] At this moment in political history, Kenya was two and a half years away from gaining independence from Britain, and, more critically, the guerrilla fighting of Mau Mau and the brutal British response had ended less than two years prior.

As Sister Millar remembered it, when she met with Mwangi she first explained to him the "facts regarding the trial" that he would be participating in. He would be required to stay in the hospital for six months, and after being discharged he would need to continue with monthly outpatient treatment for an additional six months. The experiment was a total of twelve months long, half confined in the hospital and half outside.[4] Sister Millar also told Mwangi that, if he participated, she would visit his employer (Shell) to inform them that he was part of the trial. It was implied that, with Sister Millar's help, Shell would continue paying him his salary while he was in the hospital, and that his job would be held for him until he was released—terms that were extremely unusual at this point in time. After this discussion she reported that Mwangi "agreed to enter the trial, and comply with the rules." From that point, Julius Mwangi became a human subject participating in the Medical Research Council UK's Thiacetazone Drug Trial.

The Thiacetazone Drug Trial was one of many TB drug trials being run in East Africa beginning in the 1950s and continuing through the 1970s. The Tuberculosis Research Unit (TRU) was established within the larger Medical Research Council (MRC) by 1960 and the drug therapy trials were incorporated as an East Africa High Commission research scheme. The trials were referred to as the East African/TRU Drug Trials and involved experiments at hospitals throughout the region, including the Infectious Disease Hospitals in Nairobi and Mombasa. Important discoveries were made in East Africa, including establishing that thiacetazone—a much cheaper therapy—could be substituted for para-aminosalicylic acid (PAS).

Julius Mwangi entered the MRC trial in March 1961 and remained in the hospital as a research subject for the following two to three months. At that point, one day in mid-May or mid-June, Sister Millar reported that he left his ward and came to her office within the hospital, very angry. He had received a letter from Shell stating that his salary would be cut in half for the subsequent three months of his hospital stay. Mwangi told her that "since he had allowed himself to be used by the British Government as a guinea pig" she or the MRC needed to supply him with the additional twenty pounds per month to make up the rest of his salary. After this confrontation, Sister Millar recounted that Mwangi "became very difficult," refused over several days to take his drugs, and was "insolent" about the food offered.[5] In the coming days and weeks he refused all experimental drugs and was removed from the trial.

Once removed from the MRC's drug trial, Mwangi remained at the Infectious Disease Hospital and was switched to the standard TB treatment (ordinary treatment involved a three-drug regimen of streptomycin, PAS, and isoniazid). After receiving this standard therapy for somewhere between twelve and thirty days (the accounts conflict), Mwangi was discharged in mid-July. The medical officer in charge of the Infectious Disease Hospital stated that his sputum was tested and he was found to be negative.[6] What the "negative" status refers to is not exactly clear, but it likely meant that a sputum sample had been taken, examined by microscope, and no tuberculosis bacilli were found.[7] In practical terms, this meant that Mwangi should have no longer been contagious. When Mwangi left the hospital he was told that he was "a cured case," but, without documentation provided to him, he was suspicious.

A few months after being discharged, Mwangi returned to the Infectious Disease Hospital to complain. Upon returning to work at Shell at the end of June, he was "sacked immediately." Mwangi was upset and demanded to know from the doctors why he had been fired. The medical officer in charge of the hospital claimed to have been in contact with the Shell Welfare Office and stated that the dismissal had nothing to do with pulmonary TB or the MRC trial. Rather, the patient had reported for duty late, and the company had "no alternative but to dismiss him."[8] There is no documentary record of contact between the hospital workers and Shell.

We know of Julius Mwangi's case only because he penned a note to the director of medical services for the entire Kenya colony after his unsatisfying return visit to the hospital. Mwangi's original letter, and three follow-up notes from Sister Millar, the medical officer in charge at the Infectious Disease Hospital, and the permanent secretary of health for the colony, made it through the vagaries of the colonial recordkeeping system and were ultimately deposited at the Kenya National Archives. Mwangi's case is unusual—bordering on extraordinary—in that the documents exist only because of his English literacy, palpable frustration, and persistence. The fact that multiple individuals narrate the events allows us to discern both the continuities and discrepancies in the accounts. All of the documents are in accord when it comes to describing Mwangi's admission to the Infectious Disease Hospital, his participation in the MRC's tuberculosis drug trial for three months, his withdrawal from that trial, his placement on ordinary treatment for no more than one month, and then his discharge. Beyond these areas of broad agreement, however, there are some significant and worrisome disagreements.

Based on Mwangi's recounting of events, when he first arrived at the Infectious Disease Hospital, Sister Millar told him he would undergo six months of treatment to be fully cured. But, critically, "she did not specify whether it was trial treatment or ordinary" he would be receiving. He agreed to undergo six months of "treatment," and was therefore surprised when, three months into his "treatment," Doctor Malherb and Sister Millar called him into the office and asked "whether I would like to continue with the trial treatment." Mwangi reports that he was unable to answer since "it had not been explained to me clearly which treatment I was to be having."[9]

At this point, Mwangi says that he stopped participating in the experiment. He also felt that, with his decision to quit, the general medical care he received worsened. As he put it, after withdrawing, "things were never the same again . . . they treated me most unkindly." He was discharged after only twelve days of treatment as a "cured case" but without a document showing the results of a culture/sensitivity test. Mwangi felt that this qualified as an "early discharge," and when he left the hospital he "strongly believed" he "might be a danger to my family and public, for such cases before have proved so."[10] There were two ways to establish that Mwangi was "cured" (i.e., noncontagious): by examining his sputum under a microscope and seeing no TB bacilli, or by taking a sample of his sputum and trying to culture it. A microscopic examination is inexpensive and instantaneous—as soon as the slide is read, a confirmation can be made. A culture, on the other hand, is far more precise, but would require the specimen being shipped to another laboratory in Nairobi and the sample being cultured for eight weeks before the final results would be available.[11] It appears that Mwangi was discharged after a microscopic bacterial examination, but without the more sensitive culture test that he desired.

Mwangi's suspicion of being discharged early was not unfounded, and his comment that "such cases before have proved so" was grounded in recent history. In the midst of the Mau Mau Emergency (1952–59), tens of thousands of Kikuyu were placed in British-run prison camps where many were subject to horrific conditions of famine, disease, and even torture. Thousands died as a result of outward violence and officially sanctioned neglect. When TB swept through camps between 1954 and 1956, rather than provide the detainees with standard and appropriate therapy (which would have been time-consuming and expensive and required far better public health services than were in any of the camps), the Kenya Medical Department ignored the problem. Then, as the death toll mounted inside the camps, the Medical Department opted to shift the problem to another locale. A policy was adopted of "repatriating all infectious detainees back to the reserves."[12] Men and women with active cases of TB were released from camps and forcibly returned to the Kikuyu reserves, quickly creating new epidemics. These events occurred in 1956, just five years before Mwangi's arrival at the Nairobi Infectious Disease Hospital. As a Kikuyu man living and working in the city, he had likely heard rumors or seen direct evidence of such cases of gross mistreatment in the realm of public health. Given the large percentage of the total Kikuyu population that was involved in Mau Mau in some way (through fighting, being screened, or detained in military activities such as "Operation Anvil"), it is likely that Mwangi himself had been touched by Mau Mau at some point in the past decade. Even if he had not, he surely had friends and relatives who had been forced out of Nairobi to live in the overcrowded Kikuyu reserves. That an African might receive substandard medical care and be lied to about whether he had been cured of a contagious disease was not at all a far-fetched scenario.

Mwangi's letter to the director of medical services ended with five pointed questions:

> When I returned to my job I was sacked immediately for apparently someone at the Hospital had undermined me by sending bad information. Now, this is worrying me greatly for the doctor had told me I should be eating good food, but how can I get this when out of employment?
>
> Secondly, I would like to know whether a patient is to be forced to accept a trial treatment or to be requested to do so.
>
> Thirdly I would like to know the authority which the Hospital staff have over ones [*sic*] employment.
>
> Fourthly, I would like to know the person who is responsible for paying me compensation for my lost job.
>
> Fifth, I would like to know what arrangements have been made to provide for my family in case I lose my life as a result of this half treatment.[13]

Mwangi's letter is frighteningly direct. Based on how clearly he stated his complaints and questions, it would appear there was no space for the director of medical services to skirt the issues. His letter clearly demanded a response.

But, amazingly, when the permanent secretary of health and social affairs wrote back to Mwangi on behalf of the Kenyan government, no real substantive answers were given. This final note focuses entirely on the loss of Mwangi's job at Shell, stating it was "unconnected with either your chest condition or the particular form of treatment which was given in your case."[14] No evidence was provided, and, while it was the final word, it was unlikely to have satisfied Mwangi. Entirely ignored is the most problematic claim—the accusation that Mwangi never consented to participate in the trial. The permanent secretary did not address the ethical issue of what should or could have occurred and refused to engage with Mwangi's most pointed question: "whether a patient is to be forced to accept a trial treatment or to be requested to do so."[15]

There are a few potential explanations to the knottiest issue of Mwangi's case—whether he knew he was participating in the MRC's TB drug trial, or whether his participation was a result of deception. The first explanation follows closely with what Mwangi lays out in his letter: that he was an entirely unknowing recruit into the MRC's TB drug trial. By this account, no one in the hospital ever informed him that there was a drug trial he was eligible to participate in, there was never a formal request for him to participate, nor any clarification that he was receiving anything other than ordinary treatment. He only discovered his participation months into the experiment, and then demanded to withdraw. As punishment for leaving the trial, he was discharged prematurely, while potentially still contagious, and lost his job. Nowhere in the documents was it claimed that Mwangi consented to participate. The closest anyone gets to making such a statement is when the Infectious Disease Hospital director says, "As far as I know, consents are obtained from the suitable patients before they are put into TRU/MRC [Tuberculosis Research Unit/Medical Research Council] trial."[16]

The second explanation for what Mwangi knew, and how he originally came to be enrolled in the trial, is a bit more complex, but seems more likely. Mwangi may have been initially informed about the trial by Sister Millar and agreed to participate, understanding that he was participating in an experiment. However, three months in, when he received notice that his salary at Shell would be cut by half, his willingness to participate in the experiment dissipated. It is at this moment that Mwangi contended that the British Government was using him like a "guinea pig" and that someone should compensate him for the other half of his lost salary. If Mwangi actually made this statement (only Sister Millar reports the comment), it was the fairness of the transaction that was being called into question. The demand for compensation presupposes that Mwangi had agreed to participate because the terms of the exchange were acceptable. But when his salary was reduced and the terms of the agreement changed, the transaction no longer felt equitable. He demanded that the terms of the agreement be renegotiated. His claim of being used as a "guinea pig" conveyed his outrage at being treated more like a disposable lab animal than as a human being.

Unfortunately, Julius Mwangi's case has no satisfying resolution. Based on the materials available, it is impossible to definitely say whether Mwangi was informed that he had a choice to participate in the experiment, whether he understood he was participating in an experiment, the conditions under which he quit the trial, if his TB was appropriately cared for after his withdrawal from the trial, and whether he was punished in some way for quitting. At the very least, Julius Mwangi was a highly dissatisfied participant in an international medical research trial; at the worst, he was a victim of deception and was enrolled in a drug trial without his knowledge or consent.

Modern Narrative

❖

FOCUSING ON FIELDWORKERS IN KILIFI, KENYA

On the coast of Kenya, a twenty-five-year partnership between the Kenyan Medical Research Institute (KEMRI) and the Wellcome Trust (UK) is uncovering important new findings related to modern Kenyans' understandings of medical research and the process of gaining informed consent in poor, low-literacy areas.[1] This group is part of a multidisciplinary biomedical research center that carries out medical research, but which has also been prodigious in studying the ethical issues surrounding research in resource-poor settings. The KEMRI-Wellcome Trust Research Programme is based in Nairobi and Kilifi District, with a majority of its work in Kilifi, north of Mombasa. It is an area with some of the highest rates of poverty, lowest rates of literacy, and largest gender disparities in all of Kenya. The program is housed at and around the District Hospital in Kilifi, employs nearly eight hundred people, and is internationally recognized for the biomedical and social science research that has been conducted by the group since its establishment in 1989.[2] Within this broad research agenda, a growing set of papers explores the "perceptions, understanding and appropriateness of informed consent processes."[3] In the process of conducting social science research about their own biomedical projects, they discovered that, despite their desire to share relevant information, miscommunications were rife and sometimes significant enough to jeopardize the overall integrity of consent. They also found that one of the most critical components for ensuring high quality consent were fieldworkers who were well trained and felt supported to handle the challenging and sometimes unexpected situations arising from medical research projects.

In a series of papers published over the past decade, the KEMRI-Wellcome team has captured local residents' perceptions of the research institute and general understandings of research. Focus groups held in and around Kilifi uncovered widespread misunderstanding of the biomedically accepted differences between research and therapy, and that the most commonly used Swahili word for "research" (*utafiti*) was "not widely understood to mean research."[4] Particularly striking was how rarely the KEMRI-Wellcome Trust Research Programme's work was described as "research," despite years of staff, fieldworkers, and researchers regularly visiting homes, talking people through consent forms, and explaining their work in public meetings. Many participants interviewed who were enrolled in an epidemiological study mischaracterized the research as "a community wide 'health check.'" In a malaria vaccine trial, more than half of the people interviewed believed the goal of the work was to provide "medical assistance."[5] The levels of confusion were so great that the research team ultimately came to the conclusion that "incomplete levels of understanding, or 'half knowing' are almost an inevitable accompaniment of communication efforts."[6] This is not to imply that some of the causes were not recognized (the complexity of the issues, the inevitably overlapping nature of research and treatment, the need to ensure fair benefits to participants, and linguistic challenges), nor does it imply that nothing could be done.

One aspect of the group's research agenda has focused on oft ignored but vitally important members of the consent process: fieldworkers. These are the East African men and women who I have described in other chapters as middlemen, cultural brokers, or research assistants. Historically, it was these workers who were responsible for much of the concrete work of medical research: explaining projects in local languages, gathering participants, collecting blood and bodily samples, palpating bodies, and handing out pills. There is some irony in how critical this work is but how little attention has been paid to it.[7] The role of fieldworkers is rarely, if ever, mentioned in ethical codes, research protocols, or published papers, and there is little formalization in how these individuals are trained.

The fieldworkers are tasked with a challenging set of responsibilities. They are meant to share information, recruit participants, and, perhaps most importantly, to respect the informed consent process and international ethical guidelines. But introducing another person to a medical research trial and trying to enroll them as a subject while still respecting the process of consent is not without challenges. Some of the fieldworkers employed by the KEMRI team reported a "high degree of stress" in balancing the tensions inherent in their position.[8] On the one hand, the fieldworkers believed that "rapid recruitment of large numbers of participants would reflect positively on their performance."[9] On the other hand, in accord with international ethical guidelines, they should only be enrolling subjects who had given informed, understanding, and voluntary consent. A frequent challenge is how the fieldworkers should respond when a potential subject refuses. As the KEMRI-Wellcome team framed this dilemma, it required "balancing every person's right to refuse to participate with checking that refusals were not being made on the basis of simple misunderstandings that could be easily addressed."[10] Yet a variety of community members believed that refusals to participate in research would result in suboptimal care in the future, and "several even said that they would have been chased away if they had refused." One interviewee bluntly noted that, given people's understanding of research and the perceived consequences of refusal, "It would be a miracle if anyone refused."[11]

There was a clear tension between respecting the right to refuse and gathering adequate numbers of participants. When the KEMRI-Wellcome senior researchers began to observe and speak with the fieldworkers about their work, many described their jobs as "convincing," "persuading," and "converting" parents. One man stated, "It's a matter of making them understand the importance of research. The benefits of research is what we should stress." Another fieldworker described how the work was "a form of educating the mother; making her change from where she is right now (uneducated regarding biomedicine and research) to another advanced stage of knowledge and practice. We cannot just leave her where she is now. . . . We want her to accept, not really forcing her but getting her to another level."[12] Other fieldworkers "argued strongly . . . that while they agreed with the notion of informed consent, many community members [particularly mothers with little formal education] would find it too difficult to understand and accept the information."[13] Some of these fieldworkers went so far as to argue that informed consent couldn't be gained because of the education level of the people they were working with.

There is also concern that fieldworkers themselves may be confused about the differences between research and treatment. The KEMRI-Wellcome team found a handful of cases where fieldworkers "had a low understanding of the nature of health research, often conflating treatment and research activities at the centre. . . . Of particular importance, the wider context of national and international research review processes and the existence of national and internationally agreed research ethics principles were generally unknown."[14] These findings do not preclude the fact that many fieldworkers are conscientious, aware, and respectful of international ethical codes, and are attentive to not being overly persuasive to potential subjects who choose not to enroll. Fieldworkers are critical components and there must be renewed attention to how these cultural brokers and middlemen are engaged in this work of gathering human subjects.

4 ETHICAL RECRUITMENT AND GATHERING HUMAN SUBJECTS

This chapter focuses wholly on consent as we continue our progression through the research encounter. Consent is often singled out as *the* issue on which ethical research hinges, in ways that other issues in medical research are not. In many cases, rather than the process being one of discussion between researcher and subject, it has been reduced to the mere signing of a form. A recent article described how it is "increasingly common to hear researchers describe informed consent as a task to be done, speaking of the need 'to consent' the subject. . . . 'To consent' the subject raises questions about the dynamic of power between researcher and potential subject and challenges our existing principles of autonomy and justice."[1] In order for consent to be valid, it must be informed, understanding, and voluntary. Yet we are increasingly focused on forms and formal statements rather than processes. Instead of the document being a means to an end, the signing of the form is now considered by some researchers to be the end.

This chapter presents a few different arguments. First, it argues that consent practices in the colonial era were characterized by the coercive power of chiefs acting as representatives of the state. Most of the subjects who participated in colonial-era medical research were *not* informed, *not* understanding, and *not* participating voluntarily. Many participants were recruited through practices of group consent, where chiefs agreed on behalf of others; this practice was not rooted in pre-colonial traditions, and was in fact a creation born of the colonial biomedical research enterprise. Second, it documents widespread misunderstandings of what research is, a phenomenon that has been called "therapeutic misconception" in other parts of the world. After introducing the origins of the term and explaining what it describes, I show that therapeutic misconception is an ethical concept with little explanatory power in the region, and actually obscures the therapeutic benefit being provided through participation in research.

Definitions and Requirements

Informed, understanding, and voluntary consent was one of the clearest ethical imperatives to emerge from the 1947 Nuremberg Code. The Code very clearly lays out what is required in order for ethical medical research to occur and, in the first of its ten points, specifically for the requirements of consent to be met. The document states:

> The voluntary consent of the human subject is absolutely essential. This means that the person involved should have legal capacity to give consent; should be so situated as to be able to exercise free power of choice . . . and should have sufficient knowledge and comprehension of the elements of the subject matter involved, as to enable him to make an understanding and enlightened decision. This latter element requires that . . . there should be made known to him the nature, duration, and purpose of the experiment; the method and means by which it is to be conducted; all inconveniences and hazards reasonably to be expected; and the effects upon his health or person, which may possibly come from his participation in the experiment.[2]

Modern discussions about consent are rooted in national laws and may also draw upon international treaties or international customary law. Human rights documents such as the International Covenant on Civil and Political Rights make clear that participation in research without consent is a violation of a person's rights, with the covenant stating, "No one shall be subjected without his free consent to medical or scientific experimentation."[3] In the United States, federal laws governing both research and consent procedures are referred to as the "Common Rule," and are applicable to all research carried out or funded by US agencies. There is a list of nearly two dozen specific criteria that must be included in a consent form. In language that is understandable to the subject, the form must include an explanation of the purpose of the research, a description of reasonably foreseeable risks and benefits, whom to contact for answers about subjects' rights, whom to contact in the event of a research-related injury, a statement that participation is voluntary, an explanation that refusal to participate involves no penalty, and an explanation that the subject may discontinue participation at any time.[4]

An essential element of both the Nuremberg Code and the Common Rule is that research subjects must be volunteers. This means the participant makes a decision to enroll entirely of her own accord, and is also free to leave. However, the meaning of the term "volunteer" has not remained static over time. Within the medical research community, the term was used very liberally in the 1940s and 1950s. In the early 1950s, Hilary Koprowski called the child subjects who received his experimental polio vaccine "volunteers." When it became clear that one participant had to be fed the vaccine through a stomach tube, and others were "feeble minded children," the *Lancet* published a searing editorial: "One of the reasons for the richness of the English language is that the meaning of some words is continually changing. Such a word is 'volunteer.' We may yet read in a scientific journal that an experiment was carried out with twenty volunteer mice, and that twenty other mice volunteered as controls."[5]

While the voluntary nature of some participants was questioned in the United States and Europe, in East Africa there were many examples of African volunteers demonstrating more autonomy than researchers would have liked. In 1955 a few hundred Kenyans arrived at Bondo dispensary to receive experimental outpatient treatment for onchocerciasis (river blindness). However, as the treatment results were disappointing and side effects unpleasant, the original subjects stopped attending—the voluntariness of their participation evidenced by their withdrawal en masse. As the researchers wrote when publishing the results of the failed drug trial, "We have been disappointed. The people of the Nyara valley have been discouraged. A striking testimony to this apparent failure has been the poor attendance of the original volunteers and the complete lack of any further demand for treatment by the people of the valley."[6] The example captures the components associated with a modern volunteer: that participation must be by free choice and without coercion, and that a participant may withdraw at any point and without consequence.

Based on the criteria established in the Nuremberg Code, consent relied not just on a volunteer, but also on that person making an informed and understanding decision. This meant that a researcher had a responsibility to share accurate information and to make sure it was understood by the subject prior to enrollment. The Nuremberg Code, while well known internationally, was not a document that formally governed medical research—it was not national law, nor was it the official policy at the time of any large international medical research organization. That changed in the early 1960s, when the British Medical Research Council (MRC) formalized its own policies governing research on human subjects and, in the process, clarified who could be considered a volunteer. The group's 1962–63 Annual Report included a multipage statement titled "Responsibility in Investigations on Human Subjects."[7] The statement made clear that in cases of research unlikely to result in direct benefit to the participant, "he must volunteer in the full sense of the word." They noted that "true consent" meant "consent freely given with proper understanding of the nature and consequences of what is proposed. Assumed consent or consent obtained by undue influence is valueless. . . . After adequate explanation, the consent of an adult of sound mind and understanding can be relied upon to be true consent."[8] The definition harkens back to the Nuremberg Code's emphasis on informed, understanding, and voluntary consent, and recognizes that the consent must be gained from an adult of sound mind. The MRC also wanted a signature to document consent and to have it witnessed by another person. However, they also show respect for the idea of consent as a process, reminding scientists that "written consent unaccompanied by other evidence that an explanation has been given, understood, and accepted, is of little value."[9]

The push to clarify who could be considered a volunteer in medical research was followed by more clearly defining another category of subjects: the vulnerable. While there was no discussion of this category as a distinct group until the 1970s, the bioethicist Udo Schuklenk believes that research ethics are "essentially about ways to ensure that vulnerable people are protected from exploitation and other forms of harm."[10] A well-accepted modern definition of vulnerability is "to face a significant probability of incurring an identifiable harm while substantially lacking ability and/or means to protect oneself."[11] Originally, in the US Belmont Report (1979), the term referred to a limited group of people who were physically vulnerable (pregnant women, children) or vulnerable due to their position in society (prisoners). The term has since taken on a much broader meaning, but not without significant debate. Those who advocate for a broad definition have argued that "citizens of developing countries are often in vulnerable situations because of their lack of political power, lack of education, unfamiliarity with medical interventions, extreme poverty, or dire need for health care and nutrition."[12] But with the label of "vulnerable" comes a heavy dose of paternalism through the creation of additional protections. Should these groups be the recipients (or victims) of paternalism? Are minorities, or the poor, or those who live in the global south actually more susceptible to coercion? Asked in a more pragmatic way, are there disadvantages for those groups who are considered vulnerable? The most concrete outcome of being labeled as such is that, with additional protections, less research is done. When certain groups of people are systematically ignored or excluded from medical research, diseases or health conditions that exist within those groups will also be ignored. In East Africa, many diseases are particularly damaging to pregnant women and children, such as malaria and bilharzia. When too many barriers are placed on working with vulnerable populations it may result in the further marginalization of particular health problems and populations.

Language of Consent Forms

Of the many challenges involved in ensuring informed consent, the actual language being used in forms is one often overlooked component. Consent forms are generally written and scrutinized by researchers in English or another European language, and only later are they translated into a local language. A careful translation must pay attention not only to accuracy in language, but make sure the level of detail and word choice matches local literacy levels and maximizes overall comprehension. Independent back translation, which entails taking a document that has been translated into a foreign language and then translating it back into the original language, is considered the gold standard. When read together, the three documents (the first, for example, written in English; the second translated into Swahili; and the third being the Swahili document translated back into English by a different translator) can highlight inconsistencies or areas of confusion. As Caroline Kithinji and Nancy Kass found in their review of ten different Swahili consent forms that had been used in medical research in East Africa, there was a wide range of practices when it came to translation, and no formalized rules about who was responsible for this task or how it must be done. In some cases, university lecturers and secondary school teachers translated the forms; of the ten forms reviewed, only one had been back translated.[13] Most importantly, the authors found that "a readable English-language consent form does not necessarily result in a readable form once translated in Kiswahili."[14] Thus, it is not enough to take a clear and understandable form in English and translate it word for word into Swahili.

A good translation must be able to introduce and explain an entirely new concept, and have it be well understood among the population where the research will be carried out. In the case of East Africa, this means accommodating not only varying literacy rates, but also remaining attentive to the fact that many people may not agree with biomedical conceptions of how the body functions. One must also recognize distinctions between formal vocabulary and technical terms that may not be understood. I discovered in my own work that there could sometimes be a gap between grammatically correct terms and those that were well understood. When trying to find a Swahili phrase for "medical ethics" in my own consent form, I worked with an expert Swahili speaker in Zanzibar, and he suggested the phrase *madili ya madawa,* which literally translates as follows:

madili	*ya*	*madawa*
ethics	of	medicines

My first field research trip was to Mwanza in western Tanzania, and when I used the phrase *madili ya madawa,* the young man I was speaking with looked at me strangely. When I used it again, he interrupted and asked, "What do you mean, 'Ma-deali ya madawa'?" To his ears, being someone for whom Swahili was a third language and English a fourth, the formal Swahili word for "ethics," *madili,* had been turned into a modified form of Swahili-English slang. He heard the English word "deal" being made plural by adding *ma-* at the start. He thought I was talking about a form of covert medical deals. Although my phrase was grammatically correct, it was not a functional or clear term for the people I would be interviewing, in Mwanza or in other parts of East Africa.

Other concepts routinely presented in consent forms also offer challenges in terms of translation and subjects' likely familiarity. As was discussed in the introduction, biomedical research involving large groups of people, multiple "arms" (treatment and control), and a process of random allocation, was not something that was present in East Africa prior to colonial contact. Thus, it is fair to say that while these concepts of randomization and confidentiality may have existed previously, it's only with biomedicine that they have been applied in the realm of medical research. Randomization must convey a series of potential outcomes without necessarily implying luck, or that one outcome is better than another. Yet parents believed that the decision as to which children received the experimental vaccine and which the control was reached "by means of luck, *pata potea* [get one, lose one] or *bahati nasibu* [luck/chance]."[15] This is inaccurate, since "luck" should not be understood as part of randomization. Based on such confusions, later efforts to describe randomization used the lengthy but accurate phrase "a system such that everyone has the same chance of being included in the study, without favouritism."[16] Initially, to describe confidentiality, the Swahili word *siri* (secret) was used. However, this term has negative connotations as it is typically used to refer to information that may be shameful. In this case, another phrase was substituted that more accurately described who would have access to sensitive information: "a limited number of people closely concerned with the research."[17] In 2010, a researcher attached to the KEMRI-Wellcome Trust Research Programme conducted twenty-five in-depth interviews with parents who had been asked to enroll their children in the RTS,S malaria vaccine trial (discussed in the chapter 6 narrative). The interviews highlighted the difficulty of defining even supposedly more straightforward concepts such as "compensation." The RTS,S information form used the term *fidia* to refer to compensation in case of side effects or injury. When a parent reviewed this part of the form, he asked: "So a child who has already died, what type of fidia [compensation] will you give me . . . a child like mine or what?"[18]

Given the large amount of confusion that exists—both on a global level and also as well documented in East Africa—about what constitutes research, approaches other than mechanical translation from English into Swahili are likely to yield more understandable consent forms. Studies in Kilifi showed that frequently used terms such as *utafiti* (research) and *uchunguzi* (investigation) were often not well understood by residents, and that this fact was often unknown or overlooked by investigators.[19] As an alternate approach to generating consent forms, the KEMRI-Wellcome group chose to work with native speakers and a professional translator who understood the concepts conveyed in a consent form (such as risk, benefit, compensation, and randomization), and had them describe the concepts directly in Swahili, without first producing a document in English. They found that the newly created forms were far more accurate and better understood than prior translations.[20]

Therapeutic Misconception

One consequence of this misunderstanding about the nature of research is therapeutic misconception—the situation where a subject believes that all aspects of research are designed to benefit her directly. This is a fundamental misunderstanding, since the primary goal of research is to gain generalizable knowledge, not to benefit individual subjects. Appelbaum and colleagues first addressed this phenomenon with evidence from the United States in the 1980s. They described the characteristics of therapeutic misconception thus: "Subjects appear frequently to overestimate the likely benefits of entry into research studies . . . to underestimate risks . . . to be confused about the nature of randomized assignment . . . and generally to conflate research with ordinary treatment."[21] Although research about therapeutic misconception in other parts of the world is still sparse, it is a global phenomenon and is present in East Africa.[22] It is important to note that therapeutic misconception is not necessarily the result of deception on the part of the researcher. Even when information is accurately presented, subjects may distort or ignore information that contradicts their own expectations that the research will be therapeutic.

While deception is not the sole cause of therapeutic misconception, the behaviors of researchers have often furthered its presence. During the colonial era British researchers regularly described their research activities as providing "medicine" or "treatment," or as a public health intervention. Their decision to avoid words that would have indicated a research activity was a pragmatic one meant to make the recruitment of participants, and their own work, easier. In a case in 1950, a researcher had a chief announce that all villagers should come out to receive a free inoculation against elephantiasis and hydrocele (symptoms of lymphatic filariasis). There was no such inoculation, but the research team wanted a large turnout in order to collect blood samples, especially from young children. Getting parents to voluntarily bring children for blood sampling would have been nearly impossible, hence the need for deception. As the researcher joyfully recounted, by pretending to offer free treatment, "we had all the infants in the village!"[23] In this case, convincing people to line up and receive a free treatment was much easier than convincing people to participate in an experiment with risks, or to waste any time or energy participating in something without direct benefits. It is unclear in this case whether participants were given an injection that was falsely claimed to be an effective inoculation, or whether the deceit ended once people arrived and there was no inoculation, only a blood draw.

Another clear example of deceit comes from researchers working as part of the East African Medical Survey. As they described their process of arriving in a Tanganyikan village in the 1950s and sharing information, "News was spread, through the chiefs, that a group of doctors, with a supply of medicines, was working at the dispensary, and that they were prepared to see and treat any sick people who came. Any mention of 'investigation' or 'blood samples' was carefully avoided."[24] In contrast to this outright intent to deceive, there were cases of well-intentioned researchers refusing to do experimental work without giving something tangible in return. The medical doctor-turned-researcher Hope Trant distributed proven treatment for leprosy on Ukara Island in the 1950s, while also handing out experimental drugs for lymphatic filariasis. This led some residents to insist sixty years later that Trant was a doctor and that all of the pills she gave out were effective medicine. These "benefits" were small and were often provided unofficially, but they had the unintended effect of solidifying the confusion between treatment and experimental medicine, between doctoring and researching.

Critiquing Therapeutic Misconception

While therapeutic misconception in East Africa was (and remains) widespread, there is a key element that makes this label problematic: there is a very real "therapeutic" dimension to many East Africans' participation in a medical experiment. The lack of a functioning health care system, the poverty, the high rates of preventable yet deadly diseases all mean that the meager benefits given by Western researchers to offset the risks of participation in research add up to a very real therapeutic benefit. Angeliki Kerasidou from the Ethox Center, a bioethics research group based at the University of Oxford, described therapeutic misconception in developing countries in the following way:

> In resource poor countries we often observe the following paradox. People who participate in biomedical research end up receiving better health care than 'mere' patients do. Research participants will be seen by a doctor more often than a patient in a poor hospital, they will have tests done to them that they possibly could not afford if they had not enrolled in a study, and they possibly have their minor illnesses treated. Even if the main health issue that led them to be enrolled in a study is not addressed, they will often be of better health by the time they leave the study. Therefore, one could argue that therapeutic misconception is not actually a 'misconception' when it comes to resource poor settings such as these.[25]

There is clear modern evidence indicating that East African subjects are choosing to sign up for medical research projects precisely because of these tangible benefits. Interviews conducted by the KEMRI-Wellcome Programme with parents on coastal Kenya made clear why people volunteered. One participant who had agreed to enroll his child stated, "What attracted us [was that] we knew our children will receive treatment for a whole year in every disease they suffer. If you have a problem and visit the people concerned, a call is made to the [principal investigator] he brings a vehicle and [the sick person] is carried away [to hospital]. In fact it's something we should be happy about because nobody can bring you a vehicle that easily."[26] The KEMRI-Wellcome research group has also recognized that its general and study-specific activities "offer very real clinical benefits for many," and agrees that the concept of therapeutic misconception is inappropriate for describing the conditions existing in East Africa today.[27]

The Researchers' Secret (Siri ya Watafiti)

Coupled with therapeutic misconception, there appears to be another form of deception occurring in some places in the region. It is a new type of misleading behavior that mimics the actions of colonial-era researchers but goes by a different name. During interviews in Tanzania in 2008, multiple researchers mentioned the *siri ya watifiti*—"researcher's secret" (or, "secret of researchers").[28] One man who had worked as a medical researcher for decades spelled it out: the secret was that they were conducting research. Thus, a smart researcher never tells villagers that his work is testing new drugs, because then people would believe they're only being used as "guinea pigs." If they knew they were being asked to take untested medicine, they would surely *kataa* (disagree, refuse to participate). The truth of the encounter and the risks associated with it remain a secret. Another exchange with a female nurse with decades of experience touched on the same ideas:

> MG: When you are testing a medicine, will you tell villagers, "We are checking to see if this medicine works?"
>
> Nurse: No. This is just our secret. If you told the villagers, they would not understand you.

In another interview, a male mid-level medical worker explained his success in signing up many people for projects he was in charge of.

> MG: What would you say to villagers if you have to go into a village and test if a drug works?
>
> Male Researcher: If you need to test a drug, you don't tell them this. This is a secret of the researchers [siri ya watafiti]. . . .
>
> MG: And if the medicine doesn't work, and people are vomiting and have diarrhea—what will you say?
>
> Male Researcher: You will say it's just bad luck.

These occurrences of siri ya watafiti are cause for pause. If this siri ya watafiti permeates even just a small number of research encounters, one wonders how there could be anything but widespread therapeutic misconception. Some unscrupulous researchers are actively cultivating confusion between experimental and proven therapies, since this makes their job easier in terms of recruitment. I believe that this practice is rooted in an intent to deceive, where the work of recruiting subjects is done more easily when only partial information is shared. Although it is possible that there is no bad intent, merely confusion, that seems less likely since it is termed as a siri (secret). The KEMRI-Wellcome team reports fieldworkers who also shared inaccurate information with potential subjects. However, in this case, the fieldworkers did this even when being observed by other researchers, indicating that they did not believe they were doing anything wrong, and also did not refer to their behaviors as a "secret."

Dismissing Group Consent, Reconsidering Chiefs

In past discussions about consent in Africa much has been made about the so-called practice of "group consent." A hallmark of many of these references is the lack of clarity and specificity in what is being referred to. In general, we may surmise that this supposedly traditional African practice of group consent could describe one of two things. On the one hand, the whole community could come together and decide collectively that they will participate in a project; the chief then presents this decision, as the head of the village. On the other hand, the chief has the traditional authority to consent on behalf of his subjects, regardless of their individual opinions. The first of these scenarios qualifies as voluntary—a form of collective decision making; the second does not. There is a bioethics literature that at one point suggested that individual consent could not be gained in Africa because the concept was too foreign, that individual consent would be an aberration from more communal traditions. Much of that literature about "group consent" as a traditional African practice was written decades ago and little—if any—of it was grounded in anthropological or historical data from the continent.[29] Thankfully, these claims of group consent and an African past where chiefs were making decisions on behalf of individual citizens seem to be quieting. It's become more widely understood that such claims are false. In this section I will show that the historical cases when "group consent" did occur were a result of colonial contact and were an aberration from earlier forms of chiefly authority.

Prior to colonization, within the realm of health and disease, chiefs were responsible for ensuring the overall health of a village and had authority to make sweeping decisions during times of crisis, such as epidemics of smallpox or sleeping sickness.[30] In cases of sleeping sickness epidemics, chiefs could quarantine houses with sick people, ban sick people to the outskirts of the village, or demand that the entire village be moved to a safer area.[31] In none of these examples did chiefly authority continue into the realm of dictating the care of an individual. Even shortly after European arrival in East Africa, a chief giving consent on behalf of his subjects was not the norm. In 1908, for instance, the medical missionary Albert Cook sat down at his desk in Uganda and wrote a letter to the *Lancet*. He was responding to an article in which the writer claimed it would be easier to do medical research in East Africa than in India, since doctors could rely on "complete control over the patients . . . [due to] the influence of the chief over his people."[32] In his response, Cook noted that a chief's influence "is rapidly dying out and I fear that the hold of the chiefs over their people in such a matter as periodical injections for syphilis would be extremely small."[33] The letter makes it clear that, less than a decade after substantial European contact in the area, the "hold" of the chief was perceived to be small. More likely, such an authority never existed.

Although it wasn't their duty historically, with the escalation of colonial medical research projects and the concurrent demand for human subjects and bodily samples, chiefs stepped into new roles. As a traditional authority who had been absorbed into the British governance structure through the system of indirect rule, a chief was expected to explain the research to his people, "overcome suspicion," and, most importantly, "obtain their cooperation."[34] In the context of colonial medical research projects, that meant making sure villagers were willing to participate—and often meant meeting a daily or weekly quota of human subjects. How chiefs accomplished these tasks of assuring participation and recruiting human subjects was an area where British researchers chose to maintain a cultivated ignorance. The British evidenced no preference for honey or vinegar, but they were partial to good results. In a case discussed in the chapter 5 narrative, the researcher Hope Trant was confronted with the fact that the sub-chief who was helping her recruit subjects (and had a clear quota of twenty-five to thirty people to deliver to her each day) was prepared to fine residents who refused to participate in the project. She admonished him, "fining is out of the question" even though in reality she had little authority to complain or change his behavior.[35] Archival and oral evidence shows that threatening punishment was one way a chief could "encourage" villagers to participate.[36] In some places in western Tanzania, residents recalled that chiefs would forcibly collect people from their fields and threaten physical beatings or banishment from the village during their work of recruiting people.[37] This reliance on harsh tactics and outright coercion shouldn't be too surprising. Many of the duties assigned by the colonial governments, such as collection of taxes, assistance with census, and labor requirements, were also not easy tasks to complete.[38] When chiefs were tasked with recruiting people for a medical experiment, they relied on the same techniques used to gather laborers or collect taxes: force.

Oral histories of retired Kenyans who worked for the Department of Insect Borne Diseases (DIBD) also recount cases of coercion, and the threat of force seemed to be common and not viewed as especially problematic. The DIBD was a unit within the Kenya Medical Department and carried out medical research on a number of different insect-borne diseases such as malaria, river blindness, and sleeping sickness. The Kenyan men being interviewed worked as fieldworkers during the end of the colonial era and in the decades after independence in 1963. One man recounted that, as a field researcher, it was his responsibility "to convince people." It was the chief's responsibility to explain the scientific work to people, and, if he did his job well, people would "just come." When there was "unity between the chief, the [health] workers and people," then work would progress well. However, in moments of discord, government force could be deployed through the chief. The fieldworker remembered that, if people resisted, they soon became more accepting "after we had taken some of them to the police and they were arrested and given some canes [beaten]."[39]

The irony of these cases is that the "traditional" authority expressed by the chiefs in recruiting, and forcing, villagers to participate in these medical research trials was a colonial fabrication. There was no history of a leader consenting on behalf of a whole group of people about whether or not they should receive medical care, let alone experimental medical treatment. And there certainly was no precedent for a chief deciding that villagers under his control should all provide sensitive bodily samples, like blood, to a group of foreigners. Traditional uses of blood outside the body were few, fraught, and carefully controlled in ritual exchanges meant to highlight trust and the establishment of new relationships. A decision to share blood was one made by individuals—not commanded by an authority figure. The "traditional" authority of group consent was a new authority, a product of British colonialism and the medical research enterprise. As the demand for African bodies grew, there were plenty of enterprising local leaders willing to accept this extension of chiefly authority into a new realm. These men gathered people for researchers to use, kept villagers moderately docile by threatening fines or punishment, and put researchers' minds at ease by invoking language describing "traditional" practices. The only problem with justifying such practices as traditional is that traditions change; they are mutable and malleable, and can be invented as convenient.[40]

If we are willing to accept that the "traditional" practice of group consent where a chief consented on behalf of others was actually a manufactured product of the colonial era, we may move to a more productive line of inquiry that has modern implications. Is there a space for chiefs or other community leaders to be involved in the consent process today in meaningful ways? Evidence indicates that many East Africans expect chiefs to be involved in sharing general information about medical experiments, answering questions, and providing advice about whether a particular project is in the best interest of the community. Interviews with community members on the Kenyan coast found that chiefs were perceived "as essential gatekeepers for community activities, but not necessarily as their representatives."[41] This is an accurate assessment, since in Kenya chiefs are not elected representatives, but appointed ones.[42]

Discussions on the coast also made clear that modern East Africans are comfortable with the notion of individual informed consent. During focus groups held by the KEMRI-Wellcome group, one woman pointedly told the interviewer, "[the researchers] have to ask permission from *me* before they do anything or we'll quarrel. . . . If *I* agree they can go ahead, but if they do it without asking *me* then they're in the wrong."[43] Another woman, when asked whether the chief could give permission for all people in the village, gave the following response: "It's important for the fieldworkers to get permission from the chief to move around the area, but the chief cannot decide for my child. No way!"[44] Another person described how "a community elder can organise a meeting where the parents will meet and discuss, but that elder cannot make any decision on our behalf. Elders or chiefs can agree to a piece of work but I can still disagree when you come to my home."[45] In all of these examples, it is notable how frequently the speakers refer to their own rights and decision-making capacities. There was widespread agreement among community members that chiefs could give permission for research to be carried out in an area, but that they did not have the authority to decide for specific households or individuals. Interestingly, when the KEMRI-Wellcome team then interviewed chiefs and community leaders about their own sense of their responsibilities in relation to medical research, they unanimously described themselves as providers of information. Not a single person claimed to have the authority to make decisions on behalf of another person. They considered themselves as respected individuals who could help with education and address common concerns. One leader described how it was appropriate to be involved in education, but "it would not be good for me to talk on their behalf; I don't want to act on their behalf. It's up to them to decide . . . we as leaders can tarmac the road so that you as the vehicle can do your work."[46]

As was true in the colonial era with the cases of coercion recounted earlier, there remain problems with utilizing authority figures in the process of recruiting participants and gaining consent. The KEMRI-Wellcome group noted "many positive consequences" of integrating chiefs, community health workers, and other informal local leaders into the consent process. However, this did not mean there were no problems. In one specific case it was reported

> that one of the chiefs—in his capacity as an administrator—was taking it upon himself to organize meetings about the trial, and to put significant pressure on parents with eligible children to enroll their children. In a famine prone area he had reportedly threatened to remove tickets for free food rations from eligible families who did not enroll. For this study, these threats were reported with significant

> laughter by community members, and efforts were re-doubled by the research team to emphasize the voluntary nature of trial participation.[47]

The fact that the case was publicly recounted, and that it was followed by laughter, seems to indicate that the community members were not especially intimidated by this threat, nor were they fearful of reprisals. There was no evidence that anyone who reported the chief's behavior actually felt compelled to participate. However, what counts as "compelling" will vary from person to person. Just because no one who reported the incident took it seriously does not mean that someone in a vulnerable or marginalized position may not have considered the threat legitimate. The case led the KEMRI-Wellcome researchers to the very logical and thoughtful conclusion that the incident "highlighted that engagement with community members always involves engagement with existing social relations and hierarchies, and that this can have perverse consequences."[48] Community leaders may not be easy to control or as easy to utilize as medical researchers would like. In their zeal to help with a project or accomplish a goal they may be jeopardizing the ethical principles they were enlisted to help achieve. There is no doubt that there is a role for community leaders to play when it comes to sharing information about medical experiments, answering questions, and giving permission to enter an area. However, no one should be under the illusion that anything other than individual informed consent is appropriate when it comes to enrolling subjects into a trial, or that chiefs have ever had the authority to consent on behalf of others.

There is a long list of recommendations about how consent practices may be changed to better ensure that all participants are informed and understanding, and to try to ensure the voluntariness of all subjects. These recommendations have come out of two sets of research. One set is specific to East Africa, largely written about and tested by the KEMRI-Wellcome group. Another set of recommendations have been made about how to do research in economically poor settings, in places with low or partial literacy, or where subjects may have little familiarity with biomedicine. There is largely agreement between the two literatures. Both stress the recognition of consent as a process rather than a form, less reliance on a written form and greater use of verbal presentations and visual displays, the assessing of a participant's understanding through a quiz or interview process, and the use of community leaders in education activities.[49] More specifically, being sensitive to low literacy rates may mean documenting consent through audio or videotape rather than relying on a signature.[50] Recognizing the many foreign concepts that must be captured in consent forms indicates that rigid translation from English into a local language likely will not result in the clearest document, and that efforts should be made to describe the concept directly in the local language. When translations are done, they should adhere to best practices of independent back translation, and ideally be done as a group as part of training of those who will administer it. To further check the level of understanding, potential subjects could be given a short quiz focusing on key elements such as the study purpose, risks and benefits, and their ability to refuse to participate or withdraw without consequence.[51] In general, by slowing the consent process down into a multi-day process with a series of shorter meetings, subjects will be able to more carefully weigh information, discuss with family members, and ask relevant questions.[52] Within these meetings, preference should be given toward using visual aids and delivering information verbally.

The ongoing work done by the Kilifi group has led to the creation of freely available templates that take into account the locally relevant conditions of the Kenyan coast and which are likely to be useful and appropriate across East Africa.[53] Because of the confusion around what medical research is, their consent process begins with a general explanation of what research is and how it differs from treatment. Background information is then given about the KEMRI-Wellcome group before presenting information specific to the disease being studied and the details of the experiment. The group also includes a statement on whether it's likely that the research or results will have a direct impact on the participant's health.[54] Due to the sensitivity around blood, detailed information is also given about whether the research will collect blood, and, if so, how it will be collected, how much will be taken, how it will be used, and how long it will be stored.

Even by making these changes to the consent process, a set of complicated ethical questions remain unanswered about how to protect the quality of consent in East Africa while still providing real benefit. This chapter has made clear that there are two problems associated with research practices that provide therapeutic benefits to research subjects. One of those challenges is that, as international codes have formalized the need to deliver benefit as part of medical research, these small amounts of benefit (such as free treatment or time with a doctor) contribute to therapeutic misconception. Researchers themselves switch between wearing the hats of a scientist and of a doctor, experimenting on and observing subjects, then turning around to treat the same person as a patient. Participants in medical research are increasingly confused as to what is being provided, by whom, and why. The similarities and overt overlaps between research and provision of treatment make it difficult to determine who researchers really are. The KEMRI-Wellcome researchers are fully aware of this dilemma, as they aim to do high-quality international medical research while also delivering real benefit to the communities where they have worked for the past two and a half decades. As they frame the dilemma, it is a case where "meeting one ethical requirement (for example ensuring that potential participants are given basic health care) can compromise another (for example ensuring that potential participants can distinguish clinical research from practice and thereby make an informed decision about involvement)."[55] There is no easy solution, and in the long run it may mean deciding which ethical principle must be prioritized. The second problem with providing therapeutic benefit with research when medical systems are dysfunctional is that these small benefits may quickly become coercive. In this situation, coercion does not refer to the threat of force by a chief or government authority, but to an offer that is too good to turn down, and thus inhibits truly voluntary participation. The benefits provided for participants in East Africa are small—typically only a few dollars to cover transportation costs, access to basic treatments, and time with a doctor. However, if the material conditions of East Africans continue to deteriorate—with increased disease burdens, growing levels of poverty, and even less functional government health systems—ensuring informed, understanding, and voluntary consent in the region will become even more difficult.

BALANCING RISKS AND BENEFITS

Historical Narrative

HOPE TRANT AND A COMPOUND ON FIRE IN TANGANYIKA, 1954

In April 1954, Dr. Hope Trant was employed by the East African Medical Survey to collect thousands of samples of blood, urine, and stool from residents in the districts of Kibondo and Kasulu in northwestern Tanganyika. The samples would be tested for anemia, malaria, bilharzia, intestinal worms, syphilis, yaws, and a number of other parasites. In Kasulu, they hoped to conduct four thousand physical exams, collect two thousand maternity histories, test a thousand children for tuberculosis, and conduct a dietary survey of four thousand community members.[1] The EAMS's scientists believed the massive number of samples—Kasulu and Kibondo formed only one of six sites across East Africa—would allow them to make scientifically informed recommendations about future public health and medical campaigns. They described the goal of their massive, multicountry surveying scheme as discovering "what actually are the plagues affecting the African."[2]

Hope Trant was a field researcher for the survey, and had worked as both a medical doctor and researcher across eastern and southern Africa for nearly three decades. In her prior assignments she was known for her pragmatism and strong will. After disagreeing with her superior just weeks after being hired at the survey, her boss Colonel Laurie pointed out to her, "It is simply not done . . . for a junior research officer to set herself up against a senior."[3] Her stubbornness was tolerated in part because she was better at managing community relations than many of her peers. She regularly hired the chiefs' sons, made social visits, and spent plenty of time treating sick villagers and handing out medicines.

Although Trant was a skilled researcher, like all researchers she was dependent on help from local authorities, both British and African. In the case of the survey in Kasulu District, she was given permission by the British district commissioner, the Chief *Mwami* Theresa Ntare, and assisted by

Mwami Theresa's husband, the subchief—who was responsible for delivering twenty-five to thirty people to her each day for examination.[4] Trant considered him "quite good" at bringing her a steady stream of subjects from the village of Heru Juu, but the subchief eventually ran into problems. By late July—approximately four months after arriving in the village, and after examining approximately seven hundred people—rumors began to swirl that the researchers were sucking blood. Participation dropped off and the subchief was unable to convince people to participate.

When Trant went on a short vacation and another medical researcher, Doctor Preedy, took over, conditions worsened. The project ground to a halt as the subchief was unable to recruit any participants and community members refused to volunteer. Preedy summoned the chief to discuss the situation. Trant reports their conversation in her memoir, having received the news from Preedy:

> [Mwami Teresa] was a rather haughty lady and when we asked for her help to get the survey done, she answered that her people did not want to be examined—if we really wanted them, we should pay them. Dr. Preedy explained that they would have liked to do so, but that no money was allocated for that. She then asked why blood was being taken, and he told her that it was to find out what diseases might be present among her people so that we could recommend treatment. She wanted to know whether we had any medicine. When we had to answer no, that we had none, she said that it was customary to give as well as to receive favours, and concluded the interview by saying that she could not insist that her people should come to us for nothing in return.[5]

The conversation with the chief revealed that she was perfectly clear about the source of the problem: the researchers were content to take and give nothing in return. She made clear that no one *wanted* to be examined and that they would need to be compensated—paid—if the researchers expected residents to continue participating. Mwami Theresa frankly framed the research encounter as one of exchange, which people only participated in if the price was right. Her next question was about blood. Although the researchers were collecting stool, urine, and blood samples along with maternity histories, it was the blood that people were most concerned about. Her direct question was answered indirectly. The blood would allow the researchers to see what diseases "might" be present so that the researchers could then "recommend" treatment. The vague answer about possible diseases and recommended treatment left a slew of unanswered questions in its wake. What if treatment was recommended? Where would the drugs come from? And how would local people benefit if they were forced to give blood yet weren't sick, or discovered they were sick yet had no access to drugs? The chief responded to Preedy's long-winded answer with another cutting question: Do you have any medicine? The answer was "no," and she pointed out that there was nothing customary, appropriate, polite, or desirable about always asking "favors" of people—as the researchers did with regularity by asking prying questions, taking sensitive substances, peering inside homes, and examining small children. These "favors" had a value, and needed to be reimbursed. Her conclusion was just as stark as her questioning: there was no way she could require her people to participate—they were getting nothing in return. And, the researchers shouldn't expect her to use her own political capital to force people to participate when it was clear it would only generate ill will.

A few days after the conversation with the chief, an angry group of villagers arrived at the researchers' compound, waving thick sticks and shouting *wazungu!* (white people) and *damu!* (blood). The group eventually moved on, but that same night the examination enclosure was set on fire. The district commissioner arrived in a "great state" and accused the researchers of "upsetting his people." He ordered them to stop the project immediately and move to his compound, believing their "lives might be in danger."[6] It appeared that opinions had shifted dramatically in only a few weeks, from local support—or at least grudging participation—to angry, public rioting that ended with the destruction of the researchers' compound and their hasty evacuation. In reality, it is hard to know how much the public's opinion of the project had actually changed, rather than merely being given sanction by the chief to be honestly expressed. Without records from participants, it is impossible to know exactly why people participated in the first months of the project without (noticeable) complaint, and then why suddenly there were widespread refusals. What is clear is that there was a close relationship between chiefly pronouncement and public behavior.

Trant was shaken but determined to complete the project. She addressed the rumors of bloodsucking via a public presentation where she could "explain the objects of the survey" and show local people that "nothing harmful was intended." She brought the researchers' equipment to the courtroom for a demonstration and provided a long explanation about how illnesses were discovered with blood samples and how results of the survey would benefit local people. She argued that the researchers would report the findings about disease to the health officers, who "can take steps to improve conditions."[7] It was an anticlimactic conclusion. After months of irritating and invasive procedures, it was not at all clear there were any real benefits to participating. Local residents did not find her explanations persuasive. Before the meeting even concluded, residents directly accused the researchers of sucking blood. As the villagers filed out, the British district commissioner announced that the project could not continue: the "bloodsucking" upset his people too much.

MAP 5.1. Western Tanzania/Lake Victoria region. Map by Chris Becker.

The East African Medical Survey's research in Heru Juu ended prematurely, hampered by its inflexibility to adapt to conditions on the ground and inability to prove its worth to local people or local government officials. The archival record is unclear, but if the surveying in Kasulu District was ever completed, it was on a smaller, quieter scale, sometime in early 1955.[8] The failures in Heru Juu were emblematic of the larger inability of the East African Medical Survey to provide benefits that were valued by local residents. It is very possible that after decades of frustrating interactions with the government, residents in Heru Juu were skeptical of promises to be fulfilled in the future. While the researchers in 1954 may have believed benefits would accrue to residents once samples were taken and analyzed, reports written, and government officials informed, local people were less certain. A benefit needed to be tangible and appear quickly; there was too much uncertainty to trust in anything else.

Modern Narrative

❖

A MALE CIRCUMCISION TRIAL CANCELED IN RAKAI, UGANDA, 2005

In Rakai District in southern Uganda the impact of AIDS over the past three decades has been undeniable. This was where Ugandan researchers first began to study HIV/AIDS within their country, and the subsequent epidemic has led to the deaths of tens of thousands. The deaths have had significant economic and social repercussions such as an increase in single-headed households, a growing number of orphans, and shortages of educated workers.[1] In the early 1990s, nearly 20 percent of adults in Rakai were HIV positive. In contrast to the United States and Europe, where antiretroviral therapy was prescribed, there was no treatment available in East Africa.[2] Even prevention strategies were thin, with the only options being an "A, B, C" approach emphasizing abstinence, being faithful (monogamy), and the use of condoms. Such conditions left the African continent bearing the heaviest AIDS burden in the world. In 2007, 67 percent of the 33 million HIV-positive people in the world lived on the African continent, and 75 percent of total global AIDS deaths occurred there.[3] In the midst of these dire conditions of the late 1990s and early 2000s, there was a great need to identify new methods of prevention.

In the early 2000s, medical research trials were testing whether there were protective aspects of male circumcision. Trials in South Africa and Kenya indicated circumcision could reduce the acquisition of HIV infections by 53–61 percent.[4] Observational studies had also found that circumcised men were two to three times less likely to be infected with HIV than uncircumcised men, but it was unclear whether that was the result of circumcised men engaging in fewer risky behaviors.[5] A large-scale experiment was planned in Rakai, Uganda, to take uncircumcised, HIV-negative, adult men and circumcise them to see if this resulted in lower rates of HIV acquisition. The research was a partnership between the Rakai Health Sciences

Program (RHSP) and Johns Hopkins University with funding from the US National Institutes of Health and the Ugandan Ministry of Health. The protocol for the experiment was reviewed by multiple institutional review boards (IRBs) in the United States and Uganda, in addition to being approved by the Rakai Community Advisory Board, which consisted of sixteen members who were residents in study communities.[6] A key part of the experiment design was that if circumcision was found to be effective, it would be offered to all participants—including the control group—free of charge.[7]

Midyear in 2005, the Rakai Circumcision Experiment (RCE) began enrolling upwards of five thousand (5,000) HIV-negative men in a randomized trial. Half the men were circumcised immediately; the other half would be offered circumcision at the end of the experiment, two years later. This intervention would evaluate whether men who were circumcised showed lower rates of HIV infection. All were initially screened to confirm their HIV-negative status, then retested at 6, 12, and 24 months. All trial participants were compensated for their time, including travel costs and deeper costs of absence from work. Participants who completed the entire trial and underwent circumcision earned as much as 5,190 Ugandan shillings (approximately US$30 in 2005). Participants assigned to the control group who participated in the entire trial, but opted *not* to be circumcised, earned fifteen dollars. Throughout the trial, participants could access free general health care, were given free condoms, and received counseling on HIV prevention. Participants who became HIV positive during the experiment were referred to a program providing free antiretroviral therapy.

All data from the experiment was evaluated on an ongoing basis by an outside set of specialists sitting on a Data Safety and Monitoring Board (DSMB). These boards, first mandated in the United States in the 1960s, were tasked with reviewing interim data and making recommendations about whether a trial should be continued, modified, or terminated, based on ongoing evaluation of risks and benefits.[8] The monitoring boards are generally made up of three to seven experts. These must include at least one biostatistician. Many boards include a formally trained ethicist and members knowledgeable about local conditions.[9] When the DSMB reviewed the interim data from Rakai, they recommended the trial be ended early. There were clear protective effects from the intervention: circumcision reduced HIV acquisition rates by as much as 50 to 60 percent.[10] The trial was stopped in December 2006, six months shy of the planned conclusion; the "benefit" of male circumcision was too great for it to be withheld either from the control group or the wider population.[11] As was planned in the original protocol, all of the men in the control group were offered free circumcision and 80 percent had the surgery. An additional step was then taken to deliver the intervention to the wider public: with international funding, the RHSP offered free circumcisions to all men in Rakai and neighboring districts. By 2010, over 17,000 men in the area had undergone the procedure. Since that time, additional research has established the biological basis for the effectiveness of circumcision, as well as the discovery that the procedure provides a degree of protection to female partners.[12]

It's a rare case, indeed, when research proves the efficacy of such a low-cost, simple intervention that is then quickly offered as a tangible "benefit" not only to trial participants, but also the wider community. In many ways, the Rakai experiment is a model of a well-designed trial. The project was conducting research on a disease that was locally relevant, and where new prevention strategies were desperately needed. From the very beginning, the project had budgeted to provide circumcision to all male participants should it be found effective. Results from Rakai also had far larger effects: the WHO and UNAIDS were persuaded to accept male circumcision as a global strategy for HIV prevention.[13] Collateral benefits have also accrued to the general community. A previous RHSP project that investigated the connections between sexually transmitted infections (STIs) and HIV infection did not find the expected link, but it did provide free STI treatment for all pregnant women in the community. As part of the RHSP's commitment to their research projects benefiting the wider community, the free treatment for pregnant women was continued after the trial, which has led to concrete improvements in child and maternal health outcomes.

While the commitment to delivering benefit is admirable, Rakai is currently one of the most heavily used research locations in modern sub-Saharan Africa, and possibly in the world.[14] The Rakai Health Sciences Program has been running large-scale, long-term medical research studies in the same area for the past twenty-five years, and, as of 2002, Rakai had been the subject of almost a hundred different published scientific articles.[15] The cornerstone of the RHSP's work is the Rakai Community Cohort Study, which involves repeated surveying of fifteen thousand adults in fifty communities to collect basic demographic information and health behaviors data, in addition to taking venous blood samples, urine samples, and genital swabs from all participants.[16] This makes Rakai an appropriate location to ask larger questions about "research fatigue" and whether residents are being unfairly burdened with the risks of too many research projects that ought to be more equitably spread to other communities. Even without clear answers about whether too much research is occurring in Rakai, from at least one angle this local organization's continued presence and long-term "use" of the same community is positive, since it allows for tangible benefits to accrue over time, provides a route by which medical research discoveries trickle back, and maintains and builds infrastructure so future interventions can be delivered. In Rakai, it is not outside researchers with tenuous commitments to a place, or middlemen contract research organizations, or for-profit pharmaceutical companies that are running the experiments. And, perhaps because of this sustained attention, evidence indicates that some public health indicators have steadily improved over the past decade—such as rates of voluntary HIV testing and counseling, increased use of condoms and contraceptives, and decreased HIV prevalence.[17] The large number of research projects have helped fund a new health clinic and mobile clinics. RHSP now provides free health education and HIV testing to all residents, and, during the annual surveys, all community members receive free general health care and treatment with antimalarials and antibiotics.[18] Since 2004, with funding from the US President's Emergency Plan for AIDS Relief (PEPFAR), all HIV-positive people in the district can receive free antiretroviral therapy, insecticide-treated bed nets, and water purification kits.[19] The RHSP has provided opportunities for dozens of Ugandans to pursue graduate degrees and specialized training abroad. Without a formal policy, the RHSP has shown a strong commitment to research that is locally relevant and thus likely to lead to results of import, and to working to make successful research findings available to the community.

5 FINDING AN ETHICAL BALANCE

This chapter takes up the research encounter once introductions, initial explanations, attempts to recruit participants, and consent activities were over, at that drawn-out moment when researchers were responsible for continuing projects and maintaining relationships. They had to make accurate assessments about what local people considered to be risks and benefits, while also estimating how much risk communities were willing to tolerate and what kinds of benefits would be valued. These were not straightforward activities. The main case studies come from the work of the East African Medical Survey (EAMS) in the town of Heru Juu, in Kasulu District in western Tanganyika, where insufficient benefit led to the cancellation of a project, and from the town of Makueni in the Machakos District of Kenya, where a colonial employee made a spirited defense to cancel a malaria control experiment as it grew riskier to local people. These case studies remind us not to make generalizations about colonial-era scientists and to acknowledge there was no single colonial opinion about nearly anything, let alone a topic as complicated as determining an acceptable amount of risk or benefit. This chapter builds upon the writings and ideas of Warwick Anderson, Helen Tilley, and Michael Worboys in striving to present a more nuanced picture of the history of science in colonial spaces.[1] In East African medical research circles, it is clear that differences of opinion were rife, but these disagreements often fell along predictable lines. Fieldworkers frequently advocated for more beneficial and less risky interactions than laboratory- and office-based scientists, who tended to privilege the generation of new knowledge.

The chapter is divided into three sections. The first section focuses on the "drug question" and the debate between Dr. Hope Trant and the director of the East African Medical Survey about whether participants in medical research deserved benefit. The second section highlights disagreements about acceptable levels of risk by focusing on the work of Dr. Shelly Avery Jones in the Makueni malaria experiment in Kenya in 1952–53. The chapter concludes by examining the tension between "due" and "undue" influence, the particular challenges with providing cash as benefit, and the fact that, while bioethicists are most concerned with preventing undue inducement, medical research subjects more frequently confront the problem of insufficient benefit.

❖

Although modern international ethical documents have chosen to talk about risk presuming a single definition, there is abundant evidence that risk is a subjective and culturally relative topic. From a technical perspective, risk is the combination of the probability and magnitude of some future harm. Yet, when an average person decides how "risky" something is, they typically consider *what* they are being asked to do, *who* is making the request, and the "trustworthiness and credibility" of the institutions involved.[2] In biomedicine, risk is typically presented as a discrete, quantifiable unit—something that can be assessed and measured in objective ways, holding constant over time and space. However, such a simplistic approach to defining risk ignores that risk is a social construction dependent on a variety of cultural, social, and personal factors. Trying to name, understand, or assess risk outside of a particular context would seem to be nearly impossible.[3] Recent literature on health risks comes to just this conclusion, stating, as one example, that "risk perceptions depend less on the nature of a hazard than on political, social, and cultural contexts."[4]

One need only consider the practice of blood taking to see just how broad the disagreements can be. For researchers, taking a small amount of blood is considered a low-risk procedure. Yet, for many East Africans, giving blood is risky on a physical and a spiritual level, since a loss of blood can weaken a person's physical health or lead to bewitchment, particularly given the widespread distrust of government employees involved in medical research (see chapter 2). Another example comes from Rwanda, and biomedical activities meant to reduce the risk of HIV by promoting the use of male condoms. While a condom reduces biomedically recognized risk by preventing the flow of potentially infected semen, risk is assessed differently in Rwanda. Many Rwandans recognize blocking or stopping the flow of bodily fluids as risky, as local conceptions of good health are represented by the free flow of bodily fluids such as blood, semen, and milk. In this case, paradoxically, the use of condoms can be perceived as a threat to maintaining health rather than as a strategy to reduce risk.[5] These differences in opinion about the risk of particular practices frequently led to divergent ideas about what was an acceptable amount of benefit to offer in return.

There is a similar divide when it comes to defining benefit. Biomedicine distinguishes between three different types of benefits. Direct benefits include the provision of medicine or access to a doctor, indirect benefits include the improvement of health facilities or public health activities that occur because of the research, and aspirational benefits are new treatments or improved policies that accrue only from the results of successful research. As an example, table 5.1 lists the benefits given in recent medical research projects occurring at the KEMRI/Wellcome research site on the Kenyan Coast. The KEMRI scientists note that the table indicates "the diverse range" of benefits offered in modern trials, yet the table also indicates how blurry these categories are.[6] For example, it's unclear in the malaria vaccine trial why the availability of cars for emergencies is not referred to as a "collateral" benefit, or, in the HIV trial, why free lubricant and food are not "direct" benefits. On the other hand, the chart does not list any aspirational benefits, seeming to lend credence to the idea that these are primarily hypothetical.

While there is a wide range in what is given as benefit in many modern medical research trials, very little time has been spent asking participants what they would prefer as benefit. The few articles that have been written on this subject indicate that African participants have strong opinions about "*who* should provide *what . . . how* and *why*."[7] In Rakai, Uganda, community members told interviewers that benefit should be given to the entire community and not focus narrowly on participants; in Kenya, one man invoked the notion of a "long term moral obligation to provide ART [antiretroviral therapy] to participants beyond the conclusion of a trial."[8] In Heru Juu, participants might have been vague about the exact type of benefit expected, but they were clear that the amounts offered were insufficient.

Failure in Heru Juu: Blood Sucking Rumors and Benefits

The work of the East African Medical Survey was described in chapter 1, and the specific work conducted by Dr. Hope Trant in Kibondo and Kasulu Districts of Tanganyika were described in the prior narrative. Although the EAMS may have claimed a degree of success by the sheer number of samples collected, many of their interactions were fraught. In the case of Heru

Table 5.1. Benefits provided by specific research projects at the KEMRI/Wellcome Project in Kilifi, Kenya

	Study description	*Direct benefits for participants*	*Collateral benefits*	*Transport/fares*	*Other benefits*
TB Study 2009–2011	Enrolled 1,500 severely ill children, in-patient, suspected of having TB Multiple venous blood samples taken over 2–6 months	Free management of TB, contact tracing and management of infected adults	Enhanced diagnostic and laboratory capacity of clinic	Fare for follow-up, determined by clinicians	None
Malaria Vaccine (RTS,S) 2009–2012	Testing the efficacy of a new vaccine in reducing malaria infections in children One location of a multi-sited Phase III (efficacy) trial Enrolled 1,000 children between 6 weeks and 7 months Each child received 3 doses of the experimental vaccine and had venous blood samples drawn 5 times, with follow-up appointments at home over a 3-year period	Free treatment of all acute infections including payment of hospital bills Regular screening of children for anemia, deworming Referral of chronic illnesses and those that cannot be handled at the facility Uninterrupted access to EPI & rabies vaccines, Hepatitis B, even in cases of government shortages	Physical upgrading of dispensaries and laboratory clinical support including provision of vaccines where necessary Resuscitation equipment for use by all dispensary patients Study personnel staff facility when government staff are away	Fares or lifts in KEMRI cars to hospital for medical care KEMRI cars available in emergencies	Refurbishment of facilities
HIV 2006–2016	Observational study 30 adult men who have sex with men 30-plus visits over a 2-year period with blood draws at each visit	Free screening and treatment for STIs, free condoms Free antiretroviral therapy HIV testing, disease monitoring, and support counseling Free Hepatitis B vaccine	Access to services without stigma as a result of careful training of all staff Provision of rabies and Hepatitis B vaccine	600 shillings ($1.30) fares provided at a flat rate	Free lubricants Food tickets for those on ARVs and the very poor

Source: Adapted from Molyneux et al., "Benefits and Payments," 7.

Juu, even before the project collapsed and the research compound went up in flames, there were problems afoot. Notably, there had been earlier accusations of blood theft and blood sucking. It would have been easy for Hope Trant to blame the failure of the East African Medical Survey in Heru Juu entirely on blood sucking rumors, which made sense both literally and figuratively.[9] While still carrying out the research in Heru Juu, she wrote, "The old Wazee [elders] are objecting to our taking blood from the persons examined on the grounds that we are drinking it. This is from their noticing that blood from the finger is sucked up into the pipette before diluting for estimation of haemoglobin."[10]

Blood sucking rumors have rarely been analyzed literally, but, in this case, the accusations accurately described what African participants were seeing. When performing a hemoglobin test, a finger was pricked with a needle and drops of blood accumulated, and then the blood was sucked into a pipette using the mouth before being transferred to another container.[11] In addition to the hemoglobin test, it's also possible that research subjects may have seen blood and other substances moved from one container to another using the practice of "mouth pipetting." Until the 1970s and the development of mechanical pipettes, it was common throughout the world to use one's mouth as a laboratory tool.[12]

The claims also resonated with existing idioms. Calling researchers "bloodsuckers" aligned them with colonial officials, another group of people who "sucked" resources unfairly. The historian James Brennan has shown that during the 1950s and 1960s nationalist campaigns in Tanganyika there was widespread condemnation of "sucking" (*unjojyaji*) by politicians, which was seen as a form of exploitation.[13] The idea was represented visually in a newspaper cartoon showing "non-Africans or 'exploiter' Africans standing around a poor, thin African sucking his sweat or blood with straws."[14] As chapters 2 and 3 demonstrated, many East Africans believed researchers were government employees (which they were in many cases), or worked in close concert with the colonial government (which they did).

But while Trant explored the blood sucking rumors, she ultimately settled on a different explanation for the group's failure: the difficulties arose because the researchers had nothing to give participants. While the blood taking "was already regarded with suspicion . . . it would have made our relations more friendly if we had been able to give out some slight reward for the patient's compliance (a few cigarettes, or a pinch or snuff, or sugar)."[15]

When Trant expressed her opinion to EAMS Director Laurie and made an argument for direct benefit to be given to participants, he disagreed. Laurie explained the project's failure by invoking the "suspicious attitude" of Africans who believed the surveys were "a form of witchcraft or are a Machiavellian scheme of the tax-gatherers."[16]

In seeking to explain the tense local relations, Trant could have also invoked the region's long history of interactions with the colonial government in the realm of health and disease.[17] Deadly epidemics of sleeping sickness had swept through the area repeatedly since the turn of the century, and the German colonial governments relied on policies of forced medical treatment in sleeping sickness "concentrations" with ineffective, and often dangerous, medicines.[18] Epidemiological surveys involved tens of thousands of people being physically palpated, giving blood samples, and submitting to lumbar puncture. (See figure 5.1 for a lumbar puncture being performed in the field.) None of these interventions were much valued by local populations, especially once it became apparent that the available treatments were

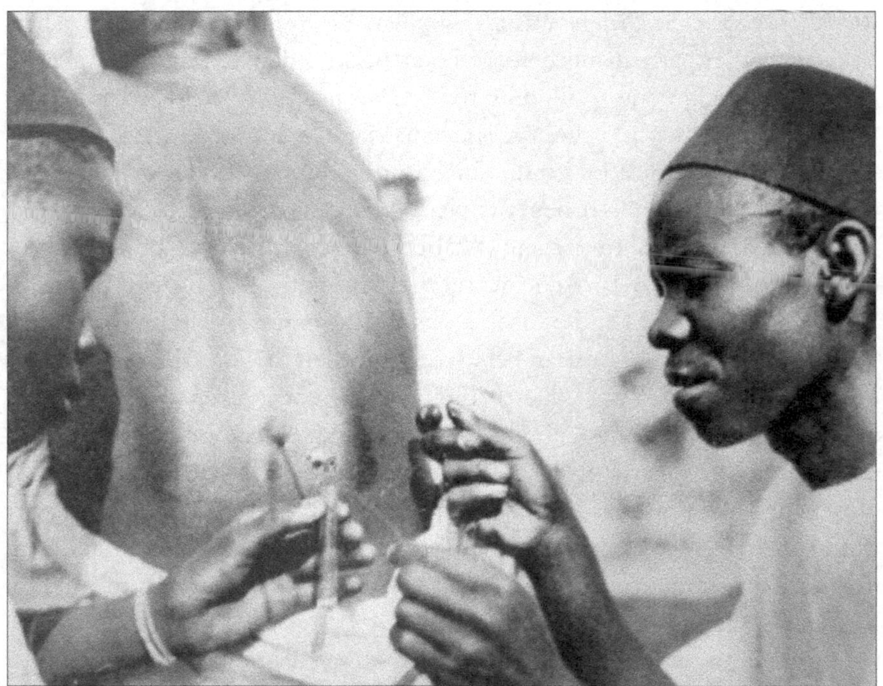

FIGURE 5.1. "Usoke African Dressers assisting at Lumbar Punctures." *Source:* Tanganyika Territory Annual Medical and Sanitary Report, 1929. Crown Copyright material is reproduced with the permission of the Controller of HMSO and the Queen's Printer for Scotland.

ineffective. The years after World War II continued these trends. In 1950, a medical officer took over nine thousand blood slides in Kibondo District—sampling over 25 percent of the total population.[19] In December 1952, there were six thousand blood slides taken from Kagunga. In February 1953, the sleeping sickness specialist returned to Kagunga to reexamine a thousand people. Mass exams occurred again in November 1954 and June 1955–July 1956.[20] An anthropologist who lived in this region in the 1950s argued that such campaigns resulted in "a strong distrust among . . . the vast majority of the population in the objectives pursued by the Government."[21] Placed against this historical backdrop, it is less surprising that residents reacted negatively to yet another medical intervention, or to the researchers' inability to deliver, or even articulate, clear benefits.

The "Drug Question"

Trant's observation that the EAMS's relations with community members would have been "more friendly" if they had been able to give out tangible benefits, and Laurie's disagreement, foreshadowed a broader argument. In 1950s East Africa, there was a disagreement among medical researchers about whether participants deserved benefits; this debate was often referred to as the "drug question," alluding to whether projects should provide drugs to participants. It was a divisive issue that circled around a set of practical and ethical questions: Did individuals and communities deserve anything for their participation? If benefits were to be given, was it because something was always due to participants, because benefits had to offset risks, or because it was likely to facilitate recruitment and ongoing participation? Then, was research in and of itself a benefit? If so, was it a sufficient benefit by any measure? Those researchers who dismissed the idea that participants deserved anything simply due to their involvement chose to engage with the "drug question" in a purely pragmatic way: what was the minimum necessary to give participants in order to allow a project to be completed? The drug question was multilayered and offered researchers the opportunity to consider ethical questions around benefit and risk in addition to meta-questions about the real purpose and likely outcomes of medical research projects in East Africa.

In the case of the EAMS, the field-based scientist (Trant) advocated for drugs to be given, while the director (Laurie) argued that treatment was an unnecessary and unjustified expense. Colonel Laurie believed that research itself was a benefit to local people, and that it was not the Survey's responsibility to provide anything more. He regularly reminded Trant that "we are doing far more for the local people than can reasonably be expected of us. We are a Research Organization."[22] A year later he wrote again to reiterate, "it was this Department who were [sic] giving services to the Native population, and not them to us."[23] While arguing that research was itself beneficial, Laurie clarified that treating sick Africans wasn't his job and, even if he was willing to treat participants, it was too expensive. He was happy to point out that "the money I have at my disposal is for research only."[24] In ongoing disagreements with Trant, missives from his office reminded her that his organization was "financed to carry out Research work and that treatment of patients is the responsibility of the Medical Department."[25] And lest she make appeals directly to the Tanganyikan Medical Department, he advised her that the national medical services had no money; asking them to provide drugs was futile.

Laurie never clearly spelled out what "services" his research was providing, but it's likely he saw his work as a necessary step on the path of delivering other forms of public health and medical benefits to communities. His research would help inform as to which diseases should be prioritized and how to best approach treatment, control, or elimination campaigns. With the benefit of hindsight, we can say unequivocally that Laurie's vision rarely materialized. Research findings were not easily translated into policy and officials in medical departments were often reluctant to accept advice from researchers. Laurie's claim that there was no money for benefit was a questionable excuse, since there's no evidence that he looked into ways of providing benefit cheaply. The most obvious and clear benefit (other than a distribution of cash) would have been to provide medicine to treat a condition such as malaria, worms, anemia, or infections. It's hard to estimate the actual costs of providing such a benefit, but even free aspirin or iron pills would have been more than most people would have had access to privately, and more tangible than what most medical research projects gave out.

Despite Laurie's general reluctance to purchase drugs, and his even greater unwillingness to recognize the need to offer benefit, there was one scenario under which he would agree to expend program funds to purchase drugs: "as a method of gaining the confidence of the people."[26] For the first eight months of the EAMS's work on Ukara Island in 1950, islanders received free medical care from Trant, who was given permission to spend her mornings working in the Bukiko government dispensary. The African dispenser and Trant distributed the small amount of drugs sent by the national medical services, and she would see an astounding twenty patients per hour.[27] Laurie approved all of this in an effort to gain the confidence of local people, but after eight months he ordered that Trant stop all treatment and focus exclusively on research. Such an approach appeared to be common. The second director of the EAMS, Colonel Bozman, advised that "no doctor should approach a primitive population without a supply of something for them."[28] It appears that, throughout Bozman's tenure, the EAMS provided limited treatment or free medical care in the early stages of a research project. The drugs and time spent away from research were justified as a practical necessity, never as an ethical obligation or as a type of compensation for the risks participants bore.

The strength of Laurie's opinion—and the magnitude of disagreement between him and Hope Trant—became clear during her posting to Ukara Island. Although some basic medicines had been provided to her for use in those first months, Trant was desperate for more. (She had iron pills, hepatex, and pencillin; she wanted hetrazan for lymphatic filariasis, ascorbic acid for scurvy, and sulpertron/sulphone for leprosy.) She wrote to request that additional drugs be purchased, and when Laurie refused, she wrote back, exasperated, "You put me here to get the people to have trust in medical treatment, and then you won't give me any more drugs . . . ! You will excuse me if I say that it just doesn't make sense. . . . They are trusting us you see and we must do our best not to let them down. At least that is how I see it."[29]

The passage begins to uncover fundamental differences of opinion about the purpose and goals of medical research, the duties of individual researchers, and the role of benefit to participants. Trant saw research as a form of humanitarian outreach where the goal was to win hearts and minds while also collecting data. Her work was an opportunity to introduce East Africans to biomedicine and to take on the responsibility of getting people to trust in the new system. Trant's desire to treat people was rooted in her experience as a medical doctor and probably her very nature as a human being, but it was also very pragmatic.[30] She recognized that unless villagers were positively disposed toward her, they would have little interest in participating in the survey. Shortly after arriving in Kibondo District, Trant pressed Laurie to return the results of stool examinations, since then she could "treat the affected ones, and this makes for good will."[31] Trant's general understanding of the local communities was that combining treatment with surveying was a surefire way to make people happy.[32]

Laurie disagreed with Trant on nearly every point, and had always insisted the goal was to identify the "plagues" of the African, not necessarily to treat them. Once he realized Trant's intention to treat and *cure* ailments, he hurriedly wrote and gave orders *not* to clear up all the cases of anemia. As he explained it from the research perspective, "we are not anxious that you clear them up as we propose a full investigation of this problem. If you do clear things up we shall not be able to find the cause! You must save life of course, but such cases will be very few. In view of this will you keep anemia treatment to an absolute minimum."[33] Trant, however, continued to badger him for additional medicines. After agreeing to send treatment for leprosy, Laurie wrote with specific instructions "that, with the exception of leprosy, please limit treatments as far as possible," and reiterated that "if you begin to clear up chronic disease you will ruin our long-laid medical survey plans!"[34] Trant was ordered to withhold the drug from Ukara Islanders. In her response to Laurie, she made a case for sending drugs to Ukerewe Island, to be distributed by those staffing the mission hospital. Trant asked pointedly if his intention was "to keep Ukerewe [Island] as a sort of preserve."[35] The word "preserve" paints Laurie as a voyeur, emphasizing his desire to observe sick Africans as a naturalist might appreciate unusual flora on a nature preserve. Laurie's view was not unusual and was part of a widespread fixation on Africans as particularly diseased—the "pathological museums" of the preface. By framing African bodies as the sites of new, unknown, and exciting diseases, scientific pleasure was achieved through the thrill of discovery, not through the banal task of treatment.

Trant seemed to believe that Laurie's refusal to purchase drugs was a product of his distance from the realities of the field. To overcome this, one week after arriving at Ukara Island she wrote him a long letter describing her work helping to save a woman who had been in labor for four days. Trant found the pregnant mother lying on the ground in a shed shared with animals, and her only tool was a stethoscope. The undelivered child was already dead. Trant borrowed a knife, cut into the dead child's scalp to pull it out, and then manually removed the placenta. The woman needed antibiotics in order to stave off infection, but Laurie had refused to purchase sulfa drugs or penicillin. Trant was reduced to begging for drugs from missionaries on nearby Ukerewe Island. She scrambled to organize a boat to Ukerewe Island,

and then had to walk seven miles each day to administer the drugs once they arrived. Trant hoped that by putting a human face to her request for drugs, Laurie might be persuaded: it was *that* woman who would die if she wasn't given basic medicines. She ended the letter with a final plea, imploring him, "they are trusting us" and "we must do our best not to let them down."[36] Laurie remained unmoved and had an administrator respond on his behalf: although "[the director] very much regrets your disappointment, he feels that in view of the work you are doing his inability to assist you may possibly not be understood. In the main he is financed to carry out Research work and that treatment of patients is the responsibility of the Medical Department, or the Territory."[37] Trant was so adamant in her beliefs that she regularly ignored Laurie's orders to stop distributing drugs, and, due to her ongoing disagreement with the policy, she was eventually demoted.[38]

As a final example of how unwilling directors were to engage with benefit as a necessary component of research, we turn to the Pare-Taveta Malaria Scheme (discussed in depth in chapter 6), which occurred in Tanganyika and Kenya starting in 1954. In the planning stages of this experiment, it was unclear whether money would be set aside to fund free health clinics for the community, which would provide basic medicines such as meparcil for malaria, cough syrup, laxatives, eye drops, and sulfa drugs in emergencies. An outside reviewer of the project noted that *five* different senior researchers familiar with the project were all "convinced" of the necessity of the health clinics, calling them "very useful" and probably an "essential service" that would allow for the smooth running of the Pare-Taveta Malaria Scheme. The outside reviewer wrote in the formal assessment that, while people in the intervention area would get the "benefit" of the experimental intervention of indoor residual spraying, the thousands of people in the control areas would receive nothing. He wrote, "It is difficult to see how the people there will continue to support the Scheme over the next 5–10 years when they are to receive nothing in return. Everyone—except those engaged on the Scheme!—I spoke to about this was convinced that something like this must be done if cooperation is to be maintained. Dr. Trant spoke from the bitter experience of their own surveys."[39] The response from the director, Bagster Wilson, was rabid, and began by noting, "I have not many criticisms to make of your criticisms," before going on to rebut nearly every point. He made clear that he was well qualified to make decisions about the provision of benefits with "seven or eight years experience of carrying out medical surveys of various sorts among various African peoples."[40] The failure of the EAMS in Heru Juu was well known to Bagster Wilson, since Hope Trant also worked on this project. Yet, he still refused to fund the health clinics or to provide any other tangible forms of benefit. Ultimately, UNICEF paid for the free clinics, although they, too, sought to "dispel the idea that the clinics are merely a free dispensary," choosing to highlight instead the pragmatic nature of the clinics, that they were sites where African mothers could be taught "elementary hygiene" and where children could receive donated clothes, dried milk, and sweets at Christmas.[41] Ultimately, Bagster Wilson used the clinics as sites of additional research, showing just how unwilling he was to allow the clinics to serve as a true benefit.

The disagreements between Trant and Laurie in the EAMS, and Bagster Wilson's reluctance to agree to medical clinics in the Pare-Taveta project, illustrate the variations and forms of the "drug question" debate. Although Trant was clearly an unusual character, her point of view was not uncommon. The perspective of colonial workers stationed in the field often differed dramatically from that of their office-bound bosses. Generally, these fieldworkers were more willing to acknowledge that research organizations had long-term responsibilities to the participants and to the communities they worked in. Colonial government officials stationed in communities (notably district commissioners and medical officers) shared many of these concerns. When planning for the EAMS in Kisii, Kenya, the locally based medical officer wrote that he considered it "most important" that, when the Survey finished, "sufficient hetrazan should be provided for the treatment of proved cases" of river blindness.[42] As the "man on the spot," he was the person local people would turn to when the researchers disappeared, yet diseases remained, no benefit was given, and no treatment was available. As the next section will make clear, it was often these officials who argued for greater benefit to communities and fewer risks.

Reducing Risks in the Kenya Malaria Experiments

Shelly Avery Jones was hired in 1951 as a parasitologist for the Kenya Medical Department to supervise experimental malaria control schemes throughout the colony.[43] Like Trant, he was an unusual character, with odd work habits that set him apart from his colleagues, such as his penchant for self-experimentation and frequent criticism of projects he felt were poorly planned.[44] After only a brief time on the job, it was clear that Avery Jones did not fit in well; communications between the director of medical services and London colonial officials implied that he did not work "placidly with his senior colleagues."[45] That allusion to working "placidly" was a reference to Avery Jones's constant questioning of authority. The historian Anna Crozier argues that East African medical bureaucracies were "conservative and hierarchical" and that until junior officers had worked their way through the ranks, they were expected "to obey their superiors and to afford fellow officers respect fitting their rank."[46] Avery Jones failed in both respects: he was neither obedient nor respectful. Within a colonial department where a premium was placed on deference to authority and maintaining a united front, his willingness to think critically, share those criticisms with his superiors, and advocate on behalf of Africans quickly alienated him. His disputes with his superiors eventually led him to leave his job.[47]

The first project Avery Jones was assigned to oversee began in August 1952 in Makueni, a native settlement ninety miles east of Nairobi, established in 1947 to relieve African overpopulation in Machakos District and make space for European settlers. The stated goal was to see what effect a single dose of daraprim (pyrimethamine) would have when given to all residents of an isolated area where malaria was "highly endemic with seasonal outbursts."[48] The experiment was premised on the idea that, in a malaria endemic area, there would always be a number of asymptomatic people with malaria parasites circulating in their blood. The hope was that a single dose of daraprim would treat many of these cases, resulting in fewer people with malaria parasites in their body, thus reducing the human reservoir for the disease. Since mosquitoes are unable to spread malaria if they don't first ingest the malaria parasites from an infected person, fewer infected people could lead to lower overall rates of malaria. If the mass administration of the drug proved successful, it would be an inexpensive way to control malaria.

Three doses of daraprim were given between September 1952 and September 1953, and between 3,700 and 4,000 men, women, and children were dosed each time. People participated voluntarily, although the "Chief had prepared the people beforehand." The documents do not say explicitly, but it is likely the drug was described as malaria treatment or prevention rather than as an experimental, short-term intervention.[49] Baseline malaria rates were collected beginning in August 1952 via blood slides from patients arriving at the dispensary who were suspected of having malaria. Rates were then monitored on a monthly basis by gathering parasite and spleen rates from school children

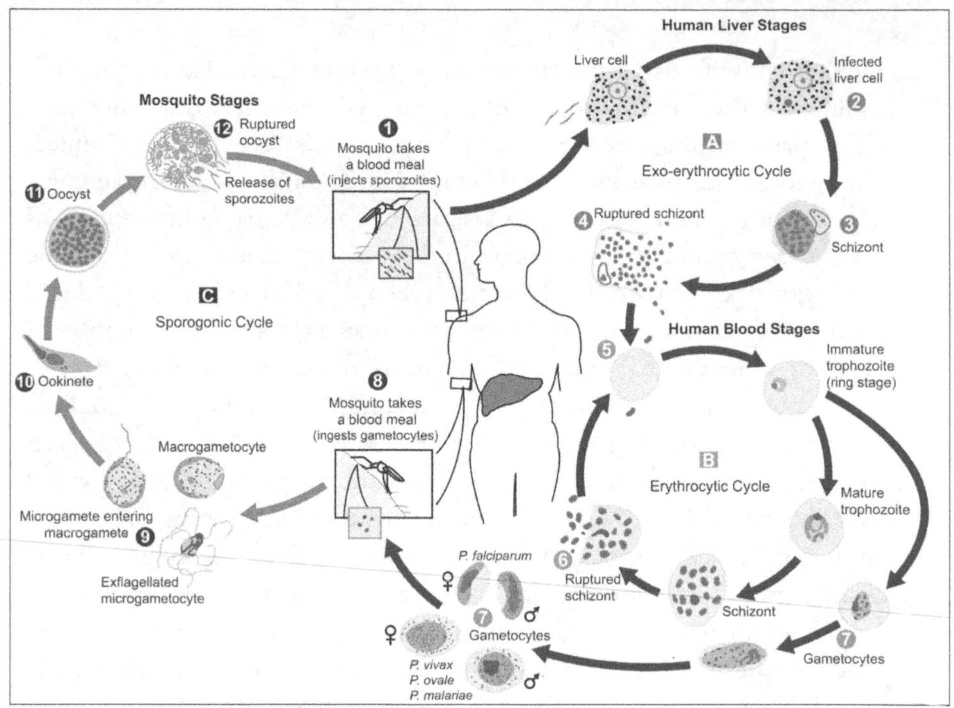

FIGURE 5.2. Malaria disease lifecycle. Produced by Chris Becker.

and random checks of 75 adults and 60 infants under the age of two. Gathering this data required blood samples and physical exams of the schoolchildren, people arriving at the clinic, and a random selection of Makueni residents. There were high expectations about what might be accomplished. Even Avery Jones was hopeful, writing in an early report that "it seems possible that malaria may be eradicated." He believed the experiment was of "great importance" and that they would want to continue it "for as long as possible."[50]

As the experiment entered its eighth month, Avery Jones grew uneasy. Initial data from multiple sources led him to believe there were drug-resistant forms of malaria present. In April 1953 he found that "resistant ST parasite rates have risen . . . and have now attained a rate of 16%." He then did a mass administration of daraprim and one week later took random blood samples from 100 adults and 100 infants. He "found asexual ST parasite rates of 6% [in adults] and 7% [in infants]," which he took "to represent resistant parasites."[51] He was worried about increased risk to local people and recommended substituting a different drug to be sure those infected with malaria would be successfully treated. Avery Jones also kept a "special malaria register" where he kept track of all malaria cases that did not respond to treatment with daraprim. By late April 1953, he noted that the "register is filling rapidly with the names of patients who have had dose after dose of Daraprim without effect on their parasitaemias."[52] As evidence mounted, Jones wrote to his supervisors to share the initial results, ask advice, and give his own (unsolicited) opinion. In a series of confidential communications in 1953, Avery Jones set off a rancorous debate over what was an acceptable amount of risk and who should decide when a project should be modified or cancelled. (Ignored by Avery Jones and his superiors was the question of when, or if, residents should be made aware of newly observed risks.) In light of the worrisome data he was collecting, Avery Jones advocated for modifying the project or stopping the drug distribution early. His concern was that if a drug-resistant form of malaria developed, or already existed naturally, the medicine being handed out would be ineffective and people would grow sicker. Furthermore, a drug-resistant parasite could be encouraged to spread through the wider community through selective pressure.

Six senior scientists all disagreed with Avery Jones's findings of resistance, his assessment of the risk, and his recommendation to end the experiment. He expressed his concerns not only to his boss, Farnworth Anderson, the director of medical services in Kenya, but also to Dr. Henry Foy of the Wellcome Research Laboratory in Nairobi, Bagster Wilson, the director of the East African Institute for Malaria and Vector-Borne Diseases, Professor P. C. C. Garnham of the London School of Tropical Medicine, Dr. A. J. Walker, deputy director of medical services in Kenya, and Dr. R. B. Heisch, a senior parasitologist and head of the Kenyan Department of Insect Borne Diseases. These were the men scientifically best qualified to consider Avery Jones's claims. It is unclear if they ignored the data entirely, or evaluated it with a strong starting assumption that the finding was incorrect. In any case, they collectively and forcefully dismissed his findings and parodied his worries. Anderson wrote that Avery Jones was "unduly obsessed with the possible dangers of the widespread transmission of so called resistant strains in the area." After Anderson and Bagster Wilson corresponded, they agreed Avery Jones had "unduly magnified the danger to public health."[53]

There were actually two disagreements simmering: first, a scientific question of whether drug-resistant malaria was actually present and what was causing it; second, whether the Makueni experiment should be modified or ended. Avery Jones was quite certain that the resistant malaria strains were present and were the result of naturally occurring foci, not the result of giving subtherapeutic doses. He was worried about the malaria parasite growing resistant and how this could change local malaria ecology for the worse. In a confidential note to Bagster Wilson, Avery Jones argued that "the rise of resistant cases . . . at a time when the mosquito population is still small justifies intervention." He went on in detail:

> My own view is that sufficient information has now been gathered about daraprim and that the risks involved in allowing the resistant infections to spread unchecked are not justified by the extra knowledge that might be obtained. The local people have already lost some immunity and such a spread would mean considerable hardship and lost working capacity. The people would see me taking bloods and spleen rates every month while their condition worsened and their confidence in our good intentions would be gravely shaken.[54]

In the passage, Avery Jones argued that the project had succeeded in gathering new scientific information about the utility of daraprim, but that a tipping point had been reached. "Sufficient information" had been gathered and there would be additional risks to continuing. He calculated that the "extra knowledge" would not be worth the pain and suffering to be borne by local people. Finally, he noted his uncomfortable position as an extractor of information while giving ineffective medicines and watching people grow increasingly sick. As he saw it, the experiment provided "a considerable amount of information about the effect of Daraprim on different malaria parasites and a very strong indication of its dangers when used by itself for mass treatment."[55]

Anderson asked Bagster Wilson to comment specifically on whether there was "undue risk" in continuing the experiment. Bagster Wilson wrote back in strong disagreement with Avery Jones. He argued that only time would tell if there was drug resistance, that the data collected to date was inconclusive. He wrote to the entire group that "the experiment has not been developed fully," and spoke for the entire group (less Avery Jones) by arguing that "it seems to us that it would be most undesirable for the original design to be abandoned at this stage simply because some undesirable public health trends appear to be resulting in this small town."[56] Heisch believed it was "too early to intervene" in Makueni, and Anderson also voted to "continue this experiment." The top men closed ranks and issued their ruling: despite some "undesirable" trends in public health, the experiment was not over and the project would continue. To add to the insult, Bagster Wilson made clear that they were "not in a position to draw any conclusions . . . nor perhaps are you."[57]

The senior scientists were unwilling to closely evaluate the collected data, likely because Avery Jones was presenting unexpected and inconvenient conclusions that jeopardized the planned experiment. This behavior struck Avery Jones as being particularly unscientific. How could malaria control projects be planned in the future if data from existing trials wasn't carefully analyzed? Although other writings have presented Bagster Wilson as a risk-averse researcher, that is not the impression one gets from his involvement in Makueni. In his advisory role, his tendency was to play down the risks raised by field workers, dismiss damning evidence, and continue on with the experiment in a quest to gather more data.[58] Foy, another well-respected scientist, dismissed the mere possibility of resistance, arguing on theoretical grounds that the resistant strains could only be the result of subtherapeutic doses despite the fact that Avery Jones claimed that this argument was "entirely unsupported by any experimental evidence." Bagster Wilson maintained into early 1954 that "there is no evidence of this as yet" and argued that a similar malaria experiment in Nandi, Kenya, should also continue as planned.[59] Their general reluctance to consider the data led Avery Jones to ultimately request that another scientist travel to Makueni to confirm the presence of "large numbers of patients who do not respond to Daraprim," collect additional data, and review his initial findings.[60] Even to this, it appears the scientists ignored his request.

We know now that Avery Jones's findings were correct. So, why did these senior scientists disagree with Avery Jones, and why were they so reluctant to evaluate the data? There are a few possible answers. First, these men were nearly all working in office-based laboratory or administrative positions, and thus may have been more tolerant of risk than a field-based scientist. Even if they believed resistant malaria existed, their interest was in privileging the generation of new knowledge rather than minimizing risk for the East Africans involved, and thus the experiment continued. Second, as these senior researchers climbed the bureaucratic ranks and escaped the messiness of field research, they became more critical of data generated outside the controlled setting of the laboratory. For the lab-based scientists, the data wasn't yet persuasive enough, especially as it ran counter to what malaria models created by senior colleagues predicted. This was an epistemological disagreement about what types of data and knowledge were valued. Avery Jones's data wasn't particularly ambiguous, but it was disrupting existing knowledge about malaria. Finally, we must consider the very real—if trite—possibility that the scientists were offended by Avery Jones's unorthodox manner and that, had he presented his information in a more traditional or humbler way, his message may have been more easily received.

There were limits to what Avery Jones's bosses could control. Although he followed instructions and continued the experiment as planned (continuing to distribute daraprim as part of the mass administration and as malaria treatment in the clinic), he pointed out that he wasn't the only one deciding whether the project would continue. He continued emphatically: "The local population is losing faith in Daraprim as case after case occurs in which the drug is ineffective. There is a strong possibility that there will be a very poor attendance for the proposed mass treatment at the height of the transmission season as by then the people will have had ample opportunity to realise that Daraprim is no longer of value to them."[61] The response was again one of smug dismissal. It seems clear in reviewing this exchange that Avery Jones had pointed out a reality on the ground that the senior officials were loath to accept: by merely observing treatment and seeing no improvements, Africans would realize what Avery Jones was documenting scientifically. With a disease like malaria, it can be fairly easy to determine whether a drug is effective or not. Since the daraprim was being given out free of charge and there was no punishment for not taking the drug, local people had little incentive to take a drug they knew had limited efficacy. Once people had made that calculation, the experiment would be ruined, whether or not Avery Jones followed instructions from his boss to continue distributing the drug. It should be clarified that there is no evidence that Avery Jones ever told Makueni residents about his suspicions of drug-resistant malaria. Although, if he ever did share this information, it's unlikely he would have been foolish enough to inform his supervisors in writing.

The Makueni experiment ended in September 1953, one year after it began. At that point, among the 4,000 people given daraprim the parasite rate had dropped dramatically (from 50 percent to 2 percent), and seven months after the project had concluded only 17 percent of the people screened had malaria parasites in their blood.[62] Blood smears taken at the time of the third dosing, in September 1953, showed that 26 percent of infants were infected with malaria strains resistant to daraprim, as were 6 percent of adults.[63] This led Avery Jones to assert that "the experiment proves conclusively that mass treatment of a population in a hyperendemic area may be followed by a resistant strain becoming predominant."[64] The Makueni experiment exposed something that no one had predicted: that there were naturally occurring drug-resistant strains of malaria, and that, following mass administration, only resistant strains were left behind, which then spread throughout the community.

Even after the Makueni experiment ended, Avery Jones continued to fight about the interpretation of the data. A year later, the Kenya Medical Department embarked on another malaria control experiment in Nandi that would replicate Makueni. Avery Jones wrote the director to reemphasize that he did "not consider the Makueni experimental results to be in any way encouraging," and that more must be known about the resistant strains prior to starting another mass treatment campaign, since it could "save much needless expense and disappointment."[65] As he interpreted the data from Makueni, it proved "that susceptible strains of ST malaria are in all probability eradicated by a single dose of daraprim but that some individuals may be infected with resistant strains from the start and they are left behind as foci from which future infections spread."[66] When Avery Jones's conclusion was circulated to Bagster Wilson, he again disagreed. As the formal advisor on both projects, Bagster Wilson reiterated his support for the new project in Nandi as planned, and told the director of Kenya medical services unequivocally, "the experiment must be continued."[67]

Colonial Science and Stifled Dissent

Avery Jones worked in a time and place that was unwilling to consider his scientific findings. He had committed multiple sins: he dared to give his opinion about a project he was hired to merely carry out, he disagreed with the recommendations of senior scientists, he was willing to jeopardize his career in order to protect local communities from risk, and he privileged minimizing risk to Africans over the benefits to be gained from producing new scientific knowledge. All of these things indicated his allegiances were wrong. Particularly among the group of scientists based in Kenya, the emphasis on group loyalty often "stifled dissenting voices."[68] As the historian Anna Crozier described the colonial medical service in East Africa, "senior officials actively pursued a policy that suppressed certain types of treachery and discord and purged undesirables. An unspoken distinction operated between acceptable and unacceptable complaints . . . individual insubordination or misconduct . . . was dealt with summarily and, often, heavy-handedly. . . . Above all else, the Colonial Service worked towards presenting an united and decorous front.[69] Such an ethos did not aid in the discovery of new data, or findings that ran counter to expectations. The handling of the Makueni malaria experiment shows just how oppressive colonial science in Kenya could be.

But while Avery Jones did not fit in 1950s Kenya, times were changing and conditions would soon favor Avery Jones's approach. After leaving the Kenya medical service, Avery Jones worked for the World Health Organization and was successfully posted to ten different countries between 1954 and 1965.[70] While gaining experience across the continent, his reputation was redeemed. Within three years of voicing his initial concerns about drug resistance, a leading malariologist with the WHO acknowledged in writing that Avery Jones's results were correct.[71] (The debate continued, however, about the cause of the resistance.[72]) By 1965, the Tanzanian Ministry of Health boasted that "it was Dr. Avery Jones himself who discovered the first signs of drug resistance . . . whilst working with the Kenyan Government in 1952."[73] While his initial research findings were validated, so too was his personal style. The same characteristics that made Avery Jones an outlier in colonial Kenya made him an asset in newly independent Tanganyika. In 1964, when deciding whether to hire Avery Jones, officials lauded his long experience working in East Africa and "his demonstrated ability to work effectively among our rural inhabitants."[74]

Conclusions: Cash, Coercion, and Undue Inducement

There was—and still is—no clear-cut right answer to the question of how much is an acceptable amount of risk and benefit to have in a medical research project. A utilitarian response would emphasize that a project must have low enough risks and great enough benefits to allow for the work to get done. An acceptable answer also had to strike a balance between what a director wanted, what a field worker was willing to do, and what subjects would agree to participate in. In the same vein, the wrong amount (whether too much risk or too few benefits) would appear to describe the instances when projects stumbled and couldn't recover.

The case studies in this chapter also presented portraits of two unusual medical researchers who might be best labeled mavericks. Although their complaints may have been shared by other colonial field workers, their behaviors were not representative. They chose to act in extreme ways, and that makes them outliers. Recounting the experiences of Dr. Hope Trant and Dr. Shelly Avery Jones hopefully provides a counterpoint to generalizations of what colonial science was. By pushing the boundaries of acceptable practice and often openly defying their bosses, these two researchers worked to make medical research encounters more fair and equitable. In doing so they endured not only personal frustration, but also suffered professionally by being demoted and fired. Although other field-based workers may have shared their concerns, Trant and Avery Jones adopted particularly strident voices, behaved as a combination of rebel and whistleblower, and were ultimately out of place in 1950s colonial East Africa.

The case studies also indicate that decades of contact with medical researchers has led East Africans to ask for their benefits early, often, and preferably in the form of cash or medicine. Heru Juu was not unusual in having a long history of government interventions promising benefits that rarely materialized. While working in Manyara, Tanganyika, prior to independence, the entomologist Alec Smith recounted how "influential villagers were often cynical of promises of benefits to be gained in the distant future."[75] It's a sentiment that is alive and well. The overload of AIDS-related epidemiological research in the 1980s led people in Kagera, Tanzania, to bluntly tell researchers, "We want help, not more research." In another HIV research project, large numbers of East Africans refused to be interviewed, complaining that they had been screened in other studies "but nothing useful was done about their results."[76]

These historical case studies provide lessons and insights regarding our modern understanding of risk and benefit. In order to recognize forms of risk and benefit valued by many participant communities, risks must be considered not just for an individual, but also for the larger community, and over a longer period of time. The effects of an experiment may still be playing out months or years after a project has formally concluded. The concept of benefit, on the other hand, must be narrowed so as to focus more on the timely and the tangible. While Western biomedicine recognizes three different types of benefits (direct and indirect to individual subjects, and collateral benefits to the wider community), there ought to be a privileging of benefit that accrues directly to the individual or community at the time of the research. Biomedicine must also recognize—and ideally find a way to incorporate the fact—that what constitutes "risk" and "benefit" cannot be simply translated across languages and cultures. We should expect slippage not just in terminology, but also in meaning. East African research subjects regularly assess risk or benefit by considering a broad range of factors that depend not just on what a participant must do or receive, but on the level of trust in the person who is asking and past interactions with that person or group.

In both the past and present, when required to enter into a research transaction, African communities will form an opinion of the likely costs—and what the pecuniary payment ought to be. Such pragmatism is not unique to East Africa. In a medical research project recently run in The Gambia, interviews with participants found that it was one of the benefits offered—access to free medicine and medical care—that led people to volunteer. One participant bluntly explained, "we now have treatment for our families . . . if there was no benefit I would not join."[77] Views from modern Kenyans about the obligations of researchers, whether compensation should be offered, and why, are equally clear. Subjects considered it researchers' "obligation" to provide benefits such as continued care and cash. These were considered to be "appropriate compensation for patients' 'helping' the investigation." As one person stated, "It's only fair that you compensate me for your research." Another, savvy to the money to be made with new drugs, noted that if the drug was effective it would be marketed, and "you are going to . . . get some money out of it." Based on that potential for economic gain, the interviewee noted that there was a "moral obligation" to continue treating subjects. She went on to admit that this obligation was "not legal, it's not binding, you cannot be taken to court. But it's only fair."[78] One Kenyan man asked rhetorically, "I have been used like a guinea pig, so how does he just leave me without compensation?"[79]

As a young Tanzanian man explained regretfully to me one day in 2008, "these days" (*siku hizi*) all people cared about was *hela*—cash. He had worked for international nongovernmental organizations running surveys and organizing community meetings on development topics throughout Tanzania. He glumly reported that no one would do anything for free. When he tried to recruit participants for even short interviews or surveys, the first question was always the same: what will you give us? The second was whether he had hela to give. (It's likely the Swahili slang *hela*—cash—is derived from a unit of currency, the *Heller*, introduced in German East Africa before World War I.[80] It's an odd historical twist on what many people see as a modern phenomenon of East Africans demanding cash compensation.) As the medical anthropologist Vinay Kamat recorded from his informants on the Swahili Coast, the common refrain was, "*Hela, hela, hela tu; kila mahali hela!*" ("Money, money, money; these days it's all about money!").[81] Which begs the question: Is it all about money? From the historical case studies and modern commentary, the demand for cash appears to be the response to a series of broken promises, benign miscommunication, and outright lying. East Africans may have initially been willing to consider the needs and risks of medical research within a reciprocal, yet informal, relationship. However, as more traditional systems of obligation and reciprocity broke down, demands were made for cash and benefit early in the process. It was a logical attempt to formalize relations and protect against unfair transactions.

Discussions among foreign workers responsible for running research and development projects imply that this demand for cash is recent. Amongst these workers, there is widespread belief in a not-so-distant past when East Africans still embraced an ethos of giving for the larger good. These imagined Africans were traditional, precapitalist, communal, and had no interest in money or accumulation of wealth. The assumption is that, because other development projects have thrown unseemly amounts of cash at East Africans, residents have now grown money-hungry. It goes without saying that these inaccuracies are easily countered with the examples in this chapter. When Mwami Teresa requested cash in 1954, a cash economy had been functioning in that remote region of Tanganyika for at least three decades, spurred by the creation and collection of colonial taxes. By the 1940s, "a money income was imperative" and cash was used to buy household goods, food, cattle, and clothing and to pay bride wealth.[82] Then and now, cash was guaranteed to be a real benefit to medical research participants.

Just as the EAMS balked at the idea of paying cash to their research subjects, distribution of cash continues to make medical researchers and sponsors working in the global south deeply uncomfortable. Although provision of money is a clear and direct benefit, there has been a general unwillingness to distribute it.[83] In many projects in East Africa, cash provided to participants is euphemistically referred to as a transportation "reimbursement" or "allowance"—it is never called a direct payment or benefit. The worry is that cash may be an "undue" inducement. (Yet, we don't worry about undue inducement among poor medical research subjects in the United States, and cash payments are widespread enough for an entire class of "professional guinea pigs" to exist.[84]) Determining what is an appropriate amount of benefit—but not too much—is a slippery subject, since inducement is okay, just not "undue" amounts that involve an "excessive offer" where people may participate against their "better judgment."[85] The Council of Europe states that financial benefit for participation would result in "inducement which compromises free consent."[86] Another worry is that money as benefit could lead us to "commercialize an altruistic endeavor."[87] But there is no evidence that most people participate in medical research because of altruism—in fact, just the opposite. As has been pointed out, "Most trial participants in both developed and developing countries join clinical trials in order to obtain medical benefits."[88] Ultimately, the distinctions between "due" and "undue" inducement, and whether cash is more "undue" than other benefits, remain unclear.[89]

Taking concrete steps toward reducing instances of undue inducement and coercion is important—no one should feel pressured to participate in medical research. Yet, the widespread refusal to distribute cash is based on a number of largely unsubstantiated assumptions: that the offer of cash would lead a person to participate in medical research they would otherwise refuse to participate in, that participants aren't already being unduly influenced by non-cash benefits, and that cash is uniquely problematic in ways that distribution of free drugs or access to free medical care are not. The small amounts of empirical work done in developing countries challenge the validity of these assumptions. Qualitative research suggests that "financial rewards do *not* distort research subjects' behavior or blind them to the risks involved with research."[90] Christine Grady, from the US National Institutes of Health (NIH), argues that money is not a unique form of inducement, and offers of medical care or treatment can be equally coercive.[91] On the other hand, fears of undue inducement—too much benefit—do not provide a sound justification for giving too little benefit. As Grady put it, "It is worthy to be concerned about possible coercion of research subjects, and, therefore, to be aware of the potential for undue inducement of trial participants. However, this concern must be balanced against the risk of exploiting subjects by failing to allocate them a fair share of the benefits of research."[92]

While undue inducement is a problem much discussed by bioethicists, lack of benefit is the problem faced by many research subjects in poor countries. In the meantime, as international research protocols refuse to provide cash as a form of benefit, participants in low-income settings are denied cash payments precisely because they are poor. Those who are cynical may infer that the risks of undue inducement have been unrealistically exaggerated, since it provides international sponsors with an "'ethical' argument for limiting the benefits they provide to trial participants."[93]

EXITS AND LONGER-TERM OBLIGATIONS

Historical Narrative

"ALMOST COMPLETELY ERADICATED"

The Pare-Taveta Malaria Scheme, 1955

For five years in the late 1950s, the World Health Organization (WHO) in partnership with the United Nations Children's Fund (UNICEF) and the East African Institute of Malaria and Vector-Borne Diseases ran a large malaria elimination experiment in the British colonies of Kenya and Tanganyika. The Pare-Taveta Malaria Scheme was called "the most comprehensive malaria survey mounted in East Africa," and the numbers support the statement.[1] The experiment's main intervention was indoor residual spraying (IRS) of 15,000 houses with the insecticide dieldrin, which was meant to kill malaria-transmitting mosquitoes. Researchers also collected 6,000 blood samples a year, vital statistics from 10,000 people, and made medical examinations of 2,500 adults and 2,000 babies.[2] The activities spread over 2,000 square miles of "lush plain and tropical forest" in the Pare District of Tanganyika and Taveta District of Kenya; within this area lived 53,000 people in treated (sprayed) areas, 79,000 people in the mountains above, and 9,000 people who came and went as part of a shifting labor force.[3] Malaria was endemic at high levels, transmission occurred year round, epidemics were uncommon, and there were two different species of anopheles mosquito (*A. funestus* and *A. gambiae*) responsible for transmitting malaria. The Pare-Taveta Malaria Scheme was part of the WHO's Global Malaria Eradication Program (GMEP).[4]

The designer of the project, Dr. Donald Bagster Wilson, was hopeful that Pare-Taveta could provide insights into the general feasibility of malaria elimination in sub-Saharan Africa. He wrote in 1951 that the goal was to study the impact of malaria "on the life and activity of people born and brought up in the most intensely affected areas . . . to define the importance, both physical and economic, of malaria under these conditions."[5] In reality,

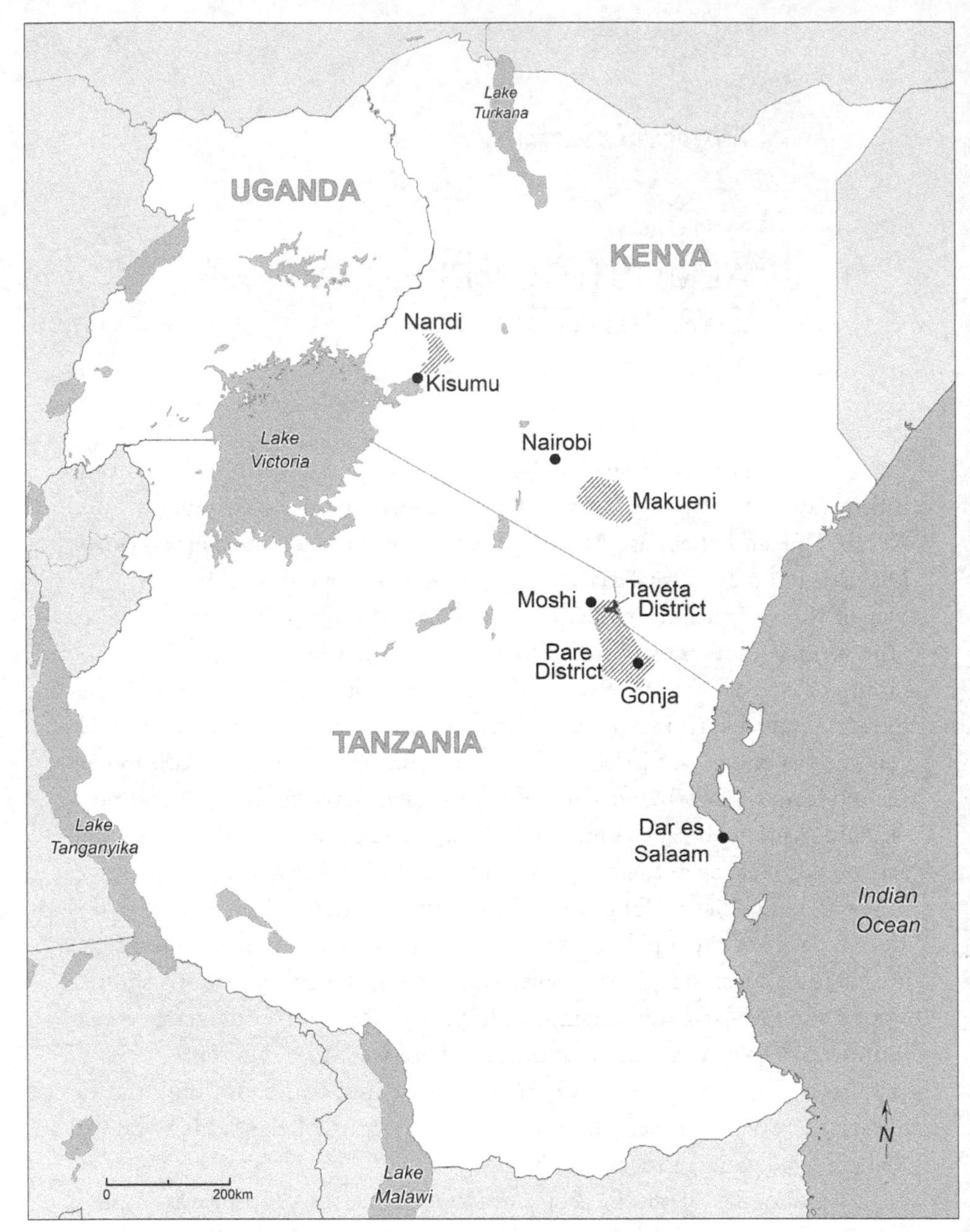

MAP 6.1. East Africa research sites. Map by Chris Becker.

the "goals" of Pare-Taveta were surprisingly malleable. In other settings, the aim of the project was described as "to find an economical means of arresting malaria transmission," "to reduce malaria and improve health," "to see if malaria transmission could be arrested by insecticides," and to determine "the importance of malaria to the 'immune' African community."[6] In addition to enumerating the physical and economic costs of malaria, they would also test the feasibility and effect of malaria elimination.[7] A 1959 newspaper article in the *East African Standard* provided a slightly different slant, stating that the purpose had been "to reduce malaria and improve the health of 120,000 people over an area of 100,000 square miles."[8]

As the Pare-Taveta experiment came to an end in 1959, at least one fact was irrefutable: malaria was still present. As a local newspaper put it, after four years of indoor residual spraying, malaria had been "*almost* completely eradicated."[9] IRS had brought malaria to very low levels of transmission, but it would be impossible to eliminate the disease using only spraying. And, while the project was officially over in 1959, there were stark realities that had to be reckoned with, most notably the risk of "rebound" or "resurgent" malaria.[10] Due to the temporary reduction in malaria transmission, local disease ecology had been modified in dangerous ways: Pare and Taveta residents had lost some of their acquired immunity, which normally provided a degree of protection against the disease. When malaria returned to the area, it could return in a more deadly, epidemic form. Rebound malaria was known to the researchers, and was more than just a hypothetical risk; nonetheless, no plans were in place to help protect community members as malaria returned.[11]

The failure to achieve elimination was due to a few different reasons. First, the spraying program effectively eliminated one vector (*A. funestus*) and reduced the levels of the second vector (*A. gambiae*) by up to 90 percent. However, the *A. gambiae* that remained grew resistant to the insecticide dieldrin *and* began to bite and rest more outdoors, away from the insecticide-coated homes—meaning no amount of IRS would kill them.[12] Second, a regular influx of malaria-infected people from neighboring regions meant the malaria parasite was continually being reintroduced; and, third, annual spraying with dieldrin was found to be too infrequent to be effective. Pare-Taveta empirically documented the difficulty, if not outright impossibility, of malaria elimination in many parts of sub-Saharan Africa. The organizations involved, however, were at pains to emphasize that Pare-Taveta was a success.

Members of the East African Council for Medical Research went on record to state, "the experiment was by no means a failure. It had attracted attention throughout the world."[13]

Beyond the world's attention and the undeniable presence of malaria, the overall results were equivocal. There were some measurable public health effects at the end of the IRS in 1959: a reduction in malaria prevalence and a 50 percent reduction in infant and child mortality (from 200/1000 to 100/1000). In 1964, five years after ending the spraying, malaria had stayed at levels lower than those prior to the intervention. (There was speculation that this was due to Africans self-treating with newly available antimalarial drugs such as chloroquine.[14]) However, when the intervention ended, malaria *transmission* rose quickly, and child mortality returned to prior levels.[15] Later papers written about Pare-Taveta through the 1970s found nearly no longer-term effects of the spraying.[16]

With the experiment's official conclusion in 1959, there were some worrisome realities that had to be reckoned with. One unintended consequence of the success of the multi-year reduction in malaria transmission was that Pare and Taveta residents had lost some of their acquired immunity, which normally provided a degree of protection against the disease. This was typically built up and maintained through regular exposure to malaria; without exposure, acquired immunity could not be maintained. When malaria did return, it would likelier be in a deadlier form of so-called resurgent, or rebound, malaria. In the 1950s, this phenomenon had already been discussed in the scientific literature—and also, notably, with consideration of the ethical questions linked to the epidemics.[17] A 1933 League of Nations report argued it would be "extremely imprudent to brutally interrupt the processes that caused immunity," and participants at the 1950 malaria conference in Kampala, Uganda, also struggled with the tension between urgently needed malaria control measures and the possibility of increasing longer-term risk.[18] Pare-Taveta scientists were well aware of the postexperiment risks of rebound malaria, as Bagster Wilson was a vocal participant in these debates.[19]

In July 1960, there was a great debate about what should be done about the returning malaria, and by whom. The only thing the involved organizations could agree about was that they faced an uncomfortable situation. Elimination had failed, malaria had returned, Pare and Taveta community members were at risk, and the African governing councils were arguing it was "morally wrong" for the researchers to end the IRS and force them to "suffer the consequences unaided."[20] Opinions about what to do ranged widely. On one end of the spectrum were those who argued that the researchers could pack up and leave ("retire gracefully over the horizon"), since the experiment was over. On the other end were those who believed researchers had inadvertently created the problem of resurgent malaria among a non-immune population and thus owned it, meaning they needed to take "responsibility for the situation."[21]

In the end, there was no more indoor residual spraying in either Pare or Taveta. The activities transitioned into "Phase 2," a short-term "continuation scheme" that distributed malaria treatment to residents who presented with fever at local clinics.[22] Phase 3 (1961–66) emerged only with a grant from the Nuffield Foundation, and studied what happened when an elimination attempt failed and malaria returned.[23] These activities were ultimately named the "Vital Statistics Survey" (VSS), and data was collected on age-specific mortality and fertility rates, in addition to physical assessments and blood slides, in two different communities where malaria had returned.[24] But the Nuffield grant expressly prohibited any malaria treatment or control activities in the area.[25] This created an awkward situation where the researchers at the Malaria Institute and the administrators at the East Africa High Commission had to dissuade communities from continuing spraying or taking up any other malaria control activities. As Gerry Pringle, the new director of the Malaria Institute, informed his field workers, when they were asked by community members to continue spraying, "Such suggestions should be side tracked as gently and as diplomatically as possible." He was clear that no more spraying paid for by the institute would happen, "whatever the justification."[26] The terms of the grant and the researchers' own line of inquiry created a situation in which communities were encouraged to stop all malaria control activities. This is what ultimately happened, although it is unclear why the local African councils stopped advocating for additional spraying. Predictably, as control measures stopped, the number of malaria infections slowly and steadily crept up in both Pare and Taveta. Luckily, the disease did not return in the deadly rebound form that scientists had feared, although no one could precisely say why.

Table 6.1. Pare-Taveta Malaria Scheme, 1954–66

	Phase 1	*Phase 2*	*Phase 3*
Dates	1954–June 1959	August 1959–60	March 1961–66
Other names	Malaria Scheme Malaria Experiment	Continuation Phase	Vital Statistics Survey (VSS)
Research questions/ goals	• Could malaria transmission be broken in a holoendemic area? • Could malaria-transmitting mosquitoes be eliminated with yearly indoor residual spraying (IRS) with dieldrin?	• Prevent a resumption of malaria transmission • "Provide for the eventual return of malaria" and "soften the impact of a wave of fresh infections" • Encourage cessation of IRS so the VSS could begin	• Research health effects when malaria returns to a place where it has been nearly eliminated
Intervention	• IRS with dieldrin of approximately 15,000 homes, four times over five years	• Distribution of anti-malarial drugs (chloroquine or neviquine) to residents presenting with fevers • House-to-house visits to treat all fever cases and acquire blood slides for testing	• Registration of 16,500-plus people and collection of births, deaths, etc. • Interviews with 1,500-plus families • Blood slides, physical examinations
Director	Donald Bagster Wilson	Donald Bagster Wilson Gerry Pringle (April 1960–)	Gerry Pringle Thomas Fletcher, acting director (July 1966–)
Researchers	Christopher Draper, malariologist, officer in charge Alec Smith, entomologist M. T. Gillies, entomologist	William Marijani (Tanzanian) Abdulrahman Salim Msangi, assistant scientific officer (Tanzanian)	Ndoloi Macharia, fieldworker (Tanzanian) Yohana Matola, fieldworker (Tanzanian; eventually became director of the institute) William Marijani, officer in charge, Pare Station
Related projects	• Nutrition study (Hope Trant) • Serum proteins of children, sickle cell (Thomas Fletcher, biochemist) • Dieldrin exposure of sprayers (Thomas Fletcher)		• Hemoglobin in children and sickle cell (Thomas Fletcher)
Funding	WHO, UNICEF, East African governments, Malaria Institute, Tropical Pesticides Research Institute	Malaria Institute	Nuffield Foundation, Malaria Institute, East African Common Services Organization

Modern Narrative

A NEW MALARIA VACCINE?

Testing the RTS,S Vaccine across Africa, 2010

The calls for malaria eradication, silenced for fifty years following the failure of the WHO's global campaign, were heard again in the new millennium. In May 2009, more than fifteen thousand African infants began participating in the first-ever phase III trial of an experimental malaria vaccine. The young children were enrolled in a randomized controlled trial to test the efficacy of the RTS,S malaria vaccine against *Plasmodium falciparum*, the deadliest of the four types of malaria. The research was sponsored by the pharmaceutical company GlaxoSmithKline and funded through an innovative public-private partnership involving the PATH Malaria Vaccine Initiative and the Bill and Melinda Gates Foundation.[1] The RTS,S vaccine is the most advanced of forty different malaria vaccine projects; it is said to be five to ten years ahead of the nearest competitor, and it could be approved for use in malarial countries as early as 2016.[2] The vaccine works by preventing the malaria parasite from infecting, maturing, and multiplying in the human liver, one of the stages vital for the disease to develop.

The experiment involved laboratories in eleven different African countries, although East Africa was most heavily represented, with five of the testing sites in Kenya and Tanzania.[3] (See maps 6.2 and 6.3 for the sites of all malaria vaccine trials in Africa.) Participating children aged 6–12 weeks and 5–7 months were randomly assigned to receive either three doses and a booster of the experimental malaria vaccine or a placebo.[4] During roughly three years, children received four intramuscular injections of vaccine over a twenty-month period, gave venous blood samples from the arm five times, and were tracked for roughly two years to document all cases of malaria, severe malaria, hospitalizations, side effects, and adverse events.[5]

The RTS,S vaccine's results through early 2014 indicate it is moderately effective at preventing malaria in young children. However, the vaccine does

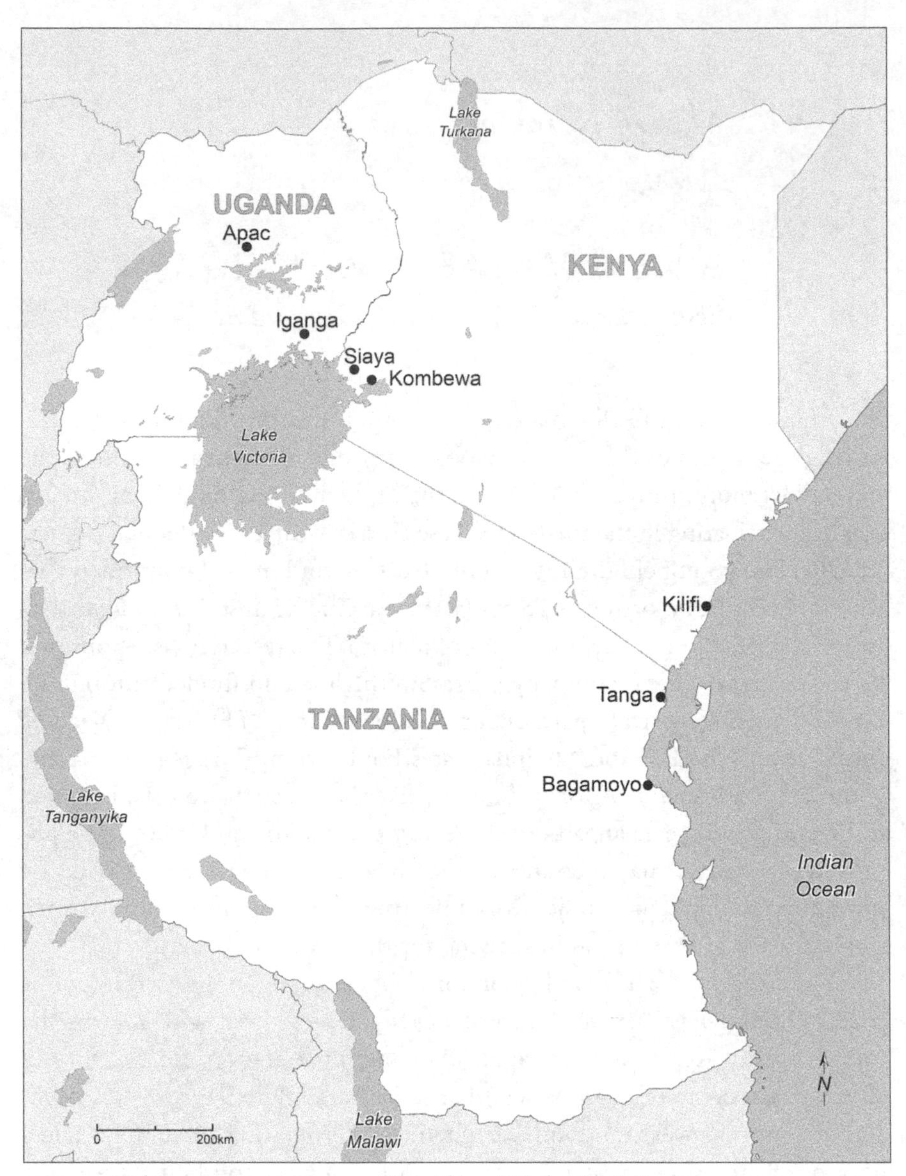

MAP 6.2. Malaria vaccine testing sites, East Africa. Map by Chris Becker.

not prevent all cases of malaria; depending on the age of the child, the vaccine reduces malaria cases by 30–55 percent. Protection was greatest in the twelve months following vaccination, and among the 5- to 7-month-old children, who experienced a 55 percent reduction in all cases, and nearly 50 percent

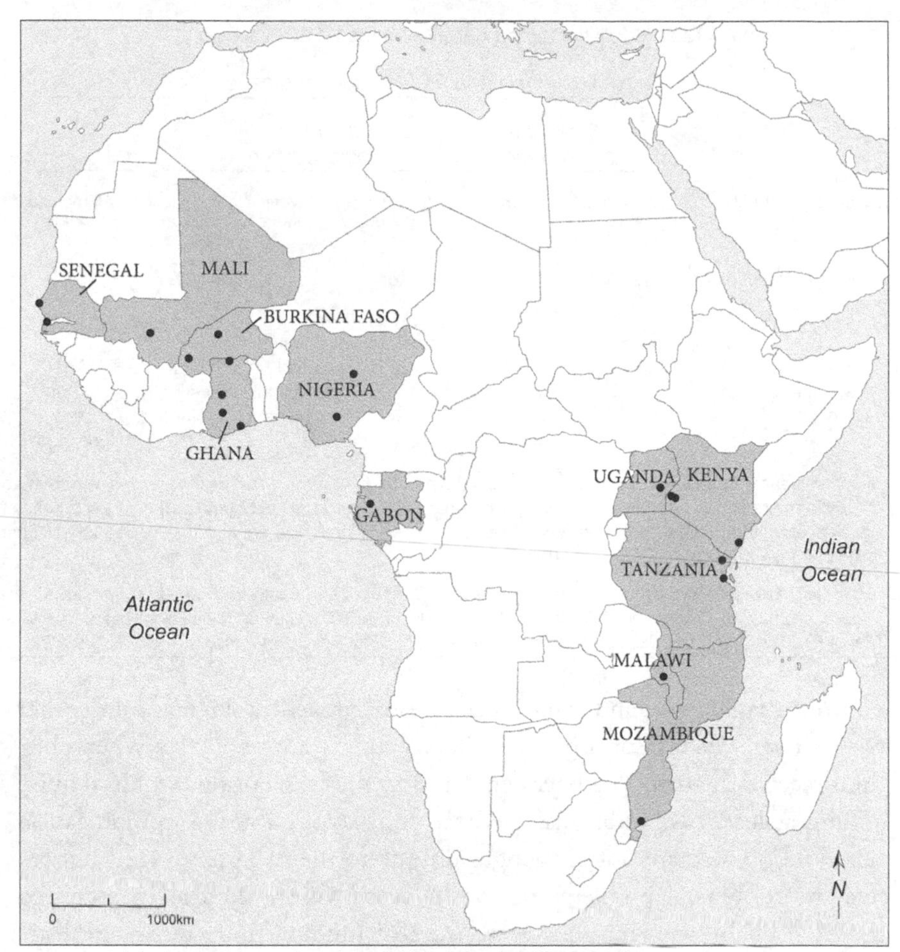

MAP 6.3. Malaria vaccine testing sites, Africa. Map by Chris Becker.

reduction in severe malaria. Infants given the experimental vaccine between 6 and 12 weeks of age had a 31 percent reduction in all cases of malaria.[6] The KEMRI/Wellcome team on the Kenyan coast tracked the vaccine's efficacy over a four-year period and found only a 16 percent reduction in cases of malaria.[7] The group documented that the vaccine's efficacy waned significantly over time, and that protection decreased most quickly in places with high levels of malaria exposure.[8] Results are summarized in table 6.2.

Although the RTS,S vaccine is closer to adoption than any malaria vaccine in history, there remain a very real set of challenges. There is the prosaic: the current vaccine requires refrigeration, which can be difficult to

Table 6.2. Effects and duration of the RTS,S vaccine

Period	Reduction of malaria cases after vaccination: *12 months post-vaccine*		*18 months post-vaccine*			*4 years post-vaccine*
Measure of morbidity	All malaria cases*	Severe malaria cases	All malaria cases	Severe malaria cases	Hospital-ization cases	All malaria cases
Vaccine administered between 5 and 17 months of age	55%	47%	46%	35.5%	42%	16.8%
Vaccine administered between 6 and 14 weeks of age	31%	37%	27%	Not statistically significant (15%)	Not statistically significant (17%)	Age group not tracked

* Clinical malaria was defined as "an illness in a child who was brought to a study facility with a temperature of 37.5°C or more and *P. falciparum* asexual parasitemia (>5000 parasites per cubic millimeter) or a case of malaria meeting the primary case definition of severe malaria." The RTS,S Clinical Trials Partnership, "First Results of Phase 3 Trial of RTS,S/AS01," 1865.

Source: Data extracted from World Health Organization, *Questions and Answers on Malaria Vaccines,* October 2013; Olotu et al., "Four-Year Efficacy of RTS,S/AS01E"; The RTS,S Clinical Trials Partnership, "A Phase 3 Trial of RTS,S/AS01"; The RTS,S Clinical Trials Partnership, "First Results of Phase 3 Trial of RTS,S/AS01."

ensure in rural areas and which has already proved problematic in earlier trials.[9] There are logistical hurdles: the vaccine is most effective when children receive the first dose between 5 and 17 months, but this would require adding at least two additional medical visits to the routine immunization schedule.[10] These are not challenges unique to the malaria vaccine, but reflect more general problems of routine vaccination, such as maintaining the cold chain, securing reliable sources of funding, and creating realistic vaccination calendars.[11] Finally, there are scientific questions where more data and analysis are needed. It remains unclear how long the vaccine is protective, and how quickly protection wanes. It is also unknown if the vaccine reduces total malaria mortality or only delays it. Malariologists and epidemiologists also disagree about how use of the vaccine could change malaria epidemiology and whether it might jeopardize the acquisition of acquired immunity.

A malaria vaccine will be a remarkable achievement, and something singularly new in the fight against malaria.[12] Yet, even with a new technology in hand, many old questions linger—questions of post-trial obligations and the particular risks posed by temporary malaria control. The longer-term epidemiological influences of the vaccine on malaria transmission patterns are unclear. Scientists have noted that there is a real risk "of changing from a stable to an unstable pattern of transmission, which would expose the population, both vaccinated and unvaccinated, to a more severe and unpredictable manifestation of the disease."[13] In the best-case scenario, vaccinated children will have a better chance of surviving their initial malaria infection because they'll be slightly older, and will go on to acquire partial immunity through subsequent infections. In a neutral scenario, the vaccine's temporary protection will merely shift the malaria mortality to a slightly older age group. Some data from the WHO already indicates this is happening: "The reduction in malaria transmission is associated with a shift in the peak age of clinical malaria to older children, as well as an increase in the median age of malaria-related hospitalization in some settings."[14]

But the greatest—known—postexperiment risk is rebound malaria. The phenomenon is explained in more detail in chapter 6, but rebound (resurgent) malaria occurs after a period of temporary malaria control, the loss of protective acquired immunity, and then the return of malaria in a more dangerous form. Results through early 2014 make clear that the vaccine's protection is not permanent. When efficacy wanes and children continue to live in a malarial zone without the benefit of acquired immunity, what will happen? One of the co-developers of the RTS,S vaccine has acknowledged that it remains unknown as to "whether or not a delayed rebound effect will be seen—a paradoxical rise in malaria incidence consequent to the loss of natural immunity due to widespread vaccine coverage."[15] Fortunately, there has been no rebound malaria to date, but that does not preclude it from developing in the future. Others writing about the malaria vaccine have urged that "researchers must continue to be vigilant" and aware of this potential risk.[16] Yet, despite the known risks of rebound malaria, there are no publicly stated plans for what will happen when the vaccine trial ends. While trials are ongoing, there are strict requirements about the search for and reporting of side effects and adverse effects. But, once a trial is technically complete, there are no mechanisms to find, report, or compensate for newly emerging risks. Conditions that develop post-trial don't necessarily qualify as "trial-related injuries" and thus do not necessarily require compensation or treatment, as is required for researchers receiving US government funds or wanting to license drugs in the United States.[17]

We are at a pivotal moment that portends great changes in how, and how effectively, we may be able to control malaria in Africa. But despite the promise and excitement of a new malaria vaccine, we remain with many of the same questions that have plagued malaria control efforts for the past sixty years. We have only a partial understanding of acquired immunity, we remain unable to explain the precise dynamics of rebound malaria, and we have yet to develop a standard method for tracking or compensating for risks once an experiment is over. We must contend with hypothetical scenarios that are not far from reality: What would happen if a malaria vaccine was adopted in East Africa but was abandoned after a few years because of donor fatigue, changing health priorities, or poor uptake rates? What would the implications be in terms of malaria epidemiology, delayed acquired immunity, and morbidity and mortality? At this point, not only do we not have answers to these questions, but we also have no systems in place to respond swiftly and adequately if these risks, or others, were to emerge.

6 COMING TO AN ETHICAL END

This chapter is focused on an important but understudied set of questions that emerge at the end of a research project. They may be framed as, "what obligations do trial sponsors and researchers have, if any, to participants at the end of a study?"[1] More specifically, as was discussed in the vaccine narrative, this chapter is concerned with questions of post-trial access to new interventions, but also with unique risks that may only manifest when a trial has concluded. This chapter moves from the narrow to the broad by first focusing on the very specific risks related to malaria control research, and then considering more general questions of post-trial obligations. I focus upon malaria because it is a singular example—an extreme case demonstrating how disease risks can be magnified through the unintended consequences of failed elimination or irresponsibly ended control activities. I consider the Pare-Taveta Malaria Scheme a case of medical research, since indoor residual spraying (IRS) with DDT modified the disease environment in ways that translated into changed human risk; this is meant to capture the longer-term risks that come with environmental interventions where risks may not appear until after the experiment. There is no other disease that behaves as malaria does, where the negative effects of temporary control are as clear, and where post-trial obligations are so obvious. But the severe nature of malaria's effects allows us to explore broader questions related to post-trial obligations and the responsibilities of researchers when other types of medical experiments end.

Through a detailed investigation of the ending of the Pare-Taveta Malaria Scheme, we explore the questions of what happens as medical research projects come to an end—which they all inevitably do. We focus on 1959–60, as phase 1 of the Pare-Taveta experiment ended, indoor residual spraying concluded, and the risk of rebound malaria appeared. It was at this moment the five groups involved in Pare-Taveta became entangled in a "malaria imbroglio" that forced colonial scientists and administrators to face a set of thorny scientific, pragmatic, and ethical questions.[2] How should a failed malaria elimination attempt be concluded? What should be done as malaria returned to the research area? When the experiment was formally concluded, were there ongoing obligations to African participants or the larger communities? If there was an obligation, what was it, and who was responsible? Were researchers compelled to consider the desires and demands of the African participants about how *they* believed they should be treated as the intervention ended? And what of the disturbing fact that most Pare and Taveta people did not even know they were participating in an experiment?

This chapter argues that the research team involved with the Pare-Taveta Malaria Scheme knew rebound malaria was likely, based on what had been observed and recorded when other malaria control activities had ended. Malaria control activities are exceptional because there are longer-term risks and new information was rapidly coming to light in the 1950s as the Pare-Taveta experiment was being designed and run. In 1951, the director of the Institute of Malaria and Vector-Borne Diseases, Dr. Donald Bagster Wilson, emphasized all that was unknown in regard to effective malaria control.[3] Yet, in the following decade, much information came to light that could have created a different ending for the Pare-Taveta Malaria Scheme. In the mid-1950s to early 1960s (the same time span during which those implementing the Pare-Taveta Malaria Scheme were spraying and debating the merits of a continuation scheme), more than twenty different WHO pilot projects were attempting to eliminate malaria in Africa.[4] The results of these experiments created new knowledge about insecticide resistance, rebound malaria, and the practical difficulties of malaria elimination in Africa. Thus, the Pare-Taveta researchers should not have been surprised when each of these issues arose. The Malaria Institute also faced questions and criticism from their own scientists and government collaborators about post-trial obligations and risks. The East African Scientific Committee and colonial government officials specifically asked about the possibility of a malaria epidemic and how it would be handled. Bagster Wilson and others at the Malaria Institute refused to create or share a concrete plan. At the beginning, they argued that it was premature to plan for the end; at the end, they did not have appropriate systems in place. The problem was never a shortage of ideas; reasonable suggestions were made early on about how to accept a degree of responsibility without making long-term financial commitments.

Community participants and researchers maintained very different ideas about what the project was meant to accomplish, which created conflict about what was an appropriate ending. Pare and Taveta community members believed they were the beneficiaries of a public health intervention, where the goal was to reduce mosquitoes (as opposed to reducing malaria). They expected the indoor residual spraying to continue indefinitely, and were thus surprised and upset to learn that it would end. Once it did, they were also alarmed that researchers had no plans to help treat the returning malaria cases. The researchers, on the other hand, being privy to the project's protocol, knew that the spraying would only last a few years and that the goal was a reduction in malaria rates, not necessarily the total number of mosquitos. Although this chapter presents the perspectives of all of the groups involved, I conclude by utilizing the few sources available that allow us to reconstruct the perspective of the local communities.

The next section presents basic information about malaria and acquired immunity, before turning to a detailed analysis of how each of the groups involved in the Pare-Taveta Malaria Scheme responded to the ending of the first phase in 1959–60. The conclusion discusses existing international regulations about postexperiment obligations, and asks larger questions about how much progress has been made since the 1960s and what an ethical ending might look like today.

Malaria, Resurgence, and Acquired Immunity

Malaria is a protozoan infection of the red blood cells, with female *Anopheles* mosquitoes serving as the vector, transmitting the disease from person to person. Mosquitoes can only spread the disease after biting a person who is infected with malaria. Once the mosquito sucks up blood containing the malaria parasite, the parasite undergoes a process of transformation inside the mosquito's gut and then migrates to the salivary glands—a process that is highly sensitive to local environmental conditions such as temperature. In warmer weather a mosquito may bite someone and then infect another individual within ten days; a slight drop in temperature can cause that process to slow down enough to prevent transmission altogether (since the mosquito may die before the process is complete).[5] Once inside a human, there are distinct developmental phases that the parasite undergoes inside the liver and blood stream.

One of the most noteworthy aspects of malaria is the dynamic nature of the disease. Historical and modern evidence documents the parasite's and the vector's abilities to adapt in impressive and frustrating ways: mosquitoes develop resistance to insecticides and also change their feeding habits to avoid control strategies such as IRS and the use of bed nets. The malaria parasite has grown resistant to every drug used since the 1950s, including to artemisin combination therapy, which has been widely available for less than a decade.[6] Malaria is also a deeply ecological disease, where transmission patterns are highly sensitive to localized conditions such as variations in temperature, rainfall patterns, soil types, and nearby flora and fauna. Perhaps most crucial in this mix of local factors is the behavior of the mosquitos in that particular place.[7] There are wide variations in mosquito behavior, including feeding and "resting" behaviors, preferred breeding sites, flight ranges, choice of blood source, and vulnerability to insecticides. Thus, one species may prefer to feed on humans indoors and then rest on a wall of the house while digesting the blood meal. Another may typically feed on animals outdoors and rest on a tree, only rarely seeking out humans. Such small differences have important ramifications when planning and running an elimination attempt, and these many variations are one of the reasons malaria is so difficult to eliminate on a large scale.

There are four different types of malaria that affect humans, but only one—*Plasmodium falciparum*—is responsible for a majority of the morbidity and mortality in East Africa and across the continent.[8] Early symptoms often involve a combination of fever, chills, headache, stomach distress, and malaise. Falciparum malaria is often referred to as "cerebral" or "deadly" malaria because of how frequently it results in cerebral infection, severe disease, and death. Since the early symptoms are rather nondescript and are often associated with other common tropical ailments, parents may not seek biomedical treatment quickly enough to prevent the risk of death. When malaria is promptly diagnosed (via a few drops of blood on a microscope slide) and properly treated, all forms of malaria, including falciparum, are fully curable.

As humans have lived with a version of the malaria parasite for the last 100,000 years, the human body has developed forms of protection.[9] Duffy negativity, the sickle cell trait, and acquired immunity provide various degrees of protection against both falciparum and vivax malaria.[10] Acquired immunity is a complex phenomenon that was not well understood in the 1950s and still is not fully understood today, but which clearly helps protect people living in areas with constant levels of malaria. When a child is born in a region where there are high levels of malaria, the child is exposed to malaria repeatedly in the first few years of life. Tragically, many of these infections will result in death.[11] However, if the child is able to survive the initial bouts of infection, a degree of immunity is conveyed and future infections are less virulent and dangerous. Thus, in regions with endemic malaria, at about five years of age malaria shifts from being a disease of mortality to one of morbidity. When this phenomenon is graphed, you see a steep drop-off of child deaths at five years of age and again at twelve.[12] In places with year-round constant levels of malaria transmission, it is unlikely that an adult would die from malaria unless they lost their immunity by no longer being exposed for a period of months or years.[13] The specifics of how long it takes for acquired immunity to develop, how easily it is lost, and the degree of protection are still unclear.[14]

Since acquired immunity can be lost if a person is not regularly exposed to malaria, temporary malaria control campaigns can create a situation ripe for rebound or resurgent malaria.[15] Rebound malaria occurs when a population has lost their acquired immunity and the disease returns. These returning epidemics typically have high mortality rates, even among adults. This phenomenon has been documented repeatedly as malaria control and elimination attempts have lapsed in various parts of Africa and the rest of the globe (a particularly devastating case occurred in Sri Lanka in the 1960s).[16] One survey of all published literature discussing resurgent malaria identified 75 cases in 61 countries between the 1930s and 2000s. They found that "almost all" of the cases were caused by the weakening of malaria control measures, most frequently "funding disruptions."[17] Within East Africa, temporary control followed by resurgence occurred in both western Kenya in the late 1950s and Zanzibar in the 1980s.[18] In Liberia in the late 1940s, as US malaria control efforts wavered, malaria swept through the formerly protected population; an epidemic in 1957 (also after control efforts lapsed) affected nearly 25 percent of all infants and children.[19] A doctor involved in the Liberia activities noted with alarm that their good intentions could lead to problems in the future. He wrote in one of the project's quarterly reports, "If the houses are sprayed, they [the children] will not get infection during this time and they might loose [*sic*] some immunity, so that the children get probably worse attacks by re-infection."[20] He articulated the tricky relationship between successful malaria control, lost acquired immunity, the ending of control measures, and the return of malaria in a resurgent form. More recent research from Senegal paints an even more complex picture of rebound malaria and emphasizes how much we still do not know about the conditions under which these rebound epidemics occur.[21] If anything, our incomplete understanding of this phenomenon should make research teams even more vigilant when it comes to longer-term planning.

As this chapter is concerned with a malaria elimination attempt that was part of a larger eradication attempt, it is worth pointing out the differences between control, elimination, and eradication. "Control" implies that efforts are being made to reduce or maintain low prevalence or incidence rates, but it does not imply a particular number or level, or even whether the activities are successful. "Elimination" refers to the removal of a disease from a particular region or country. "Eradication" refers to the entire globe, is absolute, and means no more cases of the disease exist in nature. Elimination and eradication both imply that the disease will never come back. Thus, smallpox has been *eradicated* globally, polio has been *eliminated* from all but a handful of countries, and guinea worm is being *controlled* in West Africa as a campaign for worldwide eradication is underway.

"The Malaria Imbroglio": Obligations When Ending a Project

As the Pare-Taveta Malaria Scheme transitioned into each of its three phases, there was a great deal of disagreement among the groups involved in the project. The international funders of the project, WHO and UNICEF, made clear that, as their original funding commitment ended, so did their involvement, which left decision making to the five remaining groups. Those remaining involved in the project were the East Africa High Commission (as a funder), the Kenyan and Tanganyikan governments (as funders and approvers of the activities that went on in each colony), the Malaria Institute (also a funder, and responsible for carrying out the intervention), the East African Regional Advisory Committees (responsible for reviewing the scientific design and allocating EAHC funds), and the residents of Pare and Taveta (participants). In 1960, as IRS was stopping and the risk of rebound malaria was returning, these disparate partners debated whether there was an ongoing obligation to Pare and Taveta communities once the official experiment was over, and what that obligation was. Surprisingly, there was general agreement that the research team *did* have an ongoing obligation to communities to prevent rebound malaria, and that only once the imminent threat of a malaria epidemic had passed could the Malaria Institute scale back to mere disease surveillance.

Within this agreement, however, there were shades of dissent. On one end of the spectrum were administrators at the EAHC, who claimed there was no obligation, as they were alarmed about the financial implications of continuing public health activities after the experiment. At the other end were the Pare and Taveta Governing Councils, composed of Africans from each district, who made impassioned pleas that the spraying must be continued indefinitely. An interesting, if slightly unexpected, group also agitating for increased responsibility on the Malaria Institute's part were the regional medical research oversight and advisory committees—made up of fellow scientists. These groups actually produced some of the most pointed and critical assessments of the Malaria Institute's plans to drop everything as soon as their formal project was over. One of the most notable aspects of this "malaria imbroglio" was the timeline, and that some of the toughest questions about the end of the project are asked in 1956 and 1957—years before the IRS was planned to end. Members of the Tanganyikan and Kenyan Ministries of Health and the East African regional oversight committee for medical research were raising important questions about the end as the project was just beginning. The following pages detail how each of the five groups responded in 1959–1960 when decisions had to be made about how the project would end.

East Africa High Commission: "Bury the Scheme and Retire Gracefully over the Horizon"

The EAHC was in charge of administering a number of regional research centers that were created in the late 1940s and early 1950s, one of which was the Institute of Malaria and Vector-Borne Diseases, or "Malaria Institute." The Malaria Institute designed and conducted its own projects, but a large amount of its budget was made up of EAHC funds and a smaller part of contributions from the governments of Tanganyika, Kenya, and Uganda. Although the EAHC did not advise on research protocol or design, they were involved in the institute's financial planning. Thus, when discussions about how to end Pare-Taveta turned into discussions about who was responsible for future activities, the EAHC balked. They had allocated only a limited amount of funds to be used for Pare-Taveta, and they did not want to become financially responsible for additional years. The EAHC clearly understood that the Malaria Institute would be found responsible for cleaning up after the scheme, since it was their project that had changed the disease environment, failed to eliminate malaria, and put people at risk of an epidemic. They also knew that if the Malaria Institute was roped into providing additional spraying or malaria treatment for an indefinite amount of time, the EAHC would likely have to pay for at least part of it.

In July 1960, Gerry Pringle, the newly appointed director of the Malaria Institute, accurately summarized the EAHC's position, which was one of fear that the Pare people would demand a continuation of spraying once it became obvious malaria was returning. The High Commission was afraid that, if the Pare-Taveta Malaria Scheme was still active, "the main demands for action would be directed at them rather than at the Administration and the Provincial Health authorities. The [EAHC] Administrator is trying, therefore, to shrug off responsibility for the situation that develops after the end of this year as, in his view, the Directors of the Medical Services . . . have sanctioned the High Commission to bury the scheme and retire gracefully over the horizon."[22] In his own words, the High Commission administrator was worried that "pressure of public opinion" could force the spraying to be resumed, and questioned whether the presence of Malaria Institute workers would direct "the pressure to us instead of upon those to whom it more properly belongs?"[23] It was unclear to whom the EAHC believed the pressure should be "more properly" applied, but they were clear it was not them.

Tanganyikan and Kenyan Colonial Governments: "Eliminate . . . Suffering to the Human Beings Who Are Involved"

Of all the groups involved in Pare-Taveta, it was the district-level colonial officials who asked the toughest, and earliest, questions about what would happen when the experiment reached its conclusion. The British men who worked as district officers in Pare and Taveta were responsible for keeping the peace, carrying out colonial directives, and assisting with special projects. These men often spoke Swahili and local dialects and worked hard to develop good relationships with the communities where they lived and worked. These colonial field workers were the most invested in having a clear, feasible plan for the end. In fact, from the very beginning, these officials were imagining the worst-case scenario that the research team refused to address.

In April 1956, one year into the experiment, the district officer in Taveta, Kenya, sought information about the researchers' long-term plans and their responsibilities after research. His main concern was that "it is absolutely essential that when your funds run out at the end of the scheme, the Taveta have sufficient confidence in your work to wish to continue with it, with either their own money or money subscribed through the Native Authority."[24] The note implies that the researchers' obligations were not necessarily financial, but that they did have an obligation to lay the groundwork for a new public health intervention. As the district officer understood the experiment, the goal was not necessarily to eliminate malaria, but to convince local residents of the benefits of ongoing malaria control through IRS. This questioning eventually made its way to Bagster Wilson. He was unmoved, writing in December 1956, "It is still, in my opinion, premature to formulate a plan for the continuation, or otherwise, of control in this area after mid-1959."[25]

In May 1959, another district officer in Taveta wrote to his supervisor to express his concerns about "the dangerous uncertainties" of the Vital Statistics Survey being planned. He accurately understood the main activity to be observation without any active malaria control measures. He questioned the availability of funding to resume spraying if malaria returned, voiced concern that the observational project had not been designed to minimize either risk or suffering of local people, and argued that a poorly run project would have lasting, negative effects on people's perceptions of future research projects. The Taveta district officer acknowledged the importance of the planned Vital Statistics Survey, but felt it was "equally important" that the project be run in a way that would "preclude all risk of malaria being allowed to reemerge without prompt measures being taken." Near the top of his list of worries was whether "funds [would] be available *immediately* for reintroduction of spraying if there is a resurgence of malaria here? . . . Dr. Bagster Wilson informed me that he will not have any funds available for 'contingencies.' This means that, if there were a resurgence of malaria over large areas of the Scheme, he would have to ask the territorial governments for funds to pay for spraying to be resumed."[26] The official went on to note how difficult it was to obtain government funds that hadn't been requested long in advance and wrote how he found it "most disturbing that funds have not been provided at the outset to prevent" the reemergence of malaria.[27]

Although the Taveta district officer was not a doctor, he was aware of acquired immunity and what might happen if malaria returned after a few years of control. Any delay in funds and treatment being provided "could have disastrous results on the health of the children who have grown up without developing immunity to malaria and consequently on the morale of the Wataveta and their general outlook towards Government schemes."[28] Even though money was short, he argued that, "if a scheme is to be run, it should be properly financed so as to eliminate all unnecessary risk of causing suffering to the human beings who are involved."[29] This district officer recognized the real health effects people would suffer if malaria returned—especially infants without any acquired immunity. He also recognized that people saw the Pare-Taveta Malaria Scheme as a *government* enterprise, and that they would be reluctant to participate in future government activities if the project ended poorly. He concluded by noting that, if the research was properly planned, it could "be carried out without risk to the health of the people here."[30]

During the heat of the debate in July and August 1960, government authorities in both Kenya and Tanganyika wrote in with specific questions about the plan for the future, and to clarify what they would not tolerate. In Tanganyika, the permanent secretary for health wrote to the Malaria Institute to inform them that it was no secret that malaria was returning to Pare, and that it was "causing considerable concern." He believed the Pare would press for a resumption of IRS and wanted to know whether it was best for local authorities to begin the spraying immediately.[31] Although the Tanganyikan government was not happy about the possibility of paying for the spraying, they were forced to think about picking up where the researchers left off.

Institute of Malaria and Vector-Borne Diseases: "Ameliorate the Sufferings of the Sick"

While negotiating how to end the Pare-Taveta Malaria Scheme, the Malaria Institute's new director ultimately made surprisingly broad admissions about the researchers' obligations to East African participant communities. Gerry Pringle charted a very different path from his predecessor, Bagster Wilson. Shortly after becoming director, Pringle acknowledged the specific obligation to prevent rebound malaria and the broader obligation to blend research with benefits to participants. Among his colleagues in the late 1950s and early 1960s, Pringle was an unusual example of being sensitive to both the needs of the Malaria Institute (to gather data, conduct research, and end a project without being financially responsible for years in the future) and the needs of local people (to minimize risk, receive benefits, and be left with an overall positive impression of research and government interventions). Pringle also admitted that part of the Malaria Institute's duty to reduce suffering stemmed from the fact that the suffering was a direct result of its research.

In early July 1960, Pringle wrote to respond to the advice he had received from the EAHC about how to end the project. Their advice amounted to a recommendation to get out fast and deny all responsibility. Pringle was clear in his disagreement:

> I do not think that timidity is an appropriate sentiment. . . . While I share the [East Africa High Commission] Administrator's concern at the delicacy of the situation, I think that we will come out of the position with less odium by remaining on the spot and doing all we can to ameliorate the sufferings of the sick rather than to stage a retirement from the scene. . . . I have always felt the need to combine the work of the field investigators with some sort of travelling dispensary of a very crude type. In this way the wishes of the Wapare and our own scientific requirements could be met.[32]

Pringle implies that the reputation of the researchers had suffered already, and that they needed to do the right thing by offering malaria treatment and paying attention to the "wishes of the Wapare" and not just their own scientific needs. There was an added challenge mixed in with the larger question about how to end the IRS. The researchers also had to figure out how to end the continuation scheme, which consisted of the fieldworkers treating fever cases they came across with malaria treatment. The program was "unquestionably popular" and Pringle guessed it would be "just as much a headache stopping this as it was stopping the spraying." His plan was to taper the drugs "with great care and delicacy" once the Pare had endured the next wave of malaria, which he predicted would occur in 1964. He considered withdrawing drugs any sooner to be not "possible or humane."[33]

Regional Advisory Committee: "Give All Assistance Possible"

There were two medical research oversight groups functioning in East Africa in the late 1950s. The East African Council for Medical Research (the "Council") was responsible for allocating funds and the East African Medical Research Scientific Advisory Committee (the "Committee") provided technical guidance.[34] Together, the groups dispensed advice about the design of projects, provided feedback regarding research difficulties, allocated funds, and helped distribute findings; the scientific work of the Malaria Institute was also subject to the approval of the Council.[35] The group was made up of fellow researchers and scientists, and they typically provided sensitive and nuanced advice.[36] Just like the local government officials, the advisory group thought early about worst-case scenarios and the need for contingency plans. At their 1956 meeting, the group reassured members of the Tanganyikan and Kenyan government that "in an extreme emergency such as an epidemic . . . the Institute would certainly interrupt its current programme to give all assistance possible."[37] Even at this early stage, the Council announced that the highest priority would be minimizing risk to African participants. Providing widespread malaria treatment could muddy or invalidate the Malaria Institute's research findings related to IRS, but the advisors were clear that public health would be prioritized over continuation of the experiment.

Two years later, the Committee members weighed in with a lengthy discussion about what should happen when the spraying campaign stopped. The director of the advisory committee laid out two possibilities. The first was that malaria transmission would have ceased or become very slight, and that there was no insecticide resistance, in which case it would be necessary to have a small project to keep the area free of malaria. The second possibility was that malaria would still be active and that insecticide resistance had appeared. In either case, he was certain that "there would be a research responsibility" and, if malaria was still present, "the people in the area should be protected from an upsurge of malaria . . . and the cost of this should come from research funds."[38] This stark assessment of the research team's ongoing responsibility led to one member asking whether the African communities *also* had a financial responsibility. Shouldn't the Pare and Taveta councils levy taxes or use their own funds to pay for the continued spraying? Although this idea was appealing, someone finally pointed out the obvious: that the Pare "were very poor and that it was unlikely that they would be able to afford very much."[39] The Committee ultimately recommended a small continuing research scheme whose goal would be to determine the best and cheapest methods of maintaining an area free of malaria. They believed "the results of such a scheme would be of immense value to the East African Governments."[40] However, they did not come to their decision lightly and were "seriously concerned as to whether to continue the scheme or not, on both political and humanitarian grounds."[41] They were also adamant that the costs associated with a malaria epidemic should not be the responsibility of the colonial governments, although they provided no clear advice about the shape the project should take as it moved ahead.

The Council met just a month later, in February 1960, and they also spent a great deal of time debating the Pare-Taveta Malaria Scheme and the proposed follow-up projects. They ultimately approved the continuation of the project, "but wished to emphasize that it should proceed on the basis of the search for, and treatment of malaria as a practical public health measure, and not solely as a scientific investigation."[42] The advice given by these two oversight groups indicated that other scientists and researchers working in the area understood that how the Pare-Taveta Malaria Scheme was handled would reflect not just on Pare-Taveta scientists, or on the Malaria Institute, but on future research projects.

Local Communities: It Is "Morally Wrong" to Withdraw the Spraying

Notably absent from the advisory meetings and written correspondence were the Pare and Taveta people who would have the most to lose as malaria returned. Throughout 1960, information began to emerge about what community members thought should happen as the experiment ended. Simply put, residents wanted the spraying to continue indefinitely. Throughout the years of the project, Pare and Taveta residents understood the primary intervention to be IRS, and the primary goal to be the reduction or elimination of mosquitoes. To some extent, it didn't matter how frequently the researchers restated their focus as being on malaria rather than mosquitoes, since this was a misunderstanding rooted in different systems for explaining disease. Thus, although Pare and Taveta people may have been familiar with the idea that the bite of a mosquito could lead to malaria, they were unlikely to make a distinction between different types of mosquitoes, or to believe that mosquitoes were the only way to become infected.[43] (It is also possible that participants considered malaria in adults to be a different disease than malaria in children.[44]) There was no shared understanding of malaria's etiology or pathology, which contributed to misunderstandings about the goals of the overall project how to measure the effectiveness of the interventions.

Swahili-language documents also indicate that these activities were most frequently discussed as an assumed-effective intervention to reduce malaria, rather than as an experiment with very real risks if elimination failed. This impression comes through most strongly through the researchers' and residents' reference to the insecticide as *dawa* (medicine). The problem is that dawa is assumed to be effective, and three different Swahili documents from 1957 refer to it as *dawa ya mmbu* (mosquito medicine) or *dawa ya nyumba* (house medicine).[45] In a majority of the references to the project, the researchers discuss the development (*maendeleo*), blessings (*baraka*), and benefits (*faida*) that will be delivered to people.[46] This was a calculated decision on the researchers' part, since there were other Swahili words available to explain an experiment or a test intervention. In a propaganda piece the researchers wrote for a local newspaper, they defended the potency of the insecticide by noting it had been previously "tested and investigated" (*ilipimwa na kujaribiwa*). They did, on occasion, accurately refer to their work, once calling it an "investigation" (*uchunguzi*) and in another instance describing the project as "the work of trying to reduce malaria" (*kazi ya kujaribu kuondoa ugonjwa wa malaria*) and the goal (*nia*) as "to greatly reduce and see if it's possible to eliminate entirely malaria" (*kupunguza sana na ikiwezekana kuondoa kabisa ugonjwa wa malaria*).[47] Despite these few cases of the researchers using words that indicated the experimental nature of their work (*kujaribu, uchunguzi*), the norm was to imply that the project was using medicines that had been tested, and that mosquitoes would be killed and malaria reduced as a result. In none of these documents was there any reference to potential risks, to rebound or epidemic malaria, to what would happen when the spraying stopped, or even that there would be a definitive "end" to the work.

Once it was widely known that the project would conclude, public opinion was strikingly unified. Residents wanted the spraying to be quickly restarted and continued indefinitely. The message was so clear that even the EAHC had to reluctantly admit that "there is a growing urge among the local inhabitants towards a resumption of spraying even, if they have no alternative, at their own expense."[48] The Pare Council formally stated that the spraying scheme had brought "considerable benefits which were widely appreciated and that it was morally wrong" to withdraw the spraying and let residents "suffer the consequences unaided."[49] Similar conclusions were being reached at the Taveta Council in Kenya. The field researcher stationed there wrote that the Taveta "took it for granted that they would have to put the money up themselves, that the council could not help from its present revenue and that this council would have to levy a special house tax to cover this. Everyone I spoke to seemed keen and willing that this should be done."[50] He predicted that if and when the demand fully materialized it would be "far more pressing than that made by the Pare Council."[51] A separate survey done of people in the Taveta area about their impression of the spraying found that a majority of the people interviewed said they "would be willing to contribute financially towards its continuation if this became necessary."[52] At the end of 1960, the people of Pare and Taveta had grown supportive enough of the spraying that they were willing to tax themselves so it could continue. Alas, even this was roundly discouraged by the researchers, and IRS stopped entirely.

The ending of the Pare-Taveta Malaria Scheme raises critical questions about the important—but often ignored—process of concluding a project and researchers' responsibilities upon leaving. These questions of endings and obligations are even more salient when discussing an elimination attempt for a disease like malaria, where there are great risks if the bid fails. The discussions about how to end the Pare-Taveta Malaria Scheme reveal both unsavory and uplifting aspects to the debates. On the one hand, it is abundantly clear that some participants—such as the EAHC, the WHO, UNICEF, and the Nuffield Foundation—did not want to acknowledge any long-term obligation to Pare and Taveta communities, nor did they want to engage in a discussion of researchers' responsibilities when trials end. For these groups which regularly funded field research, beginning a discussion about financial obligations *after* a trial ended appeared a fool's task. On the other hand, many of the participant organizations were tough and vocal critics of the idea that there was no obligation and that communities should be left to suffer malaria alone. Field researchers, colonial district officials, health officers, and even the director of the Malaria Institute spoke eloquently to the practical and ethical concerns involved in ending a project. Many of these men spoke humanely of the responsibility to protect people from an upsurge in malaria by eliminating unnecessary risk and ameliorating the sufferings of the sick. Those with the most foresight knew that, for research to continue productively, the wishes of scientists had to be balanced with the needs of local people, and that research ought to be conducted with great care, delicacy, and humanity.

The Pare-Taveta Malaria Scheme bridged the start and end of an era marked by attempts at global eradication—attempts that are starting anew. Privately and publicly funded projects such as the US President's Emergency Plan for AIDS Relief (PEPFAR), the Global Fund to Fight AIDS, Tuberculosis and Malaria, and the United Nations's Millennium Development Goals (MDGs) to try to "roll back" malaria have pinned their hopes on new technologies such as transgenic mosquitoes, malaria vaccines, and long-lasting insecticide-treated bed nets, in addition to drawing on older strategies such as IRS, environmental management, and mass drug administration. These efforts are partially, if not largely, attributable to the publicity, work, and funding of the Bill and Melinda Gates Foundation. Yet, most experts continue to see eradication as unrealistic. The historical record paints a mixed picture of the likelihood of elimination in the parts of sub-Saharan Africa and Asia that have the most intense transmission.[53] Sub-Saharan Africa continues to have the highest malaria morbidity and mortality rates in the world—accounting for about 81 percent of all episodes in 2010 and about 91 percent of the deaths.[54] In East Africa, malaria remains a leading cause of death for children under five.[55] Even with increased use of control strategies such as insecticide-treated bed nets, IRS with DDT, and the availability of artemisin combination drugs, malaria continues to take a huge toll in terms of human life, lost economic earnings, and slowed national growth.[56] This is not to discount very real progress that has been made since 2000: the global incidence of malaria has decreased by 17 percent, and malaria-specific mortality rates by 25 percent.[57]

While keeping track of mortality and morbidity is important, a focus on numbers can draw attention away from the ethical questions that continue to vex modern malaria control and elimination attempts. What kinds of obligations do researchers have to individual participants or the larger communities where experiments take place? Are there unique responsibilities that come with malaria control/elimination attempts because of the risks of rebound malaria? How can researchers and participants develop a common understanding of what the "end" really is? As the KEMRI/Wellcome research group on the Kenyan coast summed up these questions, "what, if anything, should be provided to research participants, and by whom, after their participation, and what, if anything, should be made available to others in the host community or country following completion of the research?"[58] These were familiar questions to the actors involved in the Pare-Taveta Malaria

Scheme, yet sixty years later we remain without good answers. No one doubts the intentions are good, but when it comes to (failed) malaria control efforts, "good intentions could cause adverse and unanticipated harms."[59]

Many of the most relevant questions around postexperiment obligations and responsibilities can be boiled down to questions of access and cost. Broadly writ, "access" means making sure the vaccine is available in the places and to the people who need it, at a price they can afford.[60] In regards to the RTS,S malaria vaccine, it remains unclear how it will be made available in malarial communities, at what cost, and whether communities where the vaccine was initially tested will have preferential access. The UK's Nuffield Council on Bioethics summarizes many of the questions that are relevant to the end of all vaccine trials, including the malaria one: "If a vaccine was found to be effective, who should provide it to the community? How many people should be treated? For how long should the vaccine be supplied? What additional costs would be involved? And most importantly, who should be responsible for meeting those costs?"[61] As the Nuffield Council makes clear, cost is a huge part of access. GlaxoSmithKline has promised to sell the vaccine "at cost plus about 5%," but there is still no actual price tag attached since the research is not yet complete.[62] It is unclear "how efficacious a malaria vaccine must be to be cost-effective."[63] The *final* calculations of whether the vaccine is cost-effective will depend on not only the final results and GlaxoSmithKline's price, but also actual (versus predicted) uptake rates, and actual (versus modeled) statistics on how the number of children vaccinated changes overall malaria transmission patterns. Once a price is set, purchasing decisions will need to account for the desires of both independent African governments and international donors, who provide a large part of East African countries' health and overall budgets. The Global Alliance for Vaccines and Immunization (GAVI) has already indicated they are considering financial support for the purchase of the malaria vaccine in the future.[64] But donor funding is a mixed blessing and raises questions about the stability of international partnerships and the long-term availability of the vaccine if it is purchased with donor funds.

Although malaria experiments are particularly problematic due to the risks of rebound malaria, questions of endings and obligations are relevant for all field research. Obligations clearly do not end when researchers declare an experiment concluded, yet there is no agreed-upon way to determine the type, scale, or duration of post-trial obligations. It does seem obvious that duties would vary and would need to take into account the type of research that was conducted, how invasive it was, the amount of time it went on, whether it introduced additional risk, and the material conditions of the research subjects. Despite the importance of the topic, modern international ethical codes and laws in the United States and Europe are vague about what should happen when experiments end.[65] Guidelines from the Council for International Organizations of Medical Sciences (CIOMS—an organization jointly created by the WHO and UNESCO in 1949) state that, "as a general rule, the sponsoring agency should agree in advance of the research that any product developed through such research will be made reasonably available to the inhabitants of the host community or country at the completion of successful testing."[66] A Nuffield Council on Bioethics report from 2002 places more of the responsibility on the national government, since if it has allowed a trial to occur it "presumably accepts some responsibility to act on the results."[67] But the calculus for why a national government would allow a research trial is far from straightforward, and, in practice, many poor nations have opted not to "act on the results" and pay for expensive new medicines. The most stringent criteria are proposed by the European Group on Ethics in Science and New Technologies, which proposes that a single standard must be used globally, even if that means "supplying the drug for a lifetime." This group argues that "the protocol of clinical trials must specify who will benefit, how and for how long," and one of their recommendations for how to deliver benefit is to "guarantee a supply of the drug at an affordable price for the community."[68]

On the other end of the spectrum are the policies of the US National Institutes of Health delineating the postexperiment obligations of researchers when HIV-positive subjects are started on antiretroviral therapy (ART). In these HIV projects, the research team is required to "make adequate plans" for how participants can continue treatment in the future. While this sounds like just the kind of responsible rule that is needed, the obligation is "to *address* the provision of post trial antiretroviral therapy," not necessarily to provide the treatment. A researcher can fulfill this obligation by merely explaining that while it would be ideal to continue treatment, none is available.[69]

The existing laws are ill-defined enough that pharmaceutical companies feel comfortable supporting the idea of providing access to newly developed drugs. GlaxoSmithKline states that it "strongly supports the goal of improving access to medicines," but goes on to clarify that such a goal does not necessarily mean "research sponsors should be routinely obliged to provide treatments to participants post trial."[70] The position paper makes clear that post-trial provision of drugs and the continued care of trial participants is the responsibility of "governments as part of national healthcare programmes" and that healthcare delivery "lies outside" the company's remit.[71] Thus, the feebleness of existing regulations allows companies to seem supportive of the idea of longer-term obligations without ever needing to financially commit to providing anything post-trial, for any amount of time.

The lack of clear national or international regulations on what should happen when an experiment ends does not mean that research subjects don't have their own ideas about what is acceptable, appropriate, or fair. When modern Kenyans were told about a hypothetical trial that would provide antiretroviral therapy and then conclude, all of the focus group participants felt it would be unfair for the drugs to be withdrawn at the end of the trial. They argued that, once started, the antiretroviral therapy ought to be continued for the duration of a research participant's life.[72] When considering a hypothetical HIV vaccine trial, a majority of the interviewed residents of Rakai, Uganda, "thought that researchers should provide some benefit to the community after a trial." They favored benefits that addressed "tangible, pressing needs in their community," such as the provision of food and basic health care.[73] When asked who should be responsible for paying for such a vaccine (if it were available), residents most frequently stated that it was a duty of "their own government or local and international aid organizations." Only 10 percent of respondents indicated that the pharmaceutical company should be responsible for paying.[74] There is clearly flexibility in determining who should be responsible for what and for how long, and these decisions must be made with the input of research participants.

Endings are inevitable, but it is time for a reimagining of what an ethical ending can and should look like. Based on what we know from the Pare-Tareva Malaria Scheme, an ethical ending relies on researchers and participants sharing an understanding of the goals and potential risks of an experiment; it means speaking early and often about the possibility of failure; it provides benefits to participants that are equivalent to both the short- and long-term risks of participation; it incorporates responsible planning for worst-case scenarios (including such known phenomena as rebound malaria); and it makes clear that research will cease when public health is at risk. Particular to HIV research that begins subjects on lifesaving ART, we may need a fundamental rethinking of how international medical research is conducted. One measure might be the creation of a continent-wide or global drug fund that all researchers must pay into, which would allow all subjects started on treatment to continue receiving it either for free or at a subsidized price. (A public-private drug fund has been in use in Thailand since 2001 with impressive results.[75]) Enacting such a proposal would dramatically change the cost and landscape of international medical research, and would be a radical but appropriate response to current failings.

A 2010 special issue of the *Lancet* addressing the issue of malaria eradication echoed the concerns of British colonial officers from the 1950s. The authors noted how much there was to lose if this latest attempt was unsuccessful. The act of "raising expectations and failing (yet again)" could set back the goal of malaria control through the innumerable people left disappointed, disillusioned, and less likely to believe the promises made about future interventions.[76] Kenyan research administrators made this point more starkly. When asked about the effects of starting HIV-positive research subjects on ART and then abruptly ending access once the trial ended, they predicted such behavior would result in "loss of trust in the doctor-patient relationship and in unwillingness by Kenyans to participate in future trials."[77] It is a gloomy warning that must be taken very seriously.

THE EXPERIMENT ENDS?

7 MODERN MEDICAL RESEARCH AND HISTORICAL RESIDUE

In this concluding chapter, I depart from the descriptive ethics that characterize most of the book and wade into the area of normative ethics—the philosophical and intellectual space where judgments are made. I turn to the question of whether any of the projects described in this book could be considered examples of "bad" science or "unethical" research, and whether we should engage in the practice of retrospective moral judgment. This requires us to recognize the particular challenges of making ethical assessments about events that occurred in the past. I then provide an overview of where the industry of international medical research is today, before returning to the small town of Gonja in Tanzania which was discussed in the book's preface.

Bad Science? Unethical Research?

What is "bad" science? Were any of the projects discussed in this book examples of bad science? There are many different ways to try to judge such a thing. One way is to consider the goal of research (to create generalizable knowledge) and whether or not the particular activity fulfilled that goal. Bad science could involve poorly designed projects that fail to generate new knowledge. Bad science may also involve a well-designed project where the knowledge generated is unimportant. Bad science may be a well-designed project that returns important results, but which, in the field of human experimentation, exposes subjects to too much risk. Bad science may be duplicative, merely reaffirming something that is already well proven and accepted. The problem with all of these descriptors is that many of them are extremely subjective and nearly all could be contested. Who decides if new knowledge is "important" or not? Who decides what is risk or what amounts to "too much" of it? There is likely to be disagreement about whether a scientific fact is "well accepted," since only through repeated testing of hypotheses do theories become firmly established. While this does not mean evaluations of science are impossible, it does mean that, in many cases, firm judgments about what is good or bad science are likely to elicit disagreement. We also have to keep in mind that bad science does not equal research that returns negative results, since an experiment showing that something does not work can be extremely valuable. A recent editorial in the *Lancet* pointed out that while US$160 billion is spent annually on medical research, there are questions about how much "good" science is going on. In a study that the editors seem to have found persuasive, it was claimed that "85% of research is wasteful or inefficient, with deficiencies in four main areas: is the research question relevant for clinicians or patients? Are design and methods appropriate? Is the full report accessible? Is it unbiased and clinically meaningful?"[1]

With the benefit of hindsight, the East African Medical Survey seems like a well-intentioned project nearly guaranteed to deliver unusable scientific conclusions. Although the project was planned on a grand scale, the design had some fatal flaws. Most fundamentally, the six locations where the EAMS collected samples were not randomly selected, and they were not geographically, ecologically, or epidemiologically representative of the larger region. This meant that no matter how many samples were collected, any attempts to generalize the findings would be problematic. Furthermore, within each of the research sites, the people surveyed would also need to be a random sample. Even though the EAMS administrators recognized the importance of the random sample and included an entire section on "randomization" in one of the project's monographs, Laurie lamented that among "the primitive peoples of Africa it is rarely possible to examine all those taken at random." The EAMS's annual report from 1954–55 described some of the barriers that the research team faced when trying to ensure a "random" sample: "The man picked may well be away, his family may not be complete, he may not want to be examined. Several times a man will say, 'Why pick on me?' While another will say, 'Why take the man next door and leave me out?' Eventually, even the headman began to lose interest and attendance dwindled off completely."[2]

The quote alludes to the EAMS's two main strategies for gathering participants: relying upon the chief to round up, coerce, or entice people to participate, and the researchers going door to door, asking every third, fifth, or tenth household to participate. While selecting households this way is a sound strategy, if too many households refuse, the sample becomes biased. The researchers' main strategy of relying upon the chief to gather participants was particularly problematic. Those who were recruited by the chief or volunteered to participate were not necessarily randomly selected. Results could be biased by over-counting more marginalized people who were more likely to be gathered by the chief, which could over-represent specific health concerns of the poor. Depending on the time of day the chief went to recruit people, only the old or infirm may have been at home rather than working in their fields, leading to a skewed demographic and disease picture. Those who appeared unbidden at the research station may have been more likely to be sick and seeking out medical care.

A counterpoint is the Makueni malaria control experiment, which was a well-designed experiment that returned clear, but negative, results. In that case, the experiment established that a single dose of a drug to treat malaria was not an effective way of reducing malaria infections. If we wanted to assess the experiment, we could argue that the problem was locally significant (people were suffering from malaria and thus effective control strategies were needed), that the intervention being testing had a reasonable likelihood of success (based on prior research), and that the project was appropriately structured. On the other hand, a case of what we may argue is bad science was the research project that *followed* Makueni.

When the Makueni scheme concluded, Shelly Avery Jones took control of another malaria control project in the Nandi Native Land Unit that was neither well designed nor minimized risk to African participants.[3] In 1953, between 80,000 and 150,000 Africans were given a single dose of the drug daraprim. The research objective was to investigate the possibility of preventing an epidemic of malaria with the use of a single administration of drugs. Avery Jones worried that the project in Nandi had been poorly planned and that the information gathered from Makueni had not been appropriately integrated. The director of medical services claimed that the project in Nandi *was* based on successful results from Makueni.[4] The director implied that the Makueni experiment had shown that malaria could be controlled with a single preventative dose of a malaria drug, but Avery Jones considered this a gross misinterpretation of the results. He wrote back, indignant:

> when the [Nandi] campaign was first suggested I thought that it was premature but was unlikely to do harm. I thought that by the time it would be likely to come into operation more experience would have been obtained from the Makueni experiment and that modifications

> might be made if necessary. Unfortunately the Makueni experiment, as you know, has shown that a rapid rise in the incidence of resistant strains of malaria can occur after mass treatment with [the drug] Daraprim but this knowledge has come too late for it to be possible to alter the arrangements made for the Nandi region. . . . The possibility, therefore, of the rise of resistant strains of malaria is quite high. As these strains are likely to be resistant both to Daraprim and to [the drug] proguanil their nuisance value may be considerable.[5]

To Avery Jones, who had been responsible for running the experiment in Makueni and for initial interpretation of the results, the idea of duplicating a failed experiment was unthinkable. What happened in Nandi—the duplication of a failed experiment, a failure to integrate past results into future research designs, and the likelihood of spreading drug-resistant forms of malaria—is clearly bad science.

We now turn toward the question of whether we can, and should, judge any of the cases discussed in this book as "unethical" research. The act of making ethical judgments about things that occurred in the past is called retrospective moral judgment, and it is a particularly fraught activity. Not only must we remember that standards and norms were different in the past, but that we are also trying to make these historical assessments in an East African context. What standard or scale should we use to judge these cases? Should it be based on what other researchers in East Africa or London did, what official policies stated, what aspirational human rights documents outlined, or what local authorities would agree to? Ethical principles are "deeply particularized" and products of their social and cultural milieu.[6] Rather than focusing on whether trials should be labeled as "unethical" or not, I believe it is far more productive to focus on the general nature of these medical research encounters and to remember the deep inequalities that characterized most of them. Individual consent was rare, and when people did consent it was often marred by threat or coercion. Researchers were often deceptive and shared only partial or misleading information with participants.

Judging projects leads us to the evaluation of individual actions and individual people. Although I am generally unwilling to make judgments about particular individuals, one example stands out among all the others I heard about in interviews and read about in the archives. Thomas Fletcher was employed by the East Africa Malaria Institute in Tanganyika between 1957 and at least 1969. He participated in a number of research projects including the multi-year Vital Statistics Survey in Pare and Taveta (described in chapter 6). Between 1962 and 1968 he was the lead researcher on a project investigating sickle cell hemoglobin. His research required the ongoing collection of blood samples from young infants (4–6 months of age) and from women of childbearing age. The project was based in the Taveta District of Kenya, which overlapped with spaces that had been regularly used for medical research in the Pare-Taveta Malaria Scheme dating back to 1954.

Between 1963 and 1965, Fletcher was in a low-level war with residents who refused to give blood and participate in his project related to sickle cell disease. In November 1963 (just a month before Kenya's independence from Britain), Fletcher first mentions that he was having "increased difficulty in persuading parents to allow me to take blood samples from infants" among residents in Kimorigo. In a letter to William Marijani, the African officer in charge of the Pare Field Station, Fletcher reported that he did "not press hard" in Kimorigo. However, five fathers in other areas *also* refused, and in those cases Fletcher was not so sanguine. He notes, "After an argument lasting about an hour in each case, these people were adamant that they were not going to cooperate." Fletcher explained the refusals as being "political" and stemming from a single man, Mr. Justice Mwokoi (or Mr. Mokoi), a representative on the local governing regional assembly. As Fletcher understood it, the dissenting fathers had visited Mwokoi after Fletcher's last blood-seeking visit and asked what action they could take. Mwokoi apparently told them, "the work is of little importance, and if they do not want to cooperate they need not."[7]

Whether it was in response to the characterization of his research as having "little importance" or the statement that they need not cooperate, Fletcher moved quickly to stamp out the dissent. In the note to Marijani he emphasized that "it is of the utmost importance that some action is taken against these people who have refused," and later referred to the group as the "six offenders."[8] Ominously, he declared that if the matter could not be solved locally, he would go to the minister of health in the central government. On the same day he wrote to Marijani about how to reapproach the fathers, Fletcher also sent a note to the district assistant in Taveta, expressing his sincere hopes that it would be possible "to approach the people who have refused, and get them to change their minds." But Fletcher was also considering what to do if the men continued to refuse. In that case, he suggested to the district assistant that the regional government could create a new policy so "the offenders can be penalised."[9]

Separate from the attempt to penalize nonparticipation, there is another troubling issue in how Fletcher described his hemoglobin research to community members. As a response to the rash of refusals, Fletcher produced a document describing the work of the Malaria Institute and going into detail explaining his own research project. In that document, which was translated into Swahili and distributed widely, he makes some troubling claims. When describing why babies (rather than adults) must give blood, he claims this is done because "it is the children who must be helped first," implying that his work is about "helping" or treating rather than research. Later in the document, he accurately states that "at present there is no cure for this disease [sickle cell] but work is being undertaken in London and elsewhere on developing a cure."[10] However, from there he makes a much more dubious claim. Implying that this "cure" was nearly discovered, he writes that, "as soon" as the cure is available, "the Malaria Institute will inform the Medical Assistant at Taveta Hospital and the children who have been found by the staff of the Institute to have SS heamoglobin can be treated."[11] Fletcher's claim that any cure would be made widely available in Taveta was a lie. There was no money set aside for such a purchase, ignoring for a moment the fact that no cure existed and that it would have been impossible to predict when a cure would be discovered or the cost and feasibility of procurement and distribution. Fletcher's statement was misleading in implying a clear plan to make a yet-undiscovered drug available in a poor, remote part of East Africa. In the same way, he argued that adult women should also give blood since SS hemoglobin could have negative effects on fertility. If it was discovered that that was the case, "steps can be taken to see if a remedy can be provided." In this case, the longed-for cure is dangled at the end of a long sentence full of modifiers implying the conditional nature of the promise. Near the end of the description of the Malaria Institute's work, Fletcher again implies benefits for participants that are not actually there. He entreats them to participate not only for their own health, but for that of people across the African continent, and concludes with a warning: "if the Taveta people refuse to cooperate . . . not only will they prevent doctors from helping them, but also prevent help being given to other people in Africa."[12] The truth was, participation or not, it was unlikely anyone was going to be directly helped by Fletcher's research.

Fletcher's sickle cell research project was still running two years later and still suffering from refusals and levels of nonparticipation he found unacceptable. That year, in 1965, Fletcher had finally convinced the District Council in Taveta to pass a law "enforcing the WaTaveta to let me pima [blood draw] them," which was being sent to Nairobi for ultimate approval or rejection.[13] When considering the possibility that the bylaw might not be passed, Fletcher admitted that, if it was not passed, "it will be extremely difficult to continue to work in the area, and I must try to persuade the Local Government authorities in Nairobi to reconsider their decision." In the meantime, with the hope that the bylaw would pass, and that the Taveta people would be *required* to let Fletcher take their blood, he compiled a list of names of past refusers "so that they can be taken to Court if the bylaw is approved."[14]

These actions are shocking and unusual. A researcher was trying to create a new law that would *require* people to participate in medical research, to give the blood of infants and women, and that would allow people who refused to participate to be taken to court. This occurred fifteen years after the Nuremberg Code established that the voluntary consent of a subject was integral to conducting ethical medical research, and that a subject must be allowed to withdraw at any time and for any reason. Fletcher made his suggestion to penalize dissenters and take nonparticipants to court just as the United Kingdom's Medical Research Council issued a statement in 1963 outlining exactly who was a volunteer and what was required to protect the voluntary participant in a research project. In this case, it seems fairly clear that not only were Fletcher's actions wrong, but so was his intent; his plan to criminalize nonparticipation was unethical. This is the only case discussed in this book where I feel comfortable saying that an individual researcher should be considered culpable for particular unethical behaviors within a medical research project he planned, led, and participated in.

The Face of Global Modern Medical Research

Globally, medical research is big business, and current estimates are that up to 40 percent of all drug trials—estimated to be "tens of thousands of studies"—are happening outside of North America and Western Europe.[15] In 2005, GlaxoSmithKline's CEO stated that a third of the company's trials took place in "low-cost" countries and he hoped to increase that number to 50 percent by 2007.[16] From what little is known about "when trials travel," the most popular destinations seem to be Eastern Europe and parts of Latin

America; no mention is made in the global literature of East Africa specifically.[17] The primary reason for the expansion of medical research in the global south is financial: research conducted in poorer countries is generally cheaper. In places where access to health care is limited, recruitment of subjects is faster, drop-out rates are lower, and compensation to participants and recruiting doctors are lower.[18] The most money is to be made on drugs that can be tested cheaply (i.e., quickly), that address chronic or lifestyle conditions that will require the drug to be taken regularly over many years, and that can be sold for high prices in the largest and richest markets (North America and Europe). This set of conditions means that many parts of the global south may be viewed as ideal testing locations for new drugs but as unlikely markets, as poor people with short life spans are unlikely consumers for expensive drugs.

However, it is difficult to be certain about these patterns since one of the very real challenges of talking about global medical research is the incompleteness of the available data. Globally, no agency collects information on the number or type of medical research trials occurring, and the picture is similarly blurry when trying to determine the scale and scope of research in East Africa. By reviewing existing databases compiled by the US Food and Drug Administration (FDA) and European Medicines Agency (EMA), one would believe that East Africa is not the site of much international research.[19] But the databases are limited. For instance, the US list includes only trials that produce a new drug that will be sold in the United States and must be approved by the FDA. Excluded are many early stage trials that don't produce a marketable drug and many other types of medical research, such as two decades worth of HIV/AIDS research conducted in Uganda related to drug regimens, drug resistance, and nutrition.[20] Surveying projects that require blood or other bodily samples (such as the Demographic and Health Survey) are also not included, nor are malaria elimination activities involving indoor residual spraying or cases of mass drug administration used to control diseases like lymphatic filariasis.

What is clear is that Kenya, Tanzania, and Uganda are sites of a large number of medical research trials related to particular diseases, such as HIV and malaria. In the most comprehensive review of published studies of HIV/AIDS research in Africa between 1987 and 2003, the East African region was by far the most heavily used.[21] More recently, Kenya and Tanzania were the site of five of the eleven African locations for the GlaxoSmithKline

RTS,S Malaria Vaccine Trial.[22] A great majority of international research funding goes toward prominent diseases such as malaria, HIV, and tuberculosis, but there remains a large class of "neglected tropical diseases" (NTDs) that could benefit from additional research funds and efforts. A 2013 study published in *Lancet Global Health* found that of the nearly 150,000 registered clinical trials for new therapeutic products in development in December 2011, only 1 percent addressed neglected diseases.[23] Research efforts into locally relevant conditions are being encouraged through the development of new research and development groups, especially public-private partnerships.[24] Some of these partnerships are explicitly targeted toward NTDs, seeking to create guaranteed markets and incentivize the development of new drugs.[25]

Today in East Africa, contract research organizations (CROs), philanthropic groups such as the Gates Foundation, and other international "partners" play an outsized role. CROs have emerged as new middlemen handling the logistics of running international research sites, recruiting subjects, and maintaining data. These activities are estimated to be part of a $10 billion enterprise employing over 100,000 people.[26] Other partners include universities and institutes based in the United States and United Kingdom that have flooded into the region since the early 1990s. In Uganda, a partnership between Makerere University and Johns Hopkins University began in 1988 with a focus on studying mother-to-child HIV transmission. Research over the next twenty-five years involved over six thousand mother and infant participants, and led to funding the construction of two new medical buildings.[27] In Kisumu, Kenya, a connection between the US Centers for Disease Control and Prevention and the Kenya Medical Research Institute, first established in 1979, grew in size and scope, with the number of research projects, staff, and annual budget increasing steadily since 2000.[28] This trend has been particularly stark in Tanzania. In 2009, a relationship with the UK Medical Research Council and the London School of Hygiene and Tropical Medicine helped construct a new laboratory at the National Institute of Medical Research site in Mwanza, Tanzania, that now handles HIV/AIDS clinical trials.[29]

These new North-South relationships are not without their critics and the anthropologist Johanna Crane has emerged as one of the most thoughtful commentators. Provocatively, she has argued that as US research universities establish "partnerships" in poor African settings, "the juggernaut of global health science is engendering a twenty-first-century academic 'scramble for Africa.'"[30] She goes on to note that the modern patterns for deciding on new medical research projects or global health priorities frequently look a lot like former colonial relationships. In regards to HIV research in Uganda,

> in the current postcolonial era, the role formerly played by the colonizing state is now partly filled by 'donors': the northern nongovernmental organizations, foundations, and governmental aid agencies that provide substantial funding and services to countries where state power has been hollowed out by structural adjustment, political unrest, and corruption. Although these providers of funding and aid can enable projects that might otherwise not be possible, they bring with them sets of expectations and priorities determined elsewhere, in much wealthier settings, which may or may not meet local scientific priorities and protocols. The result is a postcolonial science characterized by a similar 'uneasy symbiosis' of collaboration and discontent.[31]

Crane's powerful imagery of a modern-day scramble for Africa could lead one to believe the region's medical research is entirely unregulated or that equitable partnerships are impossible.

But not all international partnerships fall prey to neocolonial relations, and thoughtful programs recognize the preexisting structural inequalities. For decades, African universities, laboratories, and scientific training programs have been systematically under-resourced, which means that new programs must be conscientious and committed to funding activities meant to build both human and physical capacity.[32] There is also no denying that international medical research occupies a gray space of ethics, medicine, and the law where international regulation is not always clear and international oversight from groups such as the FDA and EMA is often weak. However, there are local actors, such as national Institutional Review Boards (IRBs), which are tasked with reviewing all proposed projects and making decisions regarding acceptability. Approval from these East Africa–based ethics committees is mandatory but not automatic. They can require changes to research protocols, informed consent forms, and practices around remuneration and follow-up, and they have the authority to shut down projects. As a way to further strengthen and empower these local groups, many international funding agencies now make the release of funds contingent upon the acquisition of ethical clearance by appropriate, competent ethics boards in the countries where the research takes place.

Unethical Medical Research Today?

If unethical research were going on today, would we necessarily know about it? It seems doubtful. We are currently relying on watchdog groups, concerned scientists, academics, and journalists to publicize cases of abuse or unethical research.[33] Their work presents a stark reminder that unethical research *is* going on globally, and that our international protections are feeble.[34]

One of the much-debated cases around the ethics of medical research in Africa occurred in 1997 with the AZT 076 drug trial in Uganda. This National Institutes of Health–funded project was testing the efficacy of a short course of the drug AZT to reduce mother-to-child transmission of HIV. Most controversially, the protocol called for a short course of the drug to be compared against a placebo, rather than a longer course of the drug, which was standard treatment in the United States. The debates were polarizing.[35] Peter Lurie and Sidney Wolfe from the watchdog group Public Citizen initially brought the trial to public attention, claiming that newborn babies were being put at unnecessary risk of contracting HIV, since preventative treatment was available, and that the use of placebos violated international ethical guidelines and US regulations.[36] The editor of the *New England Journal of Medicine,* Marcia Angell, called the trial unethical and compared it to Tuskegee.[37] The head of the NIH stood in support of the trial, defending the placebo use as a way to quickly and accurately gain important results.[38] This disagreement about whether placebos could be used when a proven therapy was available came to be referred to as the "standard of care" debate, referencing whether there should be a single standard of care globally, or whether what is given is dependent on what is locally available. In the case of AZT 076, it was argued that the local standard of care was no drug treatment at all, since a long course of AZT was too expensive to be available within Uganda, and thus that it was acceptable to provide a placebo rather than the long course of AZT available in the United States. Despite the presence of international ethical codes and regulations governing medical research, this issue remains contentious.

Other cases are more clear-cut, such as Pfizer's testing of the drug Trovan in the midst of a bacterial meningitis epidemic in Nigeria that eventually killed fifteen thousand people. The research occurred in 1996 when children in the Kano Infectious Disease Hospital were given the experimental drug. Pfizer set up their research space just across from where the humanitarian group Médecins Sans Frontières (Doctors Without Borders) was providing free, proven treatment. The experiment was meant to establish that an oral form of Trovan was as good a treatment as an intravenously delivered proven antibiotic, and two hundred children were placed in either the control or experimental group. The same number of children (four) died in each group, and it has been impossible to determine whether the deaths were a result of the experimental drug or the meningitis. It later came to light that the children in the "control" group received less than standard doses of the proven therapy, which may have increased the death rate among these children and biased the trial's data.[39] No consent forms were ever signed.

Reporters at the *Washington Post* broke the Trovan story in their "Body Hunters" series in 2000, which focused on the perils of human experimentation globally.[40] In later court cases brought against Pfizer in US courts, the parents of children who died or suffered injuries stated that they were never informed that research was occurring, were never told that the drugs their children received were experimental, and that they never consented for their child to participate. Furthermore, they were not told that free and proven treatment for the meningitis was available if they chose not to participate in the drug trial. Pfizer maintained that consent had been gained orally, that the research had been approved in both the United States and in Nigeria, and that the treatment was safe. The cases were first heard in US courts in 2001, and by 2005 all were dismissed on the grounds that wrongs committed in Nigeria could not be tried in US courts.[41] In 2009 Pfizer reached an out of court settlement worth $75 million dollars. Out of that, $35 million was placed in a fund administered by the Kano State, and allowed for payments to patients with "valid claims"; the four original families who lost children in the trial were given $175,000 each out of that fund. Ten million dollars went to cover legal costs of the Nigerian state, and an additional $30 million was for various health initiatives in Kano State.[42]

The Trovan trial technically fell under the regulatory purview of the FDA, but clearly oversight was weak. Had it not been for the reporting of journalists, the story would have likely never been widely known in the US, nor would the FDA have followed up with an investigation. The Trovan case returns us to the question, if unethical medical research was going on today, would we know about it? A US Inspector General report from 2007 found that the FDA was "unable to identify all ongoing trials and their associated trial sites and that only 1% of all trial sites was inspected between 2000 and 2005."[43] A leading American bioethicist responded by asking, "How can it be that we know how many pigs, frogs, rats and monkeys are used in research and who uses them without knowing what is going on with respect to human beings?"[44] Another report found that

> in 364 out of 1000 clinical trials checked by the US FDA the principal investigators failed to adhere themselves to the approved research plan. A stunning 140 did not bother to report adverse reactions to the investigated drug. . . . The IRB approving a major AIDS clinical trial with zidovudine required the investigators to report immediately adverse reactions by study subjects. An FDA investigation of this study revealed that none of the adverse reactions which occurred in this study were actually reported.[45]

This led another bioethicist to a logical conclusion: "It seems that we have good reasons to be worried about the implementation of research ethics standards."[46] The situation is not tangibly different with the European equivalent, the European Medicines Agency (EMA). Although they seem to be more proactive in addressing the growing number of clinical trials occurring outside of Europe, they are no better equipped to actually conduct site reviews or in-depth investigations.[47] No one should have the impression that the FDA or EMA are groups that are randomly inspecting international sites, or have the capacity to or interest in conducting independent investigations. From the information that has emerged from outside these official regulatory agencies, it seems to be not a question of *if* unethical medical research is going on, but to what degree.

The historical case studies presented in this book have shown that problems are caused not by research per se, but by how it is done: the coercion, dishonesty, and misunderstanding that characterized so many encounters past and present. It is now, at the conclusion of this book, that we can address what *ought* to be done with this knowledge. As we come to better understand cases of conflict historically, what does that mean for the present? How should this historical information inform our thinking and change practices that are still characterized by misunderstandings, injustices, and inequities?

The answer is not to stop research because there is the potential for abuse, or because there is a documented history of abuse, or even because we know abuse continues today. In this way, I agree with Harriet Washington, who has written about the historical misuse of African Americans in medical research. She argues that "acknowledging abuse and encouraging African Americans to participate in medical research are compatible goals . . . black Americans need both more research and more vigilance."[48] In the East African and larger global contexts there is no contradiction in recognizing the positive potential of medical research while still being fully aware of historical injustices and the need for stronger and stricter protections today. Potentially *because* of those historical injustices, richer countries (former colonizers), global pharmaceutical companies, and international philanthropic groups must continue research for the potential it holds to solve the health problems plaguing those who are poor and vulnerable. As the case studies showed, these "vulnerable" individuals are not people without agency, but they are people whose life options have been severely limited by structural constraints far outside their control. Medical research on diseases that plague these communities holds the potential to offer solutions and provide real benefit that may begin to offset years of inequitable encounters. But while we must continue with human experimentation, we cannot continue as we have been practicing it. Big changes must be made, the most critical of which is a commitment for real benefits to be returned to participants and their communities.

Real benefit means making newly developed medicines available in poor places, which may mean making patent exceptions for poor markets, encouraging generic drug production, or creating new public-private drug partnerships. Real benefit also means that there is a commitment to strengthening health infrastructure in poor countries—not necessarily because that is the "job" of international drug companies or NGOs, but because it is a responsibility or obligation that comes with doing research in poor places.[49] It is a way of guaranteeing that there is not a perverse incentive to keep places poor and subjects vulnerable. Real benefit may also mean providing access to drugs as a trial is running and making a commitment to provide those drugs in the future, or to contribute to larger programs that make these drugs available. When testing new combinations of ART, no HIV-positive person should be started on life-saving drugs only to be taken off them a few months later. Real benefit is exactly as it sounds: a tangible improvement in people's lives. The ultimate goal of this reimagined medical research is what Paul Farmer describes as the generation of "knowledge that can be of use in the world today" *and* that we "make sure that it is shared equitably."[50] The goal is not to end medical research, but to make it equitable and fair, and for the scientific pursuit of knowledge to be relevant and applicable to those who are in need.

Residue and Gonja

In the preface I discussed how Gonja was the first story I heard when I asked about the history of medical research in East Africa. When past and present medical researchers first told me about Gonja, it appeared to be an isolated, unexplainable act—villagers gone wild—with researchers as the unsuspecting and undeserving victims. Nearly all employees engaged in fieldwork in that part of Tanzania knew of Gonja; some version of the conflict was trotted out as evidence of how villagers lack "education" and of the conflicts caused by local residents. After hearing researchers talking so frequently about how unexplainable Gonja was, I was struck when I saw my first reference to the tiny town in the documents. The longer I spent in the archives, the more and more citations popped up.

Gonja had been used for medical research for decades, and there was a history of hostile interactions among medical researchers, government officials, and local people that put the original story in a whole new light. Researchers from the Institute of Malaria and Vector-Borne Diseases took blood samples in 1952, barging in with what the local British colonial official considered "no explanation."[51] Gonja was heavily sampled during both phases of the Pare-Taveta Malaria Scheme, which went on for five years. This meant not just that people gave blood, but that some families also allowed an observer into their homes to watch meal preparation and eating for the nutritional survey, and that the people who participated in the Vital Statistics Survey had regular physical exams and were asked to provide blood samples from their newborn children.[52] An infant welfare clinic was run in Gonja in the mid-1950s, which offered basic medical care and was meant to provide a tangible benefit to the community that was enduring the "risks" and inconveniences of the Pare-Taveta Malaria Scheme, yet the research team used the clinic as an additional site of research where they could draw blood repeatedly from sick infants and children who were brought there for treatment.[53] In 1962 and 1963, researchers worked at primary and secondary schools, collecting blood while handing out sweets.[54] In the early 1960s, Thomas Fletcher was conducting his hemoglobin research there, which required taking blood from women and babies.[55] Entomological work about mosquitoes continued into the early 1970s and human parasitological surveys, involving blood samples, were completed semiannually through 1975.[56] This was the same set of people regarding whom Bagster Wilson admitted he was afraid of "milking the same cow too often." He made this comment back in 1955, before most of the research projects listed above had even begun.[57]

Why didn't anyone mention this history of fraught medical encounters when telling the story of Gonja? Wasn't it relevant to understanding what happened? Apparently, to the researchers, it was not. During these decades of medical research in their area, the residents of Gonja had not been shy about voicing their displeasure. Long before any research vehicles were surrounded and set on fire, people had made it very clear that they didn't appreciate much of what was going on—whether it was because they discovered they were being deceived, because their participation was coerced and relied on the threat of force or fine, because there was too much risk or too few tangible benefits, or because they had seen how projects ended abruptly and with little regard for their longer-term health and welfare. During the first phase of the Pare-Taveta Malaria Scheme, which consisted primarily of indoor residual spraying of homes with chemicals, threats were made on some of the African workers' lives.[58] History indicated that feelings clearly ran hot; was it really a surprise that there was a physical conflict and conflagration around medical research?

Yet, while grappling with what *really* happened in Gonja may be interesting, it is not the most important thing. It was perhaps more important that the story had become part of the institutional memory of the new Tanzanian national research organization. A new generation of medical researchers have adopted the same rhetoric for explaining conflict and nonparticipation as colonial-era researchers. African communities and their choices to question, negotiate, or even sabotage projects were still described in much the same way they were fifty years ago. Gonja came to represent the many similarities between the past and the present. The village was a static backdrop for new researchers to experience the same difficulties and explain them with the same tired tropes: uneducated Africans and well-intentioned researchers rebuffed for no good reason. It was an old story put into service of the present.

Residue was not a concept I thought about while conducting research, but it's one that I've often returned to while writing this book. In technical terms, residue refers to the substance that remains after a process such as combustion or evaporation occurs. In general usage, residue is that which is left behind after an event has finished. Although a residue is often concrete, something that can be physically touched or observed with the naked eye, some would argue that the word "residue" may also encompass remainders that are less tangible. The residues discussed in this book are both literal and figurative. When it comes to medical research, clearly there was a palpable residue: chemical particles left behind after spraying with DDT, or molecules circulating inside people's bodies after being given antihelminthic drugs. Yet, there is also a figurative residue in these acts. Something is left behind from having people trample into one's home to spray a chemical, and when foreign experts insist that you take a pill. What's left behind is hard to predict: it may be a bitter taste of feeling coerced into participation, or gratitude for being offered a desirable intervention.

In some ways, linking the most technical definition of residue, which explicitly mentions the chemical remainders left after combustion and evaporation, with medical research is not such a stretch. In central Tanganyika in the 1950s, Hope Trant's compound really did combust. The charred compound evidenced difficult relations and inequitable research exchanges, and this conflagration affected public perception of medical research in the future. On the other end of the spectrum, some medical research projects appeared to evaporate into thin air. On the Lamu Archipelago, strong and sustained resistance from Pate Islanders meant that a planned attempt to eliminate lymphatic filariasis was canceled before it even began. Yet even the noiseless ending of this project, as H. H. Goiny's boat slipped silently back into the Indian Ocean, shaped public opinion.

As this is a history of medical research in East Africa, and a good portion of the book deals with case studies from the British colonial period, it's worth considering whether the past has a residue. Even in this realm the dictionary helps. A British dictionary lists a figurative use of the term: "the residue of the country's colonial past." Looking past the irony of the example, why does a colonial past only leave a figurative residue, and is a figurative residue any less important than a literal one? The residue left behind by history is very real, as it's made up of real events and encounters and people. This book tries to account for the residue of past encounters and show how they affect people's perceptions and behaviors. Yet, perhaps "residue" may be too benign or passive a concept to fully capture the arguments presented in this book. Past research experiences have profoundly reshaped how East Africans understand both past and present, their interactions with researchers and governments, and the ecologies they inhabit. Those experiences were the result of concrete actions and decisions made by researchers—residue doesn't merely appear out of thin air.

What the story of Gonja and the case studies in this book indicate are that local histories and past medical encounters are extremely relevant for understanding ongoing cases of disagreement. In Gonja, it wasn't important just to know that people had been asked to participate in research many different times, but to consider the nature of those experiences. Each new project—to eliminate disease, improve health, distribute drugs, or merely collect blood or bodily samples—layered atop interactions that had happened years or decades prior. After every project was completed, something remained. Every encounter left a residue. Rather than creating a smooth patina, that residue would accentuate rough spots, embalming them. There was likely to be inflammation and anger as the same mistakes and assumptions were made, time and again, and as researchers remained ignorant of the work that had come before. Residuals of past encounters lingered, and were guaranteed to be present when the next set of researchers appeared.

Appendix A

SWAHILI GLOSSARY

Swahili is a Bantu language that has been influenced by other Indian Ocean languages such as Arabic and Gujarati in addition to adopting many words from English. The words listed below are those that are either referenced in the text or which are directly relevant to the topic. A few notes:

- All verbs start with the prefix *ku–*,
- Verbs are made passive with the suffix *–isha*.
- The causative form of verbs is expressed with the suffix *–wa*.

Terms

ajili ya serikali	because of, on behalf of, the government
amri si ombi	orders not requests
bahati	luck (good or bad)
baraza	bench; informal gathering, formal public meeting
Bwana	Mister
chandarua	mosquito net
changamoto	difficulty, challenge
choo	stool (toilet)
daktari	(typically) Western-style doctor
damu	blood
dawa	medicine (good or bad)
dawa ya kienyeji	traditional medicine
dawa ya kisasa	modern medicine (Western medicine)

dawa za mitishamba	traditional medicine (medicine of the trees/ countryside)
degedege	folk illness similar to febrile malaria; convulsions
dudu	bug, insect; virus, parasite
elimu	education, knowledge
geni	foreign
gizani gizani	darkness, unknowingness
haki	rights
hatari	danger
homa	fever
imani	beliefs, faith
jambo	thing, matter
kali	fierce, sharp
kaswende	syphilis
kichocho	bilharzia
kikohozi	coughing
kipindi	season, time
kiongozi	leader
kisonono	gonorrhea
kitu	thing
kubali	to agree
kuchanja	to vaccinate
kuchanjwa	to be vaccinated
kuchoma sindano	to inject
kuchomwa sindano	to be injected
kuchukua damu	to take blood
kuchungua	to peep, to spy
kuelimisha	to educate
kufa	to die
kufanya kazi	to work, to function

kugundua	to discover
kuiba	to steal
kujaribu	to try, to test
kujenga taifa	to build the nation
kukataa	to disagree
kujificha	to hide oneself
kujitolea damu	to give blood (to give blood of oneself)
kulazimu	to be necessary
kulazimisha	to cause to be necessary (to require)
kumaliza	to finish, to die
kuumwa	to be sick, hurt, bitten
kunyonya	to suck (breast milk/blood)
kuogopa	to be afraid
kuomba	to beg (to ask, politely)
kupata chanjo	to be vaccinated
kupeleleza	to spy, pry into
kupigwa faini	to be fined
kupima	to test/check (stool, blood, urine)
kupimwa	to be tested
kupoa	to get better (to cool)
kuroga	to bewitch
kurogwa	to be bewitched
kushawishi	to persuade, coax
kutafadhali	to get better
kutafuta	to look for
kutibu	to treat
kutoa damu	to take blood
kutolewa damu	to have blood taken
kutoroka	to run away
kuugua	to get sick
kweli kweli	true
Liwali	traditional rulers, Kenyan coast (Islamic)
mabushe	hydrocele (lymphatic filariasis)
maradhi (maradi)	illness
matatizo	problems

matende	elephantiasis (lymphatic filariasis)
matibabu	treatments
mbalimbali	far away, foreign
mbu	mosquito
mchawi	witch
mchunguzi	researcher
mganga	healer
mgonjwa	sick person
miaka ya nyuma	years ago
mishipa	a medical condition (veins)
mkojo	urine
mpelelezi	researcher
mtaalamu	expert, specialist
mtafiti	researcher
Mtemi	traditional ruler, Mwanza region
Mudir	traditional ruler, Kenyan coast (Islamic)
mumiani	vampire, medicine made from blood
Mzungu	White person
ndui	smallpox
ngozi	onchocerciasis, river blindness (skin)
nguvu	strength, force
nguvu ya serikali	strength of the government
pipi	candy
porini	bush, wilderness
safura	hookworm, jaundice
sana	very
serikali	government
sharti	tool, instrument
siku hizi	these days
sindano	needle
siri	secret
siri kali	fierce, dangerous secret
sukari	sugar, diabetes
suruwa (surua)	measles

taarifa	notice, letter
tabia	personal character (unchanging)
tiba	treatment
uchafu uchafu	dirty, messy
uchawi	witchcraft
uchungu	bitter, bitterness
ugonjwa	sickness
Uhuru	Independence
ukoma	blindness
umeme	electricity
upungufu wa damu	anemia (deficiency of blood)
usaha	pus
ushauri	advice
uti wa mgongo	backbone (meningitis)
vitu	things
vyuma	rooms
wakati ule	that time, long ago
wakati wa ukoloni	time of colonialism
wasi wasi	doubts, worries
zamani	long ago
zuri	good

Appendix B

FURTHER READING ON GLOBAL MEDICAL RESEARCH

Twenty years ago Susan Lederer published a slim and erudite volume, *Subjected to Science: Human Experimentation in America before the Second World War*, an excellent introduction to the history and ethics of medical research on humans in the United States. Inspired by her brevity and organization, I've chosen to include a short essay that touches on some of the existing literature by anthropologists, historians, journalists, and public health experts that engages with medical ethics and human experimentation. This essay offers a brief overview of selected key works from this global conversation. I have chosen to highlight books, articles, and chapters that are well-researched and written in an accessible style. It is my hope that these starting points will lead interested readers deeper into the literature. Full citations for all of the works mentioned can be found in the bibliography.

Contemporary Practices

To understand how the global industry of medical research operates both in North America and the rest of the globe, start with Adriana Petryna's *When Experiments Travel: Clinical Trials and the Global Search for Human Subjects* and Jill Fisher's *Medical Research for Hire: The Political Economy of Pharmaceutical Clinical Trials.* Journalists have played an important role in bringing to light questionable practices. Although the "Body Hunters" series of articles (discussed in Chapter 7) produced by *Washington Post* reporters Mary Flaherty, Doug Stuck, Karen DeYoung, Deborah Nelson, Sharon LaFraniere, Joe Stephens, and John Pomfrey are fifteen years old, they remain moving and troubling accounts of medical research trials in India, Africa, and the United States.

Ethics in global medical research is intertwined with human rights, and authors from multiple disciplines place their work explicitly within this tradition. An outspoken critic of the unethical practices she sees plaguing the modern medical research industry, former editor of the *New England Journal of Medicine* Marcia Angell has written several provocative pieces, including "The Body Hunters" and "The Ethics of Clinical Research in the Third World." Historians David Rothman and Sheila Rothman argue in *Trust Is Not Enough: Bringing Human Rights to Medicine* that human rights must be more systematically applied to medicine, highlighting what they refer to in one chapter as "The Shame of Medical Research." Even more challenging are *Against Relativism: Cultural Diversity and the Search for Ethical Universals in Medicine* and *Double Standards in Medical Research in Developing Countries* by philosopher-bioethicist Ruth Macklin, a voracious and thoughtful critic of what she sees as double standards being used to pursue medical research in poorer parts of the globe. Physician-anthropologist and human rights champion Paul Farmer ("Rich World, Poor World: Medical Ethics and Global Inequality") is also critical of current practices but comes to a fundamentally optimistic conclusion, noting that there is an obligation to create protections that would allow for ethical research to go on, even in settings of great poverty.

As human experimentation continues in sub-Saharan Africa, scholars explore the social contexts and implications of modern research practices. In East Africa specifically, two primarily anthropological research groups based in Kenya have conducted the most sustained work on modern medical research and ethics. The first is a collaboration between the Kenya Medical Research Institute (KEMRI) and the Wellcome Trust (UK) based in Kilifi on the coast, involving Dorcas Kamuya, Geoffrey Lairumbi, Kevin Marsh, Vicki Marsh, and Catherine (Sassy) Molyneux. Their ongoing research has resulted in a prodigiously large number of articles, including a 2008 special issue (vol. 67, no. 5) of *Social Science & Medicine* on "Ethics and the Ethnography of Medical Research in Africa," and a special issue of *Developing World Bioethics* in 2013 (vol. 13, no. 1) on the topic of "Field Workers at the Interface between Research Institutions and Local Communities," which speak to some of the issues discussed in chapters 2 and 4. The second group of scholars (including P. Wenzel Geissler, Philister Adhiambo Madiega, and Ruth Prince) works primarily in western Kenya, taking advantage of the presence of a US Centers for Disease Control station. They have written about practical, theoretical, and ethical issues of medical research, including rumors of blood theft, confidentiality, cash payments, and practices of truth telling. A collaboration between these two sets of scholars resulted in the volume *Evidence, Ethos, and Experiment: The Anthropology and History of Medical Research in Africa,* edited by Geissler and Molyneux. The book is both broad and detailed in its collection of medical research accounts from across Africa.

The focused, in-depth case studies presented in medical ethnographies of Eastern and Southern Africa carefully explore many issues surrounding biomedical research. These include Stacey Langwick's *Bodies, Politics and African Healing: The Matter of Maladies in Tanzania*, and Claire Wendland's *A Heart for the Work: Journeys through an African Medical School* as well as her article, "Moral Maps and Medical Imaginaries: Clinical Tourism at Malawi's College of Medicine." I've drawn on Johanna Crane's *Scrambling for Africa: AIDS, Expertise, and the Rise of American Global Health Science* throughout this book; her article "Unequal 'Partners': AIDS, Academia, and the Rise of Global Health" is a revealing and provocative account of the ethically questionable role US academic institutions are playing in global health interventions.

In West Africa, projects in The Gambia, which has been the site of a Medical Research Council (UK) research station since 1949, have been the focus of many papers. Articles by James Fairhead, Melissa Leach, Mary Small, Ann Kelly, and Geissler have documented that local communities often differ in their understandings of medical research, the risks of giving blood, and the benefits of participation when compared to the biomedical research teams. Fairhead, Leach, and Small's article, "Where Techno-Science Meets Poverty: Medical Research and the Economy of Blood in The Gambia, West Africa," has particular resonance with themes discussed in chapter 2 of this book. Leach and Fairhead's *Vaccine Anxieties: Global Science, Child Health and Society* is a rare comparative work presenting ethnographic data about the concerns of parents in both the United Kingdom and The Gambia; particularly useful is chapter 6, which discusses experimental vaccines.

Historical Case Studies

A vast majority of the existing historical scholarship focuses primarily on cases of medical experimentation in the United States and Western Europe. Many of these works tend to focus on the twentieth century, since modern research practices were built upon the widespread acceptance of germ theory and the scientific method. Robert Baker's edited volume, *The Codification of Medical Morality*, vol. 2, *Anglo-American Medical Ethics and Medical Jurisprudence in the Nineteenth Century* addresses the early development of medical ethics in the English-speaking world, although the focus is nearly entirely on Britain and the United States. For a summary of twentieth-century developments, chapters 1–10 of the *Oxford Textbook of Clinical Research Ethics,* edited by Ezekiel Emanuel, Christine Grady, Robert Crouch, Reidar Lie, Franklin Miller, and David Wendler, are dedicated to the "History of Research with Humans." Another first-rate introduction is N. Howard-Jones's article, "Human Experimentation in Historical and Ethical Perspectives." The use of human subjects in medical research and experimentation in the United States is described in overviews such as Jonathan Moreno's *Undue Risk: Secret State Experiments on Humans*, David Rothman's *Strangers at the Bedside: A History of How Law and Bioethics Transformed Medical Decision Making*, and Susan Lederer's *Subjected to Science,* in addition to her numerous articles and book chapters.

Many studies of medical experimentation in the United States have focused on its use of vulnerable groups in institutional settings, such as children, orphans, the mentally disabled, wards of the state, and prisoners. Historians have highlighted the previously invisible use of children in medical research and established that testing on orphaned and disabled children was a widespread practice throughout the early twentieth century. Michael Grodin and Leonard Glantz's edited volume *Children as Research Subjects: Science, Ethics, and Law* provides an excellent overview, as does the more recent book by Allen Hornblum, Judith Newman, and Gregory Dober, *Against Their Will: The Secret History of Medical Experimentation on Children in Cold War America.* For an in-depth account of a particularly complex case involving the use of mentally disabled children in Hepatitis C research during the 1950s and 1960s, readers should turn to David Rothman and Sheila Rothman's *The Willowbrook Wars.* American medical researchers likewise took advantage of imprisoned populations which could be easily coerced into participation. Maurice Pappworth's *Human Guinea Pigs: Experimentation on Man,* Allen Hornblum's *Acres of Skin: Human Experiments at Holmesburg Prison*, and Jon Harkness's dissertation "Research behind Bars: A History of Nontherapeutic Research on American Prisoners" document the unjust studies carried out in prison settings and explore the many ethical questions surrounding them.

There is a particularly large and growing body of literature on the fraught history of African Americans in medical research and the modern implications of these practices. The use of African American slaves in medical research has been thoughtfully and carefully handled by Todd Savitt in "The Use of Blacks for Medical Experimentation and Demonstration in the Old South," and in Londa Schiebinger's comparative article, "Medical Experimentation and Race in the Eighteenth-Century Atlantic World."

The most well-known example of unethical medical research with African Americans is the Tuskegee Syphilis Study, and there is a large body of literature exploring this case. Susan Reverby's edited volume, *Tuskegee's Truths: Rethinking the Tuskegee Syphilis Study*, presents a well-rounded explanation of the trial, including writings by the researchers and interviews with the participants. Her later book, *Examining Tuskegee: The Infamous Syphilis Study and Its Legacy,* places the trial in a larger context and provides an excellent starting point, as do James Jones's *Bad Blood: The Tuskegee Syphilis Experiment* and Allan Brandt's article, "Racism and Research: The Case of the Tuskegee Syphilis Study."

Post-Tuskegee, African Americans have remained skeptical of biomedicine and reluctant to participate in medical research trials. Rebecca Skloot's *The Immortal Life of Henrietta Lacks* is an elegant reminder of how rumors and mistrust within minority communities are often rooted in actual misdeeds. In *Medical Apartheid: The Dark History of Medical Experimentation on Black Americans from Colonial Times to the Present,* Harriet Washington describes African Americans being subjected to unethical medical research practices for hundreds of years. Other minority groups within the United States were not immune to these injustices, although there has been less historical research. An exception is Peter Eichstaedt's *If You Poison Us: Uranium and Native Americans,* which describes how Native Americans were used to mine uranium without being clearly told of the risks, and were then subjected to observational research experiments. More recent work by the anthropologist Roberto Abadie in *The Professional Guinea Pig: Big Pharma and the Risky World of Human Subjects* charts the rise of a whole new class of individuals (many of whom could be termed vulnerable because of their poverty, education status, or ethnic background) who systematically move from one experiment to another, considering it as an occupation.

An important theme within the history of human experimentation globally is the relationship between medical research and the military, and particularly research carried out during times of war. The association was pragmatic, since research efforts were focused on topics affecting soldiers' health, and researchers also recognized soldiers as an easily controlled population where questions of consent were largely moot. The most infamous example of wartime research is that carried out by Nazi scientists during World War II. The literature related to Nazi medical research and ethics is vast, and thoughtful syntheses such as George Annas and Michael Grodin's *The Nazi Doctors and the Nuremberg Code: Human Rights in Human Experimentation*, Robert Proctor's *Racial Hygiene: Medicine under the Nazis*, and Paul Weindling's *Nazi Medicine and the Nuremberg Trials: From Medical War Crimes to Informed Consent* are ideal places to start. Multiple works have also been produced documenting the Japanese World War II experiments, including Sheldon Harris's *Factories of Death: Japanese Biological Warfare 1932–45 and the American Cover-up* and David Wallace's *Unit 731: Japan's Secret Biological Warfare Unit in World War II*. Although Africans were generally not involved in the Nazi experiments, an interesting connection is presented by Christian Bonah's chapter, "'You Should Not Use Our Senegalese Infantrymen as Guinea Pigs': Human Vaccination Experiments in the French Army," which also provides helpful comparative information.

Works that are formally comparative in nature are few and there is no book-length work examining the history of human experimentation across continents, cultures, or time periods. However, the handful of comparative studies that have been done are all of very high quality. Baader et al.'s "Pathways to Human Experimentation, 1933–1945: Germany, Japan, and the United States" recounts the different practices used by a German research organization and the American and Japanese governments. Reverby's "Ethical Failures and History Lessons: The U.S. Public Health Service Research Studies in Tuskegee and Guatemala" presents a brief introduction to both the Tuskegee Syphilis Study and the Guatemala sexually transmitted diseases inoculation research, making clear how unremarkable the practices of deception were from the 1930s throughout the 1960s. Specific to Africa, Helen Power's chapter, "'For Their Own Good': Drug Testing in Liverpool, West and East Africa, 1917–1938," finds there was little difference in how British soldiers and their African counterparts were treated during clinical drug trials. Wolfgang Eckart's "The Colony as Laboratory: German Sleeping Sickness Campaigns in German East Africa and in Togo, 1900–1914" examines early research practices in East and West Africa, documenting how dangerous drug trials worked to gather new experimental data.

In general, the history of human experimentation in Africa remains sparsely researched, although attention has grown in the past few years. Particular to East Africa, the most recent contributions can be found in a 2014 special issue (vol. 47, no. 3) of the *International Journal of African Historical Studies*, titled "Incorporating Medical Research into the History of Medicine in East Africa." Articles within the special issue include contributions from Patrick Malloy on the emergence of biomedical research in colonial Tanganyika; Jennifer Tappan on kwashiorkor research on children in Uganda; Mari Webel on German sleeping sickness camps at the turn of the century; and my more detailed exploration of the Pare-Taveta Malaria Scheme discussed in section 6. Helen Tilley's conclusion asks a series of broader questions about what constitutes medical research and the significance of place. The work of these scholars has tended to be participant-oriented in that they are concerned with African understandings of, and responses to, biomedical interventions. In this regard, the contributions of Luise White are difficult to overstate. Her innovative work, *Speaking with Vampires: Rumor and History in Colonial Africa*, on vampire and blood rumors was an early example of illuminating how medical practices took on new meanings in African settings.

Although often not directly focused on medical research, discussions of these topics are incorporated into disease-specific accounts, particularly of the efforts to treat, control, and eliminate sleeping sickness and malaria. These works capture the frequent blurring of therapeutic and experimental medical work and the scale of colonial interventions, and hint at some cases of conflict and misunderstanding. The literature on sleeping sickness in East Africa is vast, but selected works that explicitly raise questions related to human experimentation and ethics include Maryinez Lyons's *The Colonial Disease: A Social History of Sleeping Sickness in Northern Zaire, 1900–1940*, Kirk Hoppe's *Lords of the Fly: Sleeping Sickness Control in East Africa, 1900–1960*, and Deborah Neill's *Networks in Tropical Medicine: Internationalism, Colonialism, and the Rise of a Medical Specialty, 1890–1930* (particularly chapters 4 and 6).

East Africa has also been the site of dozens of malaria control programs and elimination attempts dating back more than a century. The Pare-Taveta Malaria Scheme was one of the largest, and Mary Dobson, Maureen Malowany, and Robert Snow's "Malaria Control in East Africa: The Kampala Conference and the Pare-Taveta Scheme" presents a careful discussion of the scientific debates raised around the appropriateness of attempting experimental control measures. James Webb and Randall Packard's works have documented many other experimental malaria efforts across Africa, while asking questions about the feasibility and ethics of such activities. Webb's *The Long Struggle against Malaria in Tropical Africa,* in addition to his article, "The First Large-Scale Use of Synthetic Insecticides for Malaria Control in Tropical Africa," and Packard's *The Making of a Tropical Disease* (especially chapters 6, 7, and 8), as well as his article, "'No Other Logical Choice': Global Malaria Eradication and the Politics of International Health in the Post-War Era," present superb introductions to how malaria has been constructed, controlled, and debated, and the ethical questions that arise on both local and global scales.

Another way to illuminate the history of medical research is to focus on the colonial nature of the enterprise: the organizations that funded research, the individuals who created and ran projects, and the details of the projects themselves—how they were imagined, run, and assessed. While this literature discusses projects carried out in Africa, or colonial workers who labored in Africa, and clearly describes conditions on the continent, Africans (as participants, subjects, receivers, and adapters) are not always the primary focus. Helen Tilley's *Africa as a Living Laboratory: Empire, Development, and the Problem of Scientific Knowledge, 1870–1950*, covers scientific research across much of British colonial Africa, and chapter 4 presents detailed information about medical research carried out prior to 1940. Anna Crozier's *Practising Colonial Medicine: The Colonial Medical Service in British East Africa* reconstructs the ethos and norms of the men who participated in the colonial medical service and also often worked as medical researchers. Articles by Sabine Clarke focus on the Colonial Office in London, which not only funded much of the medical research in Africa but was also instrumental in setting research priorities.

There is also a small but growing literature about the history of human experimentation in West and Francophone Africa. In French, the work of Jean-Paul Bado focuses on leprosy, sleeping sickness, and river blindness—all of which were targeted for treatment and experimental efforts. Also in French is Christian Bonah's work on the history of human experimentation in France between 1900 and 1940. Eric Silla's *People Are Not the Same: Leprosy and Identity in Twentieth-Century Mali* is explicitly oriented toward sharing sufferers' accounts (often in their own words), but chapters 5–7 also present detailed information about colonial approaches and experimental treatments given during French colonial mobile campaigns and inside leprosariums. The Pasteur Institute was also active in encouraging medical research across the continent, and innovative works by Guillaume Lachenal, Anne Marie Moulin, and Clifford Rosenberg in French and English describe the general French colonial approach while also providing a detailed history of the Pasteur Institute in places as diverse as Cameroon, Madagascar, and Algeria.

NOTES

Preface

1. East African Medical Survey Annual Report, 1952, 1.

2. Stanley George Browne, "The Indigenous Medical Evangelist in Congo," *International Review of Mission* 35, no. 1 (1946): 59–67.

3. In a survey of 650 people, 570 were anemic, 425 were malnourished, over 300 had malaria, and 159 had tuberculosis. *Parliamentary Debates*, H.C. (5th ser.) (1937), vol. 324, col. 1115, quoted in Robert Pearce, *The Turning Point in Africa: British Colonial Policy, 1938–1948* (London: Frank Cass, 1982), 15.

4. R. M. Murray-Lyon, "Important Diseases Affecting West African Native Troops," *Transactions of the Royal Society of Tropical Medicine and Hygiene* 37, no. 5 (1944): 287–96.

5. Adriana Petryna, *When Experiments Travel: Clinical Trials and the Global Search for Human Subjects* (Princeton: Princeton University Press, 2009), 18.

6. Christopher Pinney and Nicolas Peterson, eds., *Photography's Other Histories* (Durham: Duke University Press, 2003), 10.

7. Margaret Liu and Kate Davis, *A Clinical Trials Manual from the Duke Clinical Research Institute: Lessons from a Horse Named Jim*, 2nd ed. (Durham: Duke Clinical Research Institute, 2010), 17–20.

8. It's likely that Amani is the longest-used research station in the East Africa region. Andrew Zimmerman, "'What Do You Really Want in German East Africa, Herr Professor?': Counterinsurgency and the Science Effect in Colonial Tanzania," *Society for Comparative Study of Society and History* 48, no. 2 (April 2006): 437; Christopher Conte, "Imperial Science, Tropical Ecology and Indigenous History: Tropical Research Stations in Northwestern German East Africa, 1896 to the Present," in *Colonialism and the Modern World*, ed. Gregory Blue, Martin Bunton, and Ralph Crozier (Armonk, NY: ME Sharpe, 2002), 246–61.

9. A rough catalog of the materials in Amani is available for download on my website, http://pages.uoregon.edu/graboyes.

Acknowledgments

1. Melissa Graboyes, "Introduction to the Special Issue: Incorporating Medical Research into the History of Medicine in East Africa," *International Journal of African Historical Studies* 47, no. 3 (2014): 379–98. Melissa Graboyes, "'The Malaria Imbroglio': Ethics, Eradication, and Endings in Pare Taveta, East Africa, 1959–1960," *International Journal of African Historical Studies* 47, no. 3 (2014): 445–72. Melissa Graboyes, "Fines, Orders, Fear . . . and Consent? Medical Research in East Africa, c. 1950s," *Developing World Bioethics* 10, no. 1 (2010): 34–41.

THE EXPERIMENT BEGINS

Chapter 1: Medical Research Past and Present

1. East African Medical Survey, Monograph No. 1, 1954.

2. Donald Low and John Lonsdale, "Introduction: Towards the New Order, 1945–1963," in *History of East Africa, Volume III*, ed. Donald Low and Alison Smith (Oxford: Clarendon Press, 1976), 13.

3. Clarke reports that the EAMS was originally conceived as a project to study African workers involved with the ill-fated Groundnut Scheme. This idea was abandoned. Sabine Clarke, "The Research Council System and the Politics of Medical and Agricultural Research for the British Colonial Empire, 1940–52," *Medical History* 57, no. 3 (May 30, 2013): 350. The EAMS was originally based in Malya, Tanganyika, but moved to Mwanza in 1954. From the start, the Filariasis Research Unit (FRU) was intimately linked with the EAMS. It, too, was founded in 1949 as a project sponsored by the Medical Research Council and the Secretary of State, shared facilities with the EAMS in Mwanza, and was also directed by Colonel Laurie. With these separate projects sharing facilities and a director, staff also ended up being shared. In addition, there was sharing of research sites such as Ukara Island, where the EAMS did a health survey and the FRU was attempting to eradicate lymphatic filariasis and test a new set of drugs to treat filariasis. Between 1949 and May 1954, large amounts of research were conducted: in Msambweni alone, 150 villages were visited and 50,000 night blood samples were collected. The EAMS combined with the FRU in 1954 and was renamed the East African Institute for Medical Research.

4. Director Colonial Medical Research to Dr. Lewthwaite, cc Col. Davidson, "Integration of Laboratories of Medical Survey and Filariasis Unit," September 14, 1949. East African Medical Survey, NIMR, Mwanza.

5. Trant to EAMS Director, April 12, 1950, Dr. Trant Files, NIMR, Mwanza.

6. Trant to Laurie, December 4, 1953, Dr. Trant Files, NIMR, Mwanza.

7. Laurie, "Field Medical Staff East African Medical Survey, Work 1952," n.d. [July–August 1952, Dr. Trant Files, NIMR, Mwanza.

8. Vaughan to Anderson, April 16, 1952, MOH/22/15, KNA.

9. East African Medical Survey Annual Report, 1953.

10. Unknown author and recipient, "Personal and Confidential," August 25, 1955, NIMR, Mwanza, Emphasis added.

11. Helen Tilley, *Africa as a Living Laboratory: Empire, Development, and the Problem of Scientific Knowledge, 1870–1950* (Chicago: University of Chicago Press, 2011).

12. John Iliffe, *East African Doctors: A History of the Modern Profession* (Cambridge: Cambridge University Press, 1988); Johanna Crane, *Scrambling for Africa: AIDS, Expertise, and the Rise of American Global Health Science* (Ithaca: Cornell University Press, 2013).

13. During the project's final years, it was discovered in experiments outside of Africa that low, continuous doses of chloroquine could lead to blindness; it does not appear this information was ever shared with Mto wa Mbu community members. Tanganyika Medical Department Annual Reports, 1961, 1962; East African Institute of Malaria and Vector-Borne Diseases Annual Reports 1968, 1969, 1971, 1972–73.

14. W. S. Haynes, "The Treatment of Pulmonary Tuberculosis in Kenya Africans with Thiacetazone," *East African Medical Journal* 29, no. 9 (September 1952): 339–55.

15. Mwangi to Director of Medical Services, Kenya, "King George VI Hospital," August 14, 1961, BY/22/49, KNA.

16. Udo Schuklenk, "Protecting the Vulnerable: Testing Times for Clinical Research Ethics," *Social Science & Medicine* 51, no. 6 (2000): 969.

17. N. A. Christakis, "Ethics Are Local: Engaging Cross-Cultural Variation in the Ethics for Clinical Research," *Social Science & Medicine* 35, no. 9 (1982): 1080.

18. Maryinez Lyons, *The Colonial Disease: A Social History of Sleeping Sickness in Northern Zaire, 1900–1940* (Cambridge: Cambridge University Press, 1992): 110. Emphasis in original.

19. Itesio Leprosarium Patients [about six, signatures unreadable] to Director of Medical Services, Kenya, "Petition for the Better Treatment are Required by the Sick-Patients of Leprosy," November 13, 1957, BY/42/8, KNA.

20. This definition of medical research is in accord with that given by groups such as the National Institute of Health (US), the General Medical Council (UK), and the Medical Research Council (UK). One of the weaknesses of this definition is that it does not acknowledge the vast middle space between curative medicine (proven therapies given by doctors to patients with the goal of improving individual health) and medical research on humans (experiments run by researchers on subjects with the goal of collecting data and generating generalizable knowledge). In between treating patients and experimenting on human subjects are activities that can be harder to define, which some scholars refer to as "experimental treatment," "innovative therapy," "non-validated innovative treatment," or "therapeutic research." These terms emphasize the fact that research does sometimes benefit an individual participant. But, on the whole, I believe these terms—and this general category—obscure more than they reveal, since the goals of doctoring and researching are in tension.

21. This is a low estimate based only on the references to trials I've personally seen. I have only counted trials that were clear about the number of subjects they included, and have used the researchers' own numbers. An Excel file with the quantitative data is available online at: http://pages.uoregon.edu/graboyes.

22. East African Medical Survey and Research Institute Annual Report, 1954–55; Tanganyika Medical Department Annual Report, 1950.

23. These consisted of 5,900 blood samples in 1951 and 14,000 samples from the general population and 5,000 from hospital patients in 1952. East African Medical Survey Annual Reports 1951, 1952.

24. Colonial Medical Research Committee Tenth Annual Report, 1954–55. The Filariasis Research Unit was subsequently absorbed into the East African Medical Survey.

25. Tanganyika Medical Department Annual Reports, 1941, 1944, 1949, 1950.

26. Trant is quoting Dr. Donald Bagster Wilson. Trant to Bozman and Holmes, May 22, 1955, Dr. Hope Trant Files, NIMR Mwanza.

27. "Sleeping Sickness and Reclamation Kigoma and Kasulu Districts," 523A/M.11/2, TNA. Thanks to Julie Weiskopf for sharing this citation.

28. Tanganyika Medical Department, Western Region Annual Report, 1950, 450/1501/4, TNA. Thanks to Julie Weiskopf for sharing this citation.

29. Vicki Marsh et al., "Working with Concepts: The Role of Community in International Collaborative Biomedical Research," *Public Health Ethics* 4, no. 1 (2011): 37.

30. Ibid., 36.

31. Ibid., 31.

32. Melissa Graboyes, "From *Mumiani* to Transfusions: Economies of Blood in East Africa" (paper presented at University of Oregon, May 2014).

33. Per Marx, in a commodity economy, persons and things can be divided, a commodity is an object, and a commodity has a value. That commodity can then be exchanged between people. Karl Marx, *Capital: Volume 1: A Critique of Political Economy* (London: Vintage Books, 1976), esp. "Part I: Commodities and Money."

34. Not just a modern economy in blood, but also in albino body parts. These are believed to be potent as yet another form of medicine, and thus command high cash prices and lead to the killing and mutilation of albinos. This has been reported from in western Tanzania.

35. Parker Shipton, *The Nature of Entrustment: Intimacy, Exchange, and the Sacred in Africa* (New Haven: Yale University Press, 2007), 28.

36. Ruth Prince, "HIV and the Moral Economy of Survival in an East African City," *Medical Anthropology Quarterly* 26, no. 4 (2012): 536. She is drawing upon Angelique Haugerud, *The Culture of Politics in Modern Kenya* (Cambridge: Cambridge University Press, 1993); John Lonsdale, "The Moral Economy of Mau Mau: Wealth, Poverty and Civic Virtue in Kikuyu Political Thought," in *Unhappy Valley: Conflict in Kenya and Africa,* Book 2, *Violence and Ethnicity*, ed. Bruce Berman and John Lonsdale (London: James Currey, 1992), 315–468; J. P. Olivier de Sardan, "A Moral Economy of Corruption in Africa?" *Journal of Modern African Studies* 37, no. 1 (1995): 25–52.

37. A great example of this economy in bodily tissues and the moral economy that sprung up around it is chronicled in Warwick Anderson, *The Collectors of Lost Souls: Turning Kuru Scientists into Whitemen* (Baltimore: Johns Hopkins University Press, 2008). Over multiple decades, there were near-constant requests for blood, tissue, and brains by foreign scientists. The Fore were generally willing to give blood in the early years of the work; they saw this as a way "to enter into, or cement, a relationship with the scientist; they sought to make him indebted to them or to reciprocate for his donations of cargo." Anderson, *Collectors of Lost Souls*, 95.

38. James Fairhead, Melissa Leach, and Mary Small, "Where Techno-Science Meets Poverty: Medical Research and the Economy of Blood in The Gambia, West Africa," *Social Science & Medicine* 63, no. 4 (2006): 1119.

39. Sherry Ortner, "Resistance and the Problem of Ethnographic Refusal," *Comparative Studies in Society and History* 37, no. 1 (1995): 173–93.

40. Arthur Kleinman, *Writing at the Margin: Discourse between Anthropology and Medicine* (Berkeley: University of California Press, 1995), 67.

41. Wenzel Geissler et al., "'He Is Now Like a Brother, I Can Even Give Him Some Blood'—Relational Ethics and Material Exchanges in a Malaria Vaccine 'Trial Community' in The Gambia," *Social Science & Medicine* 67, no. 5 (2008): 696–707.

42. Ibid., 700.

43. Ibid., 697; P. Wenzel Geissler, "Public Secrets in Public Health: Knowing Not to Know While Making Scientific Knowledge," *American Ethnologist* 40, no. 1 (2013): 18.

44. Iliffe, *East African Doctors*, 8.

45. Steven Feierman and John Janzen, eds., *The Social Basis of Health and Healing in Africa* (Berkeley: University of California Press, 1992); Steven Feierman, "Struggles for Control: The Social Roots of Health and Healing in Modern Africa," *African Studies Review* 28, no. 2 (1985): 73–147; Randall Packard, *White Plague, Black Labor: Tuberculosis and the Political Economy of Health and Disease in South Africa* (Berkeley: University of California Press, 1989); Megan Vaughan, *Curing Their Ills: Colonial Power and African Illness* (Stanford: Stanford University Press, 1991).

46. Crane, *Scrambling for Africa*; Stacey Langwick, "Devils, Parasites, and Fierce Needles: Healing and the Politics of Translation in Southern Tanzania," *Science, Technology & Human Values* 32, no. 1 (2007): 88–117; Stacey Langwick, "Articulate(d) Bodies: Traditional Medicine in a Tanzanian Hospital," *American Ethnologist* 35, no. 3 (2008): 428–39; Julie Livingston, *Debility and the Moral Imagination in Botswana* (Bloomington: Indiana University Press, 2005); Claire Wendland, *A Heart for the Work: Journeys through an African Medical School* (Chicago: University of Chicago Press, 2010); Susan Reynolds Whyte, *Questioning Misfortune: The Pragmatics of Uncertainty in Eastern Uganda* (Cambridge: Cambridge University Press, 1997).

PERCEPTIONS

Chapter 2: East African Perceptions of Medical Research

1. Bilharzia (Schistosomiasis, species *Schistosoma mansoni* and *S. haematobium* present in East Africa) is a parasitic infection transmitted by infected freshwater snails living in all of the Rift Valley lakes, including Lake Victoria. A fluke can attach within the urinary or gastrointestinal tract and cause irritation and bleeding in addition to longer-term problems. Once individuals are infected, their urine or feces contain the flukes which, when they enter a body of water, complete the cycle of infection. Bilharzia disproportionately affects children, probably because of bathing and swimming in infected water. Gordon Cook and Alimuddin Zumla, eds., *Manson's Tropical Diseases*, 22nd ed. (Philadelphia: Saunders, 2009), 1425–60.

2. Interview with Mama Christine Nzito, August 8, 2008, Bukumbi, Tanzania.

3. Ibid.

4. Ministry of Health Annual Report, Mwanza Region, Tanzania, 1965.

5. Interestingly, in a joint WHO and Tanzanian project to address bilharzia in this region during the 1970s, an early project report seemed to reference the earlier deaths by noting that "utmost care must be taken to select the most appropriate drug for mass chemotherapy, and also to determine the most satisfactory method of its administration to a scattered population. A single death attributable to treatment could jeopardize the project irrevocably." East African Institute for Medical Research Annual Report, 1969, 42.

6. Even today, when there are at least two different effective treatments for bilharzia (metrifonate and praziquantel), there is still research being done on the ideal treatment regimen. East African Institute for Medical Research Report, 1963–64.

7. Interview with Bibi Teresa Wilburt and Mzee Edward Bunduki, August 8–9, 2008, Bukumbi, Tanzania.

8. The information presented in this paragraph is compiled from about two dozen formal interviews and dozens of informal conversations with people in Mwanza Region, Tanzania, in July and August 2008.

9. *Bwana* followed by a descriptor of a person's work is not an uncommon title. Pels reports how the cattle and game officer was often called Bwana Nyama (Mr. Meat), and Alec Smith, an entomologist, was referred to as Bwana Dudu—Mr. Insect (or, as he translated it, Insect Man). Peter Pels, "Mumiani: The White Vampire: A Neo-Diffusionist Analysis of Rumour," *Etnofoor* 5, nos. 1–2 (1992): 168; Alec Smith, *Insect Man: A Fight against Malaria in Africa* (New York: Palgrave Macmillan, 1993).

10. Luise White also makes a distinction between European-styled vampires who drink blood and East African vampires who desire blood as a commodity but don't drink it or need it to live. Luise White, *Speaking with Vampires: Rumor and History in Colonial Africa* (Berkeley: University of California Press, 2000).

11. During daytime hours, microfilariae retreat deeper into the body and samples of blood will give false negatives. This behavior of becoming active in the peripheral blood at night is the result of symbiotic evolutionary development with the main vector. Since mosquitoes typically bite during the night, if the microfilariae are active at night in the peripheral blood they increase their chances of being sucked up by a mosquito and passed on to another person.

12. The information presented in this paragraph is compiled from about two dozen formal interviews and dozens of informal conversations with people in Mwanza Region, Tanzania, in July and August 2008.

13. This is also a comment on untrustworthy state institutions, such as the police, that pillage and terrorize local people more than they protect them. Interview with Mzee Thomas Inyassi, August 8, 2008, Kigongo, Tanzania.

14. The information presented in this paragraph is compiled from about two dozen formal interviews and dozens of informal conversations with people in Mwanza Region, Tanzania, in July and August 2008.

15. Interview with Mzee Thomas Inyassi, August 8, 2008, Kigongo, Tanzania.

16. After a person is bitten by an infected mosquito, microfilariae (tiny worms) lodge in the lymphatic system. Damage to this system causes swelling of the testicles

and scrotum (hydrocele) and legs (elephantiasis). The disease takes many years, often decades, of receiving infected mosquito bites for a person to develop obvious symptoms. The disease is not fatal on its own, although it can lead to complications such as gangrene and sepsis, which can cause death. Cook and Zumla, *Manson's Tropical Diseases*, 1477–91.

17. By the end of June 1954, a filarial survey conducted by the East African Medical Survey and Research Institute had collected 50,000 night blood slides from Tanganyika alone. East African Medical Survey and Research Institute Annual Report, 1954–55.

18. East African Medical Survey Annual Report, 1951.

19. Interview with Mzee Thomas Inyassi, August 7, 2008, Kigongo, Tanzania.

20. Interview with Mama Christine Nzito, August 8, 2008, Bukumbi, Tanzania.

21. Anne Marie Moulin, "Defenseless Bodies and Violent Afflictions in the Global World: Blood, Iatrogenesis, and Hepatitis C Transmission in Egypt," in *Global Health in Africa: Historical Perspectives on Disease Control*, ed. Tamara Giles-Vernick and James L. A. Webb, Jr. (Athens: Ohio University Press, 2013), 138–58; G. T. Strickland, "Liver Disease in Egypt: Hepatitis C Superseded Schistosomiasis as a Result of Iatrogenic and Biological Factors," *Hepatology* 43, no. 5 (May 2006): 915–22.

22. Guillaume Lachenal, "The Doctor Who Would be King," *Lancet* 376, no. 9748 (9 October 2010): 1216–17.

23. Jacques Pépin and A. C. Labbé, "Noble Goals, Unforeseen Consequences: Control of Tropical Diseases in Colonial Central Africa and the Iatrogenic Transmission of Blood-Borne Viruses," *Tropical Medicine & International Health* 13, no. 6 (June 2008): 744–53; Jacques Pépin et al., "Risk Factors for Hepatitis C Virus Transmission in Colonial Cameroon," *Clinical Infectious Diseases* 51, no. 7 (October 2010): 768–76; Richard Njouom et al., "The Hepatitis C Virus Epidemic in Cameroon: Genetic Evidence for Rapid Transmission between 1920 and 1960," *Infection, Genetics and Evolution* 7, no. 3 (June 2007): 361–67; G. T. Strickland, "An Epidemic of Hepatitis C Virus Infection While Treating Endemic Infectious Diseases in Equatorial Africa More Than a Half Century Ago: Did It Also Jump-Start the AIDS Pandemic?" *Clinical Infectious Diseases* 51, no. 7 (October 2010): 785–87.

24. Interview with Bibi Agnes Kagera, August 8, 2008, Bukumbi, Tanzania.

25. Interview with Father Mwalimu, August 11, 2008, Mwanza Parish, Tanzania.

26. The mistaken idea that research is meant to help the individual is not confined to Africa or the developing world. Many Americans, when interviewed about what medical research is, erroneously state that medical research is being done to help treat them. This idea is commonly referred to as "therapeutic misconception."

27. Interview with Bibi Agnes Kagera, August 8, 2008, Bukumbi, Tanzania.

28. Interview with Mzee Masatu Ndege and Mzee Manyama Mtaki, August 26, 2008, Bwisya, Ukara Island, Tanzania.

29. Interview with Mzee Thomas Inyassi, August 7, 2008, Kigongo, Tanzania. There is slippage in terms. *Mganga* is not always used exclusively to discuss a traditional healer or herbalist, and *daktari* does not always refer to a Western-style doctor.

The terms were often used interchangeably and for many of the interviewees Swahili was not their first language.

30. This chart is based on Frederick Johnson, *A Standard Swahili-English Dictionary* (Oxford: Oxford University Press, 1939).

31. In my own interviews I usually used *utafiti*. Many of the people who work as medical researchers also chose to call themselves *mtafiti*. People would also use *uchunguzi* at times. No one ever used any of the *kupeleleza* words around me.

32. Patrick Thomas Malloy, "'Holding [Tanganyika] by the Sindano': Networks of Medicine in Colonial Tanganyika" (PhD diss., UCLA, 2003), 406, citing an interview with Charles Chambika, August 21, 1996, Tengeni.

33. Interview with Mzee Thomas Inyassi, August 7, 2008, Kigongo, Tanzania.

34. The Swahili word *mtaalum* is probably related to the Arabic words *muta'leem* (literate) and *aalim* (scholar). The original Arabic has likely been modified in Swahili with the addition of the *m* at the beginning to signify the *m-/wa-* noun class. Thanks to Natalie Mettler for help translating and tracing the roots. Munir Baalbaki and Rohi Baalbaki, *Al-Mawrid Al-Quareeb Pocket Dictionary, English-Arabic Arabic-English*, 20th ed. (Beirut: Dar El-Ilm Lilmalayin, 2007).

35. Some might argue that I was told more frequently about the "good" parts of research because interviewees didn't want to offend me, they thought I wanted to hear about positive things, or they thought they might get something if they talked about the positives. I don't think any of these reasons are true, mostly because I am convinced through talking with people and observing rural conditions in East Africa that medical research actually *is* beneficial. Individuals often benefited from visits with doctors, simple treatments, or experimental drugs for conditions that would have otherwise remained untreated. The one trend that I did notice running through my data was that Christians were much more likely to link Western medicine and medical research with their local mission station or individual Fathers/Sisters and presented a much more benevolent view of research.

36. This is another case of people commenting on reality, since international researchers increasingly resemble doctors. In a quest to make research beneficial to communities, researchers distribute drugs, making it harder to distinguish between treatment and research, doctoring and experimenting.

37. Interview with Mama Christine Nzito, August 8, 2008, Bukumbi, Tanzania.

38. Simeon Mesaki, "Witchcraft and Witch-Killings in Tanzania: Paradox and Dilemma" (PhD diss., University of Minnesota, 1993), 89. Other works report similar findings. Malloy, "'Holding [Tanganyika] by the Sindano,'" makes this point in chapter 2, discussing the similarities between "Healers and Harmers."

39. Marja-Liisa Swantz, *Blood, Milk and Death: Blood Symbols and the Power of Regeneration among the Zaramo of Tanzania* (London: Bergin & Garvey, 1995), 112; Susan Beckerleg, "Maintaining Order, Creating Chaos: Swahili Medicine in Kenya" (PhD diss., University of London, 1989), 203.

40. T. O. Beidelman, "Witchcraft in Ukaguru," in *Witchcraft and Sorcery in East Africa,* ed. J. Middleton and E. H. Winter (London: Routledge & Kegan Paul, 1963), 62.

41. M. Gelfand, *Witch Doctor: Traditional Medicine Man of Rhodesia* (London: Harvill Press, 1964), 55.

42. Ludwig Krapf, ed., *A Dictionary of the Suahili Language: With Introduction Containing an Outline of Suahili Grammar* (London: Tauber, 1882), 224–25.

43. Maureen Malowany, "Medical Pluralism: Disease, Health and Healing on the Coast of Kenya, 1840–1940" (PhD diss., McGill University, 1997), 45.

44. Lloyd Swantz, *The Medicine Man: Among the Zaramo of Dar es Salaam* (Uppsala, Sweden: Nordic Africa Institute, 1990), 28.

45. Ibid., 25; Ole Bjorn Rekdal, "Cross-Cultural Healing in East African Ethnography," *Medical Anthropology Quarterly* 13, no. 4 (1999): 466. There are also the "collective therapeutic rites" discussed by Feierman and Janzen in *The Social Basis of Health and Healing in Africa,* ed. Steven Feierman and John Janzen (Berkeley: University of California Press, 1992), 171.

46. Pels, "Mumiani: The White Vampire," 176.

47. Interview with Dr. Malenganisho, August 1, 2008, Mwanza, Tanzania.

48. Interview with Bibi Agnes Kagera, August 8, 2008, Bukumbi, Tanzania.

49. Pels argues something different. He recognized that many "researchers" who were veterinary workers, surveyors, and engineers were often accused of being bloodsuckers because their work took them into the "bush" outside of human society. Pels, "Mumiani: The White Vampire," 175, 183n29.

50. The Pare-Taveta Malaria Scheme, discussed in chapter 6, had an ever-changing set of goals that medical department officials couldn't keep straight, and countless "stray" researchers who attached their individual projects to the larger one about malaria.

51. The Aptitude Testing Project was organized by the Kenyan government during the Mau Mau years and was headed by the infamous J. H. Vint, a British doctor based in Nairobi who was active in eugenics debates in the 1920s and 1930s. Much of his "data" came from autopsies performed on African bodies coming from the King George VI Hospital. District Officer, Taveta, to Aptitude Testing Staff, "The Aptitude Testing Unit: A Progress Report Covering the Period from March 1953–June 1957," Confidential, NIMR, Amani.

52. District Commissioner, Pare, to Wilson, "Medical—General—Malaria Research," June 7, 1952, "Special Investigations in Man/Immunity Pare General," NIMR, Amani.

53. Brayne-Nicholls to Goiny, June 20, 1956, DC/LAMU/2/23/29, KNA.

54. Tappan, referencing White, *Speaking with Vampires*, when interviewees tell her about a "sucking rubber tube" and "rubber sucking tubes." Jennifer Tappan, "Blood Work and 'Rumors' of Blood: Nutritional Research and Insurrection in Buganda, 1935–1970," *International Journal of African Historical Studies* 47, no. 3 (2014): 473–94.

55. White, *Speaking with Vampires*, 106.

56. Interview with Mzee Thomas Inyassi, August 8, 2008, Kigongo, Tanzania.

57. *Nguvu* can also be used to describe the ineptness of the modern Tanzanian state and question its overall authority. As one local government official commented,

"It hasn't any strength [*nguvu*]. It is government on paper." Maia Green, "After Ujamaa? Cultures of Governance and the Representation of Power in Tanzania," *Social Analysis* 54, no. 1 (2010): 26.

58. Interview with Bibi Agnes Kagera, August 8, 2008, Bukumbi, Tanzania.

59. Interview with Bibi Teresa Wilburt, August 8, 2008, Bukumbi, Tanzania.

60. Interview with Mzee Donat, August 27, 2008, Namanga, Ukara, Tanzania; interview with Mzee Damian Mumwi, August 27, 2008, Bwisya, Ukara, Tanzania.

61. Interview with Mzee Donat, August 27, 2008, Namanga, Ukara, Tanzania.

62. Interview with Mzee Edward Bunduki and Bibi Teresa Wilburt, August 8, 2008, Bukumbi, Tanzania; interview with Mzee Thomas Inyassi, August 7, 2008, Kigongo, Tanzania.

63. Interview with Mzee Masatu Ndege and Mzee Manyama Mtaki, August 26, 2008, Bwisya, Ukara, Tanzania.

64. There's an interesting parallel here. If East Africans saw medical research as government work, it supports academic claims of medicine being a "tool" of colonialism or government.

65. The Swahili original being, "serikali yetu imeamua kufanyia majaribio hii dawa ya SP kwetu ili waone kama watu wangapi watakufa." Stephen E. D. Nsimba, "How Sulfadoxine-Pyrimethamine (SP) Was Perceived in Some Rural Communities After Phasing Out Chloroquine (CQ) as a First-Line Drug for Uncomplicated Malaria in Tanzania: Lessons to Learn towards Moving from Monotherapy to Fixed Combination Therapy," *Journal of Ethnobiology and Ethnomedicine* 2, no. 5 (2006): 6.

66. Interview with Mzee Mwendadi, October 17, 2008, Amani, Tanzania.

67. The Medical Research Council UK (MRC) has been working in The Gambia since 1949. See James Fairhead, Melissa Leach, and Mary Small, "Where Techno-Science Meets Poverty: Medical Research and the Economy of Blood in The Gambia, West Africa," *Social Science & Medicine* 63, no. 4 (2006): 1109–20; P. Wenzel Geissler et al., "'He Is Now Like a Brother, I Can Even Give Him Some Blood'—Relational Ethics and Material Exchanges in a Malaria Vaccine 'Trial Community' in The Gambia," *Social Science & Medicine* 67, no. 5 (2008): 696–707.

68. Geissler et al., "He Is Now Like a Brother," 700.

69. P. Wenzel Geissler, "'*Kachinja* Are Coming!': Encounters Around Medical Research Work in a Kenyan Village," *Africa* 75, no. 2 (2005): 179.

70. Interview with Bibi Agnes Kagera, August 8, 2008, Bukumbi, Tanzania.

71. Geissler, "'*Kachinja* Are Coming!,'" 179.

72. Johnson, *A Standard Swahili-English Dictionary*; Krapf, *A Dictionary of the Suahili Language*; A. C. Madan, *Swahili-English Dictionary* (Oxford: Clarendon Press, 1903); Charles Sacleux, *Dictionnaire Swahili-Français* (Paris: Institut d'Ethnologie, 1939); *Kamusi ya KiSwahili Sanifu* (Dar es Salaam: Oxford University Press, 1981).

73. Pels, "Mumiani: The White Vampire," 79–80; Swantz, *Blood, Milk and Death,* 68; Thomas Buckley and Alma Gottlieb, eds., *Blood Magic: The Anthropology of Menstruation* (Berkeley: University of California Press, 1988).

74. An individual's ability to reproduce and the productive capabilities of their descendants were seen as markers of the entire community's vitality. Swantz, *Blood, Milk and Death.*

75. Brad Weiss, "Electric Vampires: Haya Rumours of the Commodified Body," in *Bodies and Persons: Comparative Perspectives from Africa and Melanesia*, ed. M. Lambek and A. Strathern (Cambridge: Cambridge University Press, 1998), 172–94.

76. Vinay Kamat, "Negotiating Illness and Misfortune in Post-Socialist Tanzania: An Ethnographic Study in Temeke District, Dar es Salaam" (PhD diss., Emory University, 2004), 270–71.

77. Malowany, "Medical Pluralism," 43.

78. Rekdal, "Cross-Cultural Healing," 472.

79. Susan Reynolds Whyte, "Penicillin, Battery Acid and Sacrifice: Cures and Causes in Nyole Medicine," *Social Science & Medicine* 16, no. 23 (1982): 2056, 2060.

80. Vinay Kamat, "Dying Under the Bird's Shadow: Narrative Representations of *Degedege* and Child Survival among the Zaramo of Tanzania," *Medical Anthropology Quarterly* 22, no. 1 (2008): 73.

81. I am not the first person to consider East African medical beliefs as syncretic. Susanna Hausmann Muela et al., "Medical Syncretism with Reference to Malaria in a Tanzanian Community," *Social Science & Medicine* 55, no. 3 (2002): 403–13. The concept of syncretism is often used to discuss the effect of new religions appearing in Africa after contact with Christianity and Islam. An interesting discussion of that religious context can be seen in John Thornton, "Perspectives on African Christianity," in *Race, Discourse, and the Making of the Americas,* ed. Vera Hyatt and Rex Nettleford (Washington, DC: Smithsonian Institution, 1994), 169–98; John Thornton, "On the Trail of Voodoo: African Christianity in Africa and the Americas," *Americas* 44, no. 3 (1988): 261–78. Many scholars now argue against the theory of syncretism, positing instead that Africans maintained parallel religions. James Sweet, *Recreating Africa: Culture, Kinship, and Religion in the African-Portuguese World, 1441–1770* (Chapel Hill: University of North Carolina Press, 2003).

82. Johnson, *A Standard Swahili-English Dictionary*, 20.

83. J. W. T. Allen, ed. and trans., *The Customs of the Swahili People: The Desturi Za WaSwahili of Mtoro Bin Mwinyi Bakari and Other Swahili Persons* (Berkeley: University of California Press, 1981).

84. Ibid.

85. Beckerleg, "Maintaining Order, Creating Chaos," 165.

86. Wayne Melrose, *Lymphatic Filariasis: A Review 1862–2002* (Killarney, Australia: Warwick Educational Publishing, Inc., 2004), 17.

87. Beckerleg, "Maintaining Order, Creating Chaos," 166.

88. Muela et al., "Medical Syncretism," 407.

89. Susanna Hausmann Muela, Joan Muela Ribera, and Marcel Tanner, "Fake Malaria and Hidden Parasites—the Ambiguity of Malaria," *Anthropology & Medicine* 5, no. 1 (1998): 45, 47.

90. Brad Weiss, *The Making and Unmaking of the Haya Lived World: Consumption, Commoditization and Everyday Practice* (Durham: Duke University Press, 1996), 208.

91. Muela et al., "Medical Syncretism," 408; William Schneider and Ernest Drucker, "Blood Transfusions in the Early Years of AIDS in Sub-Saharan Africa," *American Journal of Public Health* 96, no. 6 (2006): 989–90, citing K. S. Dewhurst, "Observations on East African Blood Donors," *East African Medical Journal* 22 (1945): 276–78.

92. David L. Schoenbrun, "Conjuring the Modern in Africa: Durability and Rupture in Histories of Public Healing between the Great Lakes of East Africa," *American Historical Review* 111, no. 5 (2006): 1410; Luise White, "Blood Brotherhood Revisited: Kinship, Relationship, and the Body in East and Central Africa," *Africa* 64, no. 3 (1994): 359–72; Steven Feierman, "Explanation and Uncertainty in the Medical World of Ghaambo," *Bulletin of the History of Medicine* 74, no. 2 (2000): 330.

93. The idea of blood as medicine doesn't strike all English speakers as logical. But blood is a vital component of many medicines. Today blood is separated out into component parts: plasma (the liquid of blood), red blood cells, white blood cells, and platelets. The ability to separate blood and use the plasma to generate new treatments led directly to new medicines such as Factor VIII, which is now the standard treatment for the majority of hemophiliac patients.

94. See *Kamusi ya KiSwahili Sanifu.*

95. Krapf, *A Dictionary of the Suahili Language,* 266–67.

96. Johnson, *A Standard Swahili-English Dictionary,* 314.

97. For an extensive discussion of the word's etymology, see Malloy, "'Holding [Tanganyika] by the Sindano,'" 360–75.

98. Mesaki, "Witchcraft and Witch-Killings in Tanzania," 180.

99. Peter Lienhardt, introduction to *The Medicine Man: Swifa ya Nguvumali*, by Hasani Bin Ismail, ed. and trans. Peter Lienhardt (Oxford: Clarendon Press, 1968), 59.

100. Geissler, "'*Kachinja* Are Coming!,'" 179.

101. Schneider and Drucker, "Blood Transfusions," 989.

102. B. Jacobs and Z. A. Berege, "Attitudes and Beliefs about Blood Donation among Adults in Mwanza Region, Tanzania," *East African Medical Journal* 72, no. 6 (1995): 346, 348. People also mentioned fear of contracting diseases such as HIV through unsterilized equipment. Difficulty in recruiting blood donors is not unique to East Africa, but has been reported in other parts of the continent. For just a few examples, see O. U. J. Umeora, S. O. Onuh, and M. C. Umeora, "Socio-Cultural Barriers to Voluntary Blood Donation for Obstetric Use in a Rural Nigerian Village," *African Journal of Reproductive Health* 9, no. 6 (2005): 72–76, for Nigeria; and Mamady Cham, "Maternal Mortality in the Gambia: Contributing Factors and What Can be Done to Reduce Them" (PhD diss., University of Oslo, 2003).

103. Maia Green, "Medicines and the Embodiment of Substances among Pogoro Catholics, Southern Tanzania," *Journal of the Royal Anthropological Institute* 2, no. 3 (1996): 488.

104. Green notes that among the Pogoro of Tanzania, *mtera* refers to medicines made from plants and trees. Ibid., 489. Other Bantu languages follow the general pattern of plant medicines using the word for "tree." See Jean Comaroff, *Body of Power, Spirit of Resistance: The Culture and History of a South African People* (Chicago: University of Chicago Press, 1985), 66; Harriet Ngubane, *Body and Mind in Zulu Medicine: An Ethnography of Health and Disease in Nyuswa-Zulu Thought and Practice* (London: Academic Press, 1977), 22; and Audrey Richards, *Chisungu: A Girl's Initiation Ceremony among the Bemba* (Edison, NJ: Tavistock, 1982), 27.

105. Green, "Medicines and the Embodiment," 488–89.

106. Ibid., 488.

107. Ibid., 489.

108. Douglas Edwin Ferguson, "The Political Economy of Health and Medicine in Colonial Tanganyika," in *Tanzania under Colonial Rule*, ed. Martin H. Y. Kaniki (London: Longman, 1980), 311.

109. Some of the phrases I used to capture this idea (created with help from native speakers) included "tutaonana kama dawa hii inafanyakazi?" and "tujaribu dawa hii kuona . . ." Others have used phrases such as "utafiti wa tiba" or "utafiti wa dawa."

110. There is another argument that can be made about the presence of therapeutic misconception in East Africa: that by merely offering access to a doctor and treatment for basic ailments, people really are receiving a "therapy." Additionally, in an environment where there may be no effective therapy geographically available or within a person's financial reach, one could argue that access to an experimental drug that *may* be effective is a rational decision, and, thus, there is no "misconception" since participation in the project truly is therapeutic.

111. Paul S. Appelbaum et al., "False Hopes and Best Data: Consent to Research and the Therapeutic Misconception," *Hastings Center Report* 17, no. 21 (April 1987): 20–24; "Investigating Parents' Understanding and Decision Making Process for Children's Participation in a Malaria Vaccine Trial, in Kenya: Qualitative Data Analysis" (MSc thesis, London School of Hygiene and Tropical Medicine, 2012), 32. For more on therapeutic misconception in African contexts, see Geissler et al., "'He Is Now Like a Brother,'" 699; Vicki Marsh et al., "Beginning Community Engagement at a Busy Biomedical Research Programme: Experiences from the KEMRI CGMRC-Wellcome Trust Research Programme, Kilifi, Kenya," *Social Science & Medicine* 67, no. 5 (September 2008): 721–22; Molyneux et al., "'Even If They Ask You to Stand by a Tree All Day, You Will Have to Do It (Laughter) . . . !': Community Voices on the Notion and Practice of Informed Consent for Biomedical Research in Developing Countries," *Social Science & Medicine* 61, no. 2 (2005): 447, 452; Udo Schuklenk, "Protecting the Vulnerable: Testing Times for Clinical Research Ethics," *Social Science & Medicine* 51, no. 6 (2000): 973.

112. Appelbaum et al., "False Hopes and Best Data," 20–24.

113. Melissa Graboyes, "Fines, Orders, Fear . . . and Consent? Medical Research in East Africa, c. 1950s," *Developing World Bioethics* 10, no. 1 (2010): 34–41.

114. Jean Comaroff and John Comaroff, "Occult Economies and the Violence of Abstraction: Notes from the South African Postcolony, " *American Ethnologist* 26, no. 2 (1999): 279–303; Clifton Crais, *The Politics of Evil: Magic, State Power and the Political Imagination in South Africa* (Cambridge: Cambridge University Press, 2002); Weiss, "Electric Vampires"; White, *Speaking with Vampires*; James Brennan, "Destroying *Mumiani*: Cause, Context and Violence in Late Colonial Dar es Salaam," *Journal of Eastern African Studies* 2, no. 1 (2008): 96; Geissler, "'*Kachinja* are Coming!'"; P. Wenzel Geissler and Robert Pool, "Editorial: Popular Concerns about Medical Research Projects in Sub-Saharan Africa—a Critical Voice in Debates about Medical Research Ethics," *Tropical Medicine & International Health* 11, no. 7 (2006): 975–82.

115. Brennan, "Destroying *Mumiani*," 96; Jonathan Stadler and Eirik Saethre, "Rumours about Blood and Reimbursements in a Microbicide Gel Trial," *African Journal of AIDS Research* 9 no. 4 (2010): 345–53.

116. White, *Speaking with Vampires*. A thoughtful critique of White can be found in Gregory Mann, "An Africanist's Apostasy: On Luise White's Speaking with Vampires," *International Journal of African Historical Studies* 41 (2008): 117–22.

117. Melissa Leach and James Fairhead, *Vaccine Anxieties: Global Science, Child Health and Society* (London: Earthscan, 2007), 171.

118. United National Children's Fund, Eastern and Southern Regional Office, "Combatting Antivaccination Rumours: Lessons Learned from Case Studies in East Africa." 2001.

119. Alan Irwin and Brian Wynne, eds. *Misunderstanding Science? The Public Reconstruction of Science and Technology* (Cambridge: Cambridge University Press, 1996), 6.

120. Green, "After Ujamaa?," 26, referencing Matthew Costello, "Administration Triumphs Over Politics: The Transformation of the Tanzanian State," *African Studies Review* 39, no. 1 (1996): 123–48; and Joel Samoff, "The Bureaucracy and the Bourgeoisie: Decentralization and Class Structure in Tanzania," *Comparative Studies in Society and History* 21, no. 1 (1979): 30–62.

121. Leach and Fairhead, *Vaccine Anxieties*, 33.

122. Umeora et al., "Socio-Cultural Barriers," 75.

123. Jacobs and Berege, "Attitudes and Beliefs," 345.

124. Mzee D____, 2008.

125. Anna Crozier, *Practising Colonial Medicine: The Colonial Medical Service in British East Africa* (London: I. B. Tauris, 2007), 93.

126. Rekdal, "Cross-Cultural Healing," 471.

RESEARCHERS ARRIVE

Historical Narrative: "Inspeakable Entomologists": H. H. Goiny and a Failed Attempt to Eliminate Lymphatic Filariasis, Pate Island, Kenya, 1956

1. As map 3.1 shows, Pate Island is made up of a number of towns including Pate, Siu ("Siyu"), and Faza, in addition to smaller villages.

2. Goiny describes the men as "Arab" although the families of these men had likely lived on Pate Island for many generations. The category was one regularly used by British colonial officials, although the men may not have identified themselves that way. This is not to ignore the very real ethnic diversity that existed on Pate Island as a result of centuries of contact with the Middle East and larger Indian Ocean world, which did often result in immigration, intermarriage, and permanent resettlement.

3. Goiny to Senior Parasitologist, "Activities during Jan 1956—Lamu Pit Latrine Campaign, Pate Island Filariasis Investigations," n.d. [late January 1956], DC/LAMU/2/23/29, KNA.

4. Goiny to Senior Parasitologist. "Pate Island," February 9, 1956, DC/LAMU/2/23/29, KNA.

5. It is not entirely surprising that the letter was written in Arabic. Pate Island was the site of an Islamic college, had long been literate and Islamic, and writing in Arabic may have emphasized the writers' education and cosmopolitanism, as well as the formality and seriousness of the content.

6. Their research team was stationed in two separate villages. Goiny and Henry Gigiri were in the village of Siu and two other Africans, Michael Ikata and Faros Enos, were stationed in Faza village at the tip of the island. Ikata and Enos also had a hard time. Although they limited themselves to walking outdoors and doing daytime mosquito catches, they were ominously warned not to enter houses unless they were "prepared to take the consequences." Goiny to Senior Parasitologist. "Pate Island," February 9, 1956, DC/LAMU/2/23/29, KNA.

7. Professor P. C. C. Garnham of the London School of Tropical Medicine was involved in the original attempt. Newspaper Clipping, "Campaign to Wipe out Elephantiasis: Hope of Success on Island," December 29, 1956, DC/LAMU/2/23/29, KNA.

8. Goiny to Senior Parasitologist, "Activities during Jan 1956—Lamu Pit Latrine Campaign, Pate Island Filariasis Investigations," DC/LAMU/2/23/29, KNA.

9. Maureen Malowany, "Medical Pluralism: Disease, Health and Healing on the Coast of Kenya, 1840–1940" (PhD diss., McGill University, 1997), 190–91, 200, 202.

Modern Narrative: A "Remarkable Achievement"? A Lymphatic Filariasis Elimination, Zanzibar, 2001

1. The unprecedented donation of drugs led to the creation of the Global Alliance to Eliminate Lymphatic Filariasis (GAELF) in 2000. The group is a public-private partnership tasked with advocacy and mobilization of resources.

2. Mass drug administration against LF relies upon a single annual dose of 400 mg albendazole plus either 6 mg/kg diethylcarbamazine (DEC) or 150–200 mcg/kg ivermectin. World Health Organization, *Progress Report 2000–2009 and Strategic Plan 2010–2020 of the Global Programme to Eliminate Lymphatic Filariasis: Halfway towards Eliminating Lymphatic Filariasis* (Geneva: World Health Organization Press, 2010).

3. World Health Organization, *The Global Elimination of Lymphatic Filariasis: The Story of Zanzibar* (Geneva: World Health Organization Press, 2010), 4, 34.

4. The activities in Zanzibar were part of a larger attempt to eradicate malaria globally, which ran from 1954 to 1969 and failed to achieve its greater goal, although malaria was eliminated in certain countries. The global program—and other attempts based in East Africa—are discussed in chapter 6. World Health Organization, *Progress Report 2000–2009*, 40.

5. Rebound malaria is discussed more fully in chapter 6.

6. World Health Organization, *Progress Report 2000–2009*, 41.

7. Ibid., 13, 34.

8. Ibid., 12.

9. Ibid., 11.

10. Ibid., 12.

11. Ibid., 32.

12. Ibid., 39–40.

13. Khalfan A. Mohammed et al., "Progress towards Eliminating Lymphatic Filariasis in Zanzibar: A Model Programme," *Trends in Parasitology* 22, no. 7 (July 2006): 342.

14. Paul E. Simonsen et al., *Lymphatic Filariasis: Research and Control in Eastern and Southern Africa* (Frederiksberg, Denmark: DBL – Centre for Health Research and Development, 2008), 161.

15. Ibid., 27.

16. Ibid., 4, 157.

17. John Gyapong and Nana Twum-Danso, "Editorial: Global Elimination of Lymphatic Filariasis: Fact Or Fantasy?" *Tropical Medicine & International Health* 11, no. 2 (2006): 125.

18. World Health Organization, *Progress Report 2000–2009*, 11.

19. Information presented in different parts of the WHO report appears to be contradictory. On page 347, it states that, "Of the 73 countries where LF is currently considered endemic, 53 are implementing MDA to interrupt transmission, of which 12 countries have moved to a post- MDA surveillance phase." Yet, on pages 14 and 47, it indicates that only five countries have entered that phase. Ibid., 14, 47, 347.

20. Two countries have been declared LF free: Korea and China, although China embarked on its own program using DEC-treated salt, which fell outside of the WHO recommendations. The WHO's own documents indicate that the Pacific Islands only account for 3 percent of the global burden. Ibid., 46.

21. Ibid., 351.

22. Mohammed et al., "Progress towards Eliminating," 341.

Chapter 3: Arrivals, First Encounters, First Impressions

1. Formal, published research on LF prevalence and incidence levels in East Africa dates back to 1910. Earlier references to conditions such as hydrocele and elephantiasis, which are often symptoms of LF, appear in written accounts dating

back to the 1890s. C. N. Wamae et al., "Lymphatic Filariasis in Kenya since 1910, and the Prospects for Its Elimination: A Review," *East African Medical Journal* 78, no. 11 (1970): 595–603; J. W. T. Allen, ed. and trans., *The Customs of the Swahili People: The Desturi Za WaSwahili of Mtoro Bin Mwinyi Bakari and Other Swahili Persons* (Berkeley: University of California Press, 1981), 138.

2. Gordon Cook and Alimuddin Zumla, eds., *Manson's Tropical Diseases*, 22nd ed. (Philadelphia: Saunders, 2009), 1484.

3. Filariasis Research Unit Annual Report, 1950.

4. Division of Insect Borne Diseases Annual Report, 1956.

5. Another example of a singular biomedical condition being viewed by African sufferers as two distinct diseases (based on two dominant sets of symptoms) is TB and the Tswana condition *thibamo*. Livingston discusses how the diagnoses were mutually reinforcing even though they were based on different etiologies and had "different moral consequences." Julie Livingston, "Productive Misunderstandings and the Dynamism of Plural Medicine in Mid-Century Bechuanaland," *Journal of Southern African Studies* 33, no. 4 (December 2007): 801–10.

6. Brayne-Nicholls to Provincial Commissioner, Coast, "Elephantiasis Survey: Pate Island," March 2, 1956. DC/LAMU/2/23/29, KNA.

7. Interview with Mzee Chiloli, August 27, 2008, Nyang'ombe, Ukara Island, Tanzania.

8. Allen, *Customs of the Swahili People*, 138.

9. Ibid., 135–36; East African Institute of Malaria and Vector Borne Diseases Annual Report, 1974–75.

10. The same sort of disagreements continue into the present. During the 2000s, WHO and Ministry of Health workers disagreed with Zanzibaris about how bad lymphatic filariasis really is, and whether it was worth so much attention and effort. The WHO described LF as a "stigmatizing and incurable disease," but most Zanzibaris did not feel at risk. World Health Organization, *The Global Elimination of Lymphatic Filariasis: The Story of Zanzibar* (Geneva: World Health Organization Press, 2010).

11. "Report on work at Ukara during the months of November and December, 1950," Dr. Trant Files, NIMR, Mwanza.

12. "Faza People, Tunda, Bajumale, Kizingitini, Moyaboge" to Mudir of Faza, February 25, 1956, DC/LAMU/2/23/29, KNA.

13. Brayne-Nicholls to Senior Parasitologist, "Elephantiasis Survey Pate Island," March 13, 1956, DC/LAMU/2/23/29, KNA.

14. Brayne-Nicholls to Provincial Commissioner, Coast, "Elephantiasis Survey: Pate Island," March 2, 1956, DC/LAMU/2/23/29, KNA.

15. Wayne Melrose, *Lymphatic Filariasis: A Review 1862–2002* (Killarney, Australia: Warwick Educational Publishing, 2004); Eric Ottesen, "Lymphatic Filariasis: Treatment, Control and Elimination," *Advances in Parasitology* 61 (2006): 395–441.

16. Not much has been written about Ukara and Ukerewe Islands, and nothing specific to medicine and disease. Other sources include Aniceti Kitereza, *Mr.*

Myombekere and His Wife Bugonoka, Their Son Ntulanolwo and Daughter Bulihwali (Dar es Salaam: Mkuki na Nyota Publishers, 2002); Hans Paasche, *The Journey of Lukanga Mukara into the Innermost of Germany*, last revised September 3, 1998, http://wp.cs.ucl.ac.uk/anthonysteed/lukanga. Thanks to Michael Paasche-Orlowe for the reference.

17. In a modern medical research trial conducted by the Medical Research Council (UK) in The Gambia, African research assistants were hosted by community members, which seemed to improve researcher-community relations. Historically, on nearby Lamu Island, the DIBD originally intended for their African staff members to be hosted by community members, but the offers of housing were rescinded shortly after Goiny's arrival. James Fairhead, Melissa Leach, and Mary Small, "Where Techno-Science Meets Poverty: Medical Research and the Economy of Blood in The Gambia, West Africa," *Social Science & Medicine* 63, no. 4 (2006): 1109–20.

18. There was some discussion about providing indoor residual spraying of houses as a way to incentivize residents to take their pills, but it doesn't appear to have been done on a large scale. The drug DEC was donated by the British drug company Burroughs Wellcome. East African Medical Survey and Research Institute Annual Report, 1956–57.

19. It appears that much of the LF research quieted down in the 1960s and possibly into the 1970s. A fifteen-year follow-up to the original study was prepared for in 1971 and 1972, but it does not appear the project actually happened. East African Institute for Medical Research Reports, 1958–59, 1960–61.

20. Alec Smith, *Insect Man: A Fight against Malaria in Africa* (New York: Palgrave Macmillan, 1993), 34–35.

21. Ibid., 36.

22. Trant's situation of spending so much time providing medical care was unusual since the directors of research projects were typically unwilling to allow such arrivals focused on getting to know communities rather than starting the research right away. In this case, much of this disapproval was a result of Trant's personality and her openly disobeying orders. The FRU was so shortsighted about all the benefits Trant brought them that she was actually demoted and temporarily suspended because she wouldn't stop treating patients. None of this was mentioned in her autobiography or the organization's official publications, but it is meticulously documented in near-daily letters between Trant and her boss, Col. Laurie. More about the East African Medical Survey and Hope Trant is in chapter 5.

23. Interview with Mzee Masatu Ndege, August 26, 2008, Bwisya, Ukara Island, Tanzania; Interview with Mzee Manyama Mtaki, August 26, 2008, Bwisya, Ukara Island, Tanzania.

24. It is worth clarifying that despite the presence of traditional healing systems on Ukara Island, residents were eager to receive Western therapies such as pills and injections—especially those that proved their efficacy quickly. Thus, pills to treat malaria, fevers, and infections were in high demand, despite or because of little access prior to Trant's arrival. Dr. Trant Papers, NIMR Mwanza.

25. Smith, *Insect Man,* 41.

26. Ibid.

27. Regional Medical Department to Trant, November 14, 1950, Dr. Trant Papers, NIMR, Mwanza.

28. Nelson to Director of Medical Services, Kenya, "Filariasis Investigation Lamu," September 19, 1958, DC/LAMU/2/23/29, KNA.

29. Brayne-Nicholls to Senior Parasitologist, "Elephantiasis Survey Pate Island," March 13, 1956, DC/LAMU/2/23/29, KNA; Provincial Commissioner, Coast, to District Commissioner, Coast, "Elephantiasis Survey: Pate Island," August 22, 1956, DC/LAMU/2/23/29, KNA.

30. Brayne-Nicholls to Senior Parasitologist. "Elephantiasis Survey: Pate Island," February 15, 1956, DC/LAMU/2/23/29, KNA.

31. Maureen Malowany, "Medical Pluralism: Disease, Health and Healing on the Coast of Kenya, 1840–1940" (PhD diss., McGill University, 1997), 1.

32. Patricia Romero, *Lamu: History, Society and Family in an East African Port City* (Princeton: Markus Wiener, 1997), 86, 156.

33. Ibid., 207.

34. Fears about government hospitals were already well established, and did not disappear in future decades. In March 1960, an angry mob attacked the Mombasa Provincial General Hospital, trying to prevent a post mortem from being performed. Although hospital officials claimed the family had consented to the autopsy, relatives either changed their minds or agreed under dubious conditions, since they participated in the protest. "Angry Mob Besiege Hospital," March 20, 1960, MOH/42/7, KNA; ADMS to Chief Medical Officer, March 23, 1960, MOH/42/7, KNA.

35. Ahmed Salim, *The Swahili-Speaking Peoples of Kenya's Coast, 1895–1965* (Nairobi: East African Pub. House, 1973), 155.

36. Susan Beckerleg, "Maintaining Order, Creating Chaos: Swahili Medicine in Kenya" (PhD diss., University of London, 1989), 306; Peter Lienhardt, "The Mosque College of Lamu and Its Social Background," *Tanganyika Notes and Records* 53 (1959): 228–42; Susan Beckerleg, "Medical Pluralism and Islam in Swahili Communities in Kenya," *Medical Anthropology Quarterly* 8, no. 3 (1994): 309.

37. Rebecca Gearhart and Said Abdulrehman Munib, "Purity, Balance and Wellness among the Swahili of Lamu, Kenya," *Journal of Global Health* (December 2013), 3.

38. My emphasis. Rebecca Gearhart and Munib Said Abdulrehman, "Concepts of Illness among the Swahili of Lamu, Kenya," *Journal of Transcultural Nursing* (December 31, 2013): 2, doi:10.1177/1043659613515713; Beckerleg, "Medical Pluralism and Islam," 300.

39. Gearhart and Abdulrehman, "Concepts of Illness," 2, 4.

40. Marguerite Ylvisaker, "Lamu in the Nineteenth Century: Land, Trade and Politics" (PhD diss., Boston University, 1979), 18.

41. Ylvisaker, "Lamu in the Nineteenth Century," 83.

42. Lienhardt, "Mosque College of Lamu," 240.

43. Salim, *Swahili-Speaking Peoples*, 153–54.

44. Brayne-Nicholls to Provinicial Commissioner, Coast, "Elephantiasis Survey: Pate Island," March 2, 1956, DC/LAMU/2/23/29, KNA.

45. "Traditional" is in quotation marks since there's plenty of documentation to show that traditional positions of authority changed dramatically over time. Eric Hobsbawm and Terence Ranger, eds., *The Invention of Tradition* (Cambridge: Cambridge University Press, 1992); Thomas Spear, "Neo-Traditionalism and the Limits of Invention in British Colonial Africa," *Journal of African History* 44, no. 1 (2003): 3–27.

46. The concept of middlemen is particularly apt on the Swahili coast, harkening back to earlier Indian Ocean connections where coastal residents worked as "middleman traders" and "economic and cultural brokers" between the African mainland and foreigners from the Middle East and India. Gearhart and Abdulrehman, "Concepts of Illness," 1.

47. Surprisingly little has been written about the real demands placed on chiefs and the limitations of their power. Benjamin Lawrance, Emily Osborn, and Richard Roberts, eds., *Intermediaries, Interpreters, and Clerks: African Employees in the Making of Colonial Africa* (Madison: University of Wisconsin Press, 2006); Frederick Cooper and Anne Stoler, "Introduction Tensions of Empire: Colonial Control and Visions of Rule," *American Ethnologist* 16, no. 4 (1989): 609–21; P. Wenzel Geissler, "Parasite Lost: Remembering Modern Times with Kenyan Government Medical Scientists," in *Evidence, Ethos and Experiment: The Anthropology and History of Medical Research in Africa* (New York: Berghahn Books, 2011), 309.

48. Roland Moore and A. Roberts, "An Investigation of the Pattern of Disease Prevalent in Parts of the Rufiji District," East African Medical Survey and Research Institute, n.d.

49. Filariasis Research Unit Annual Report, 1951.

50. Interview with Mzee Salvatory Nambiza, August 25, 2008, Nansio, Ukerewe, Tanzania.

51. Interview with Mzee Masatu Ndege and Mzee Manyama Mtaki, August 26, 2008, Bwisya, Ukara Island, Tanzania.

52. Interview with Mzee Salvatory Nambiza, August 25, 2008, Nansio, Ukerewe, Tanzania; interview with Mzee Donat, August 27, 2008, Namanga, Ukara Island, Tanzania. A similar sentiment was expressed about the DIBD men in Kenya, where Geissler notes there was a "clear 'chain-of-command' from the ministry via the chief, to 'the people' (*wananchi*)." Geissler, "Parasite Lost," 309.

53. Goiny to Senior Parasitologist, "Activities during Jan 1956—Lamu Pit Latrine Campaign, Patte Island Filariasis Investigations," n.d. [late January 1956], DC/LAMU/2/23/29, KNA.

54. Salim, *Swahili-Speaking Peoples*, 146.

55. Mudir to District Commissioner, Lamu, "Entomological Research—Faza," February 28, 1956, DC/LAMU/2/23/29, KNA.

56. Goiny to Brayne-Nicholls, June 25, 1956, DC/LAMU/2/23/29, KNA.

57. Provincial Commissioner, Coast, to District Commissioner, Coast, "Elephantiasis Survey: Pate Island," August 22, 1956, DC/LAMU/2/23/29, KNA.

58. Medical Officer in Charge, District Hospital Lamu, to District Commissioner Lamu, "Filariasis Mass Treatment at Faza," January 12, 1957, DC/LAMU/2/23/29, KNA.

59. Goiny to Senior Parasitologist, "Patte Island," February 9, 1956, DC/LAMU/2/23/29, KNA.

60. Unknown author [Heisch?] to Brayne-Nicholls, February 20, 1956, DC/LAMU/2/23/29, KNA.

61. Goiny to Senior Parasitologist, "Activities during Jan 1956—Lamu Pit Latrine Campaign, Patte Island Filariasis Investigations," n.d. [late January 1956], DC/LAMU/2/23/29, KNA.

62. Michaek Ikata was also referred to as "Michael Okata" in some documents. Goiny to Senior Parasitologist, "Patte Island," February 9, 1956, DC/LAMU/2/23/29, KNA.

63. Goiny to Senior Parasitologist, CC DC Lamu, "Patte Island," February 9, 1956, DC/LAMU/2/23/29, KNA.

64. Goiny to Senior Parasitologist, "Activities during Jan 1956—Lamu Pit Latrine Campaign, Patte Island Filariasis Investigations," n.d. [late January 1956]. DC/LAMU/2/23/29, KNA.

65. Jordan, Rhodes-Jones, and Goatly, "Parasitology," East African Medical Survey and Research Institute Annual Report, 1956–57.

66. Goiny to Senior Parasitologist, CC DC Lamu, "Patte Island," February 9, 1956, DC/LAMU/2/23/29, KNA.

67. Anthropological research has begun to report on a new type of middlemen appearing in East Africa. "Village reporters" are paid volunteers "who act as an interface between villagers" and international research groups. They are described as being both "with the community" and with the research organization. Although these workers are intended to have dual allegiances, research has found that their actual allegiances clearly skew toward supporting the work of the research institutes that pay them. Chantler et al., "Ethical Challenges That Arise at the Community Interface of Health Research: Village Reporters' Experiences in Western Kenya," *Developing World Bioethics* 13, no. 1 (March 22, 2013): 30–31.

68. In the available materials chronicling the Pate Islanders' refusals, there is no information about internal community divisions. Organized meetings and protests emphasized *all* residents' opposition to the plan in an outward show of unity. If there were factions within the community—the rich, women, or the well-educated, for example—who supported the work of the DIBD, that remains to be documented.

69. Alan Irwin and Brian Wynne, eds., *Misunderstanding Science? The Public Reconstruction of Science and Technology* (Cambridge: Cambridge University Press, 1996), 42. Wylie has discussed the subjectivity of science in a South African context: Diana Wylie, *Starving on a Full Stomach: Hunger and the Triumph of Cultural Racism in Modern South Africa* (Charlottesville: University of Virginia Press, 2001).

70. Romero, *Lamu: History, Society and Family*, 70–80.

71. Malowany, "Medical Pluralism," 199.

72. A state's public health powers are bound to a state's police powers, which meant there was a good chance measures could be oppressive. Ibid., 121–22.

73. Ibid., 190.

74. Romero describes how, "In the nineteenth century bathrooms, pottery jars locally made held large quantities of water brought in from town wells that served as breeding grounds for mosquitoes. . . . Some houses and mosques contained cisterns that held stagnant and impure water. These cisterns drew mosquitoes." Romero, *Lamu: History, Society and Family*, 79–80.

75. Brayne-Nicholls to Provincial Commissioner, Coast, "Elephantiasis Survey: Pate Island," March 2, 1956, DC/LAMU/2/23/29, KNA.

76. "The Scout's and King's African Rifles Uniforms [Photographs]," Children and Youth in History, item #101, accessed December 5, 2009, http://chnm.gmu.edu/cyh/primary- sources/101.

77. P. Wenzel Geissler, "Stuck in Ruins, or Up and Coming? The Shifting Geography of Urban Public Health Research in Kisumu, Kenya," *Africa* 83, no. 4 (2013): 541.

78. Fairhead, Leach, and Small, "Where Techno-Science Meets Poverty," 1118.

79. Mau Mau has drawn large amounts of scholarly attention. Colonial-era authors such as J. C. Carothers played up the savagery and even psychoses of African behavior. J. C. Carothers, *The Psychology of Mau Mau* (Nairobi: Government Printer, 1954). Shortly after independence, nationalist approaches abounded; a wave of Marxist-inspired scholars in the 1970s and 1980s framed Mau Mau as primarily a struggle for land and living wages. More recently, the violence of the uprising and the oppressiveness of the British response have been highlighted by Caroline Elkins, *Imperial Reckoning: The Untold Story of Britain's Gulag in Kenya* (New York: Owl Books, 2005) and David Anderson, *Histories of the Hanged: The Dirty War in Kenya and the End of Empire* (New York: W. W. Norton, 2005).

80. Brayne-Nicholls to Heisch, February 20, 1956, DC/LAMU/2/23/29, KNA.

81. Goiny to Senior Parasitologist, "Patte Island," February 9, 1956, DC/LAMU/2/23/29, KNA.

82. Ibid.

83. Brayne-Nicholls to Heisch, February 27, 1956, DC/LAMU/2/23/29, KNA.

84. Heisch to Brayne-Nicholls, February 15, 1956, DC/LAMU/2/23/29, KNA. Pate Island was often referred to as "Patte" by the British officials, and I have retained that spelling in direct quotes.

85. Brayne-Nicholls to Heisch, February 20, 1956, DC/LAMU/2/23/29, KNA.

86. Heisch to Brayne-Nicholls, March 5, 1956, DC/LAMU/2/23/29, KNA.

87. This literature spanned the 1980s and early 1990s. Headrick's was the most deterministic account of the effects of quinine on "opening up" the African continent for colonialism. However, later works by Arnold, Turshen, and Vaughan branched into related but distinct categories. Turshen took a political economy

approach and was heavily influenced by Marxism. Vaughan, on the other hand, integrated Foucault's ideas of social control and bio-power into an African context. All seem to have overestimated the authority and oppressiveness of the state. Michael Adas, *Machines as the Measure of Men: Science, Technology, and Ideologies of Western Dominance* (London: Cornell University Press, 1989); Daniel Headrick, *The Tools of Empire: Technology and European Imperialism in the Nineteenth Century* (New York: Oxford University Press, 1981); Meredeth Turshen, *The Political Ecology of Disease in Tanzania* (Piscataway: Rutgers University Press, 1984); David Arnold, ed., *Imperial Medicine and Indigenous Societies* (Manchester: Manchester University Press, 1988); David Arnold, *Colonizing the Body: State Medicine and Epidemic Disease in Nineteenth-Century India* (Berkeley: University of California Press, 1993); Roy MacLeod and Milton Lewis, eds., *Disease, Medicine and Empire: Perspectives on Western Medicine and the Experience of European Expansion* (New York: Routledge, 1988); Megan Vaughan, *Curing Their Ills: Colonial Power and African Illness* (Stanford: Stanford University Press, 1991).

88. Sara Berry, "Hegemony on a Shoestring: Indirect Rule and Access to Agricultural Land," *Africa* 62, no. 3 (1992): 327–55; Bruce Berman, *Control and Crisis in Colonial Kenya: The Dialectic of Domination* (London: James Currey, 1980); A. H. M. Kirk-Greene, "The Thin White Line: The Size of the British Colonial Service in Africa," *African Affairs* 79, no. 314 (1980): 25–44.

89. Brayne-Nicholls to Provincial Commissioner, Coast, "Elephantiasis Survey: Pate Island," March 2, 1956.

90. Unknown author [Heisch?] to Brayne-Nicholls, March 5, 1956, DC/LAMU/2/23/29, KNA.

91. Data for 1920 from G. Dunderdale, "Notes on the Incidence of Filarial Infection on the Neighbourhood of Lamu, British East Africa," *Transactions of the Royal Society of Tropical Medicine and Hygiene* 15 (1921): 190–97. Data for 1950s from R. B. Heisch, G. S. Nelson, and M. Furlong, "Studies in Filariasis in East Africa: 1. Filariasis in the Island of Pate, Kenya," *Transactions of the Royal Society of Tropical Medicine and Hygiene* 53, no. 1 (1959): 41–53. Data for 1980s from C. N. Wamae, W. Nderitu, and F. M. Kiliku, "*Brugia Patei* in a Domestic Cat from Lamu Island, Kenya," *Filaria Links* 2 (1997): 7; C. S. Mwandawiro, "Studies on Filarial Infection in Lamu and Tana River Districts" (MS thesis, University of Nairobi, 1990); C. N. Wamae et al., "Lymphatic Filariasis in Kenya since 1910, and the Prospects for Its Elimination: A Review," *East African Medical Journal* 78, no. 11 (2001): 595–603.

CONSENT OR COERCION?

Historical Narrative: "Forced to Accept Trial Treatment?" A Tuberculosis Drug Trial, Nairobi, Kenya, 1961

1. The name "Mwangi" was a name of the Kikuyu generational ruling class from the late 1800s. Godfrey Muriuki, *History of the Kikuyu, 1500–1900* (Oxford:

Oxford University Press, 1975). Thanks to Lynsey Farrell for help confirming this fact.

2. Gordon Cook and Alimuddin Zumla, eds., *Manson's Tropical Diseases*, 22nd ed. (Philadelphia: Saunders, 2009), 985, 987, 992. "Cavitary Tuberculosis," accessed May 5, 2014, http://granuloma.homestead.com/TB_cavitary_gross.html.

3. Sister Margaret Millar was the senior resident nurse at the hospital.

4. Margaret Millar to unknown recipient, "Mwangi Julius," August 22, 1961, BY/22/49, KNA.

5. Ibid.

6. The doctors maintain Mwangi remained in the hospital receiving "ordinary" care for a month, but Mwangi claimed it was only twelve days.

7. A WHO tuberculosis survey was going on in 1960 in Kenya, with results published in 1961. The article makes clear that the standard for diagnosing cases of TB involved a Mantoux tuberculin test, X-ray, and a microscopic exam of a sputum sample. When a positive sample was identified, another sample was collected and cultured at the WHO/UNICEF Tuberculosis Diagnostic Laboratory in Nairobi, with a final reading being done after eight weeks. E. Roelsgaard and J. Nyboe, "A Tuberculosis Survey in Kenya," *Bulletin of the World Health Organization* 25 (1961): 851, 854.

8. Medical Officer in Charge, Infectious Disease Hospital, Nairobi, to Medical Superintendent, King George, "Julius Mwangi," August 23, 1961, BY/22/49, KNA.

9. Julius Mwangi to Director of Medical Services, Kenya, "King George VI Hospital," August 14, 1961, BY/22/49, KNA.

10. Ibid.

11. Roelsgaard and Nyboe, "Tuberculosis Survey in Kenya," 851, 854.

12. Caroline Elkins, *Imperial Reckoning: The Untold Story of Britain's Gulag in Kenya* (New York: Owl Books, 2005), 144.

13. Julius Mwangi to Director of Medical Services, Kenya, "King George VI Hospital," August 14, 1961, BY/22/49, KNA.

14. Permanent Secretary Health and Social Affairs, Kenya to Julius Mwangi, "King George VI Hospital," September 20, 1961, BY/22/49, KNA.

15. Julius Mwangi to Director of Medical Services, Kenya, "King George VI Hospital," August 14, 1961, BY/22/49, KNA.

16. Medical Officer in Charge, Infectious Disease Hospital Nairobi to Medical Superintendent, King George, "Julius Mwangi," August 23, 1961, BY/22/49, KNA.

Modern Narrative: Focusing on Fieldworkers in Kilifi, Kenya

1. Sassy Molyneux, Dorcas Kamuya, and Vicki Marsh, who are members of the KEMRI-Wellcome Trust Research Programme, reviewed the contents of section 4 and provided feedback and comments. I am grateful to them for taking the time to edit carefully and present some new information and perspectives I had not considered. What is presented in the following modern narrative and analytic chapter is my own interpretation and representation of this group's work.

2. C. S. Molyneux, N. Peshu, and K. Marsh, "Understanding of Informed Consent in a Low-Income Setting: Three Case Studies from the Kenyan Coast," *Social Science & Medicine* 59, no. 12 (2004): 2548.

3. Ibid., 2547.

4. C. S. Molyneux, D. R. Wassenaar, N. Peshu, and K. Marsh, "'Even If They Ask You to Stand by a Tree All Day, You Will Have to Do It (Laughter)…!': Community Voices on the Notion and Practice of Informed Consent for Biomedical Research in Developing Countries," *Social Science & Medicine* 61, no. 2 (2005): 452.

5. "Investigating Parents' Understanding and Decision Making Process for Children's Participation in a Malaria Vaccine Trial, in Kenya: Qualitative Data Analysis" (MSc thesis, London School of Hygiene and Tropical Medicine, 2012), 19, 31; Vicki Marsh et al., "Beginning Community Engagement at a Busy Biomedical Research Programme: Experiences from the KEMRI CGMRC-Wellcome Trust Research Programme, Kilifi, Kenya," *Social Science & Medicine* 67, no. 5 (September 2008): 721; Vicki Marsh et al., "Experiences with Community Engagement and Informed Consent in a Genetic Cohort Study of Severe Childhood Diseases in Kenya," *BMC Medical Ethics* 11, no. 13 (2010): 9; Molyneux et al., "'Even if They Ask You," 443; C. S. Molyneux, N. Peshu, and K. Marsh, "Trust and Informed Consent: Insights from Community Members on the Kenyan Coast," *Social Science & Medicine* 61, no. 7 (October 2005): 1463–73.

6. Vicki Marsh et al., "Working with Concepts: The Role of Community in International Collaborative Biomedical Research," *Public Health Ethics* 4, no. 1 (2011): 36.

7. An entire special issue of *Developing World Bioethics* was devoted to this very topic in 2013. Sassy Molyneux et al., "Editorial: Field Workers at the Interface," *Developing World Bioethics* 13, no. 1 (2013): ii-iv; Caroline Gikonyo et al., "Feedback of Research Findings for Vaccine Trials: Experiences from Two Malaria Vaccine Trials Involving Healthy Children on the Kenyan Coast," *Developing World Bioethics* 13, no. 1 (2013): 48–56; Vibian Angwenyi et al., "Working with Community Health Workers as 'Volunteers' in a Vaccine Trial: Practical and Ethical Experiences and Implications," *Developing World Bioethics* 13, no. 1 (2013): 38–47; Dorcas M. Kamuya et al., "Engaging Communities to Strengthen Research Ethics in Low-Income Settings: Selection and Perceptions of Members of a Network of Representatives in Coastal Kenya," *Developing World Bioethics* 13, no. 1 (2013): 10–20; Dorcas M. Kamuya et al., "Evolving Friendships and Shifting Ethical Dilemmas: Fieldworkers' Experiences in a Short Term Community Based Study in Kenya," *Developing World Bioethics* 13, no. 1 (2013): 1–9.

8. Marsh et al., "Experiences with Community Engagement," 4.

9. Ibid., 4.

10. Ibid.; see also Molyneux, Peshu, and Marsh, "Understanding of Informed Consent," 2555.

11. Ibid., 2551.

12. Ibid., 2554. This vocabulary is very much in keeping with words I heard Tanzanian fieldworkers use to describe their work in interviews in 2008. Frequently used was the Swahili verb *kushawishi*, which means to persuade, coax, entice, tempt, or allure. Frederick Johnson, *A Standard Swahili-English Dictionary* (Oxford: Oxford University Press, 1939), 418.

13. Molyneux, Peshu, and Marsh, "Understanding of Informed Consent," 2554.

14. Marsh et al., "Experiences with Community Engagement," 4.

Chapter 4: Ethical Recruitment and Gathering Human Subjects

1. Sandra Crouse Quinn, "Protecting Human Subjects: The Role of Community Advisory Boards," *American Journal of Public Health* 94, no. 6 (2004): 919.

2. Nuremberg Code (1947), available at http://www.hhs.gov/ohrp/archive/nurcode.html.

3. International Covenant on Civil and Political Rights, Article 7 (adopted 1966), available at http://www.ohchr.org/EN/ProfessionalInterest/Pages/CCPR.aspx.

4. "General Requirements for Informed Consent," 45 C.F.R. 46:116 (July 14, 2009), available at http://www.hhs.gov/ohrp/humansubjects/guidance/45cfr46.html. For the Federal Policy for the Protection of Human Subjects, or "Common Rule" (1991), see http://www.hhs.gov/ohrp/humansubjects/commonrule/index.html.

5. "Poliomyelitis: A New Approach," *Lancet* 1 (1952): 552; David Smith and Alison Mitchell, "Sacrifices for the Miracle: The Polio Vaccine Research and Children with Mental Retardation," *Mental Retardation* 39, no. 5 (2001): 405–9.

6. George S. Nelson, "A Preliminary Report on the Out-Patient Treatment of Onchocerciasis with Antrypol in the West Nile District of Uganda," *East African Medical Journal* 32, no. 11 (November 1955): 413–29.

7. Medical Research Council, "Responsibility in Investigations on Human Subjects," Annual Report 1962–63, reprinted in *British Medical Journal* 2 (July 18, 1964): 178–80.

8. Ibid., 179.

9. Ibid.

10. Udo Schuklenk, "Protecting the Vulnerable: Testing Times for Clinical Research Ethics," *Social Science & Medicine* 51, no. 6 (2000): 969.

11. D. Schroeder and E. Gefenas, "Vulnerability: Too Vague and Too Broad?" *Cambridge Quarterly of Healthcare Ethics* 18, no. 2 (2009): 113–21.

12. Leonard Glantz et al., "Research in Developing Countries: Taking 'Benefit' Seriously," *Hastings Center Report* 28, no. 6 (1998): 38–42.

13. Caroline Kithinji and Nancy E. Kass, "Assessing the Readability of Non-English-Language Consent Forms: The Case of Kiswahili for Research Conducted in Kenya," *IRB: Ethics & Human Research* 32, no. 4 (2010): 10.

14. Ibid.

15. "Investigating Parents' Understanding and Decision Making Process for Children's Participation in a Malaria Vaccine Trial, in Kenya: Qualitative Data

Analysis" (MSc thesis, London School of Hygiene and Tropical Medicine, 2012), 16, 30.

16. Boga et al., "Strengthening the Informed Consent Process in International Health Research through Community Engagement: The KEMRI-Wellcome Trust Research Programme Experience," *PLoS Medicine* 8, no. 9 (2011): 3.

17. Ibid.

18. Vibian Angwenyi et al., "Complex Realities: Community Engagement for a Paediatric Randomized Controlled Malaria Vaccine Trial in Kilifi, Kenya," *Trials* 15, no. 1 (2014): 11.

19. Molyneux, Peshu, and Marsh, "Understanding of Informed Consent," 2553.

20. Boga et al., "Strengthening the Informed Consent Process," 2.

21. Charles Lidz et al., "Therapeutic Misconception and the Appreciation of Risks in Clinical Trials," *Social Science & Medicine* 58, no. 9 (2004): 1691; Paul S. Appelbaum et al., "False Hopes and Best Data: Consent to Research and the Therapeutic Misconception," *Hastings Center Report* 17, no. 21 (1987): 20–24.

22. V. M. Lema, "Therapeutic Misconception and Clinical Trials in Sub-Saharan Africa: A Review," *East African Medical Journal* 86, no. 6 (2009): 291–99.

23. Filariasis Research Unit Annual Report, 1951.

24. Roland Moore and A. Roberts, "An Investigation of the Pattern of Disease Prevalent in Parts of the Rufiji District," East African Medical Survey and Research Institute, n.d.

25. Angeliki Kerasidou, "Therapeutic Misconception in Research in Developing Countries," *Ethox Blog*, accessed May 5, 2014, http://www.ethox.org.uk/ethox-blog/therapeutic-misconception-in-research-in-developing-countries.

26. Vicki Marsh et al., "Working with Concepts: The Role of Community in International Collaborative Biomedical Research," *Public Health Ethics* 4, no. 1 (2011): 34.

27. Molyneux, Peshu, and Marsh, "Understanding of Informed Consent," 2557.

28. I was told about the *siri ya watafiti* by a number of different Tanzanian medical researchers who admitted to working in this way and introduced me to the phrase and concept, which I had never heard about before. Even with my questioning, and a discussion about international medical research and ethical regulations, none of them claimed to be doing anything wrong. No one in Nairobi or Kisumu mentioned this topic independently, but when I asked about the siri ya watifiti, I had multiple male Kenyan researchers laugh and say that, while they had never heard the phrase, the practice was not foreign to them.

29. M. Barry, "Ethical Considerations of Human Investigation in Developing Countries: the AIDS Dilemma," *New England Journal of Medicine* 319, no. 16 (1988): 1083–86; N. A. Christakis, "The Ethical Design of an AIDS Vaccine Trial in Africa," *Hastings Centers Report* 18, no. 3 (1988): 31–37; C. E. Taylor, "Clinical Trials and International Health Research," *American Journal of Public Health* 69, no. 10 (1979): 981–83. Ijsselmuiden and Faden are an exception to this rule, and present a thoughtful analysis of the inappropriateness of the concept. C. Ijsselmuiden

and R. Faden, "Research and Informed Consent in Africa—Another Look," in *Health and Human Rights: A Reader,* ed. Mann et al. (New York: Routledge, 1999), 363–72.

30. Steven Feierman, *Peasant Intellectuals: Anthropology and History in Tanzania* (Madison: University of Wisconsin Press, 1990).

31. Helge Kjekshus, *Ecology Control and Economic Development in East African History: The Case of Tanganyika, 1850–1950* (Berkeley: University of California Press, 1977); James Giblin, *The Politics of Environmental Control in Northeastern Tanzania, 1840–1940* (Philadelphia: University of Pennsylvania Press, 1992).

32. A. R. Cook, "Syphilis in Uganda," *Lancet* (December 12, 1908): 1771.

33. Ibid.

34. District Commissioner, Lamu to Mudir of Faza and Witu, "Filariasis Investigation," September 5, 1958, DC/LAMU/2/23/29, KNA.

35. Trant to Senior Medical Officer, Tabora, August 17, 1954, Dr. Trant Papers, NIMR, Mwanza.

36. Interviews in Mwanza, Ukara Island, and Ukerewe Island. Further evidence that the practice of fining was used by chiefs during the colonial era comes by looking at the Swahili phrase for levying a fine. *Kupiga faini* is derived from the English word "fine," which is added to the multipurpose Swahili verb *kupiga* (to do or make).

37. When conducting interviews about the authority of the government in forcing people to participate in medical research, people often spoke generally of the *serikali* (government). When I asked for clarification about *which* government, people were emphatic that all governments behaved the same. Interviews in Mwanza, Ukara Island, and Ukerewe Island, July and August 2008. Melissa Graboyes, "Fines, Orders, Fear . . . and Consent? Medical Research in East Africa, c. 1950s," *Developing World Bioethics* 10, no. 1 (2010): 34–41.

38. David Killingray, "The Maintenance of Law and Order in British Colonial Africa," *African Affairs* 85, no. 340 (July 1986): 417; John Tosh, *Clan Leaders and Colonial Chiefs in Lango: The Political History of an East African Stateless Society c. 1800–1939* (Oxford: Oxford University Press, 1978), 188.

39. P. Wenzel Geissler, "Stuck in Ruins, or Up and Coming? The Shifting Geography of Urban Public Health Research in Kisumu, Kenya." *Africa* 83, no. 4 (2013): 544.

40. Eric Hobsbawm and Terence Ranger, eds., *The Invention of Tradition* (Cambridge: Cambridge University Press, 1992); Thomas Spear, "Neo-Traditionalism and the Limits of Invention in British Colonial Africa," *Journal of African History* 44, no. 1 (2003): 3–27.

41. Marsh et al., "Working with Concepts," 29.

42. National Council for Law Reporting, *Laws of Kenya,: Chiefs' Act: Chapter 128,* rev. ed. 2012 [1998], http://www.icnl.org/research/library/files/Kenya/chief.pdf.

43. Molyneux et al., "'Even If They Ask You to Stand by a Tree All Day, You Will Have to Do It (Laughter) . . . !': Community Voices on the Notion and Practice

of Informed Consent for Biomedical Research in Developing Countries," *Social Science & Medicine* 61, no. 2 (2005): 446. Emphasis added.

44. Ibid.

45. Ibid.

46. Ibid., 450.

47. Marsh et al., "Working with Concepts," 36.

48. Ibid.

49. C. Gikonyo et al., "Taking Social Relationships Seriously: Lessons Learned from the Informed Consent Practices of a Vaccine Trial on the Kenyan Coast," *Social Science & Medicine* 67, no. 5 (2008): 708.

50. Oscar Benitez, D. Devauz, and J. Dausset, "Audiovisual Documentation of Oral Consent: A New Method of Informed Consent for Illiterate Populations," *Lancet* 359, no. 9315 (April 20, 2002): 1406–7.

51. Daniel Fitzgerald et al., "Comprehension during Informed Consent in a Less-Developed Country," *Lancet* 360, no. 9342 (October 26, 2002): 1301–2.

52. Ibid.

53. Available at the Global Health Trials website, https://globalhealthtrials.tghn.org.

54. Boga et al., "Strengthening the Informed Consent Process," 3.

55. Molyneux, Peshu, and Marsh, "Trust and Informed Consent," 1470.

BALANCING RISKS AND BENEFITS

Historical Narrative: Hope Trant and a Compound on Fire in Tanganyika, 1954

1. Governor to Secretary of State for the Colonies, "Health Survey Buha," August 17, 1953, 523 M 10/5, TNA. Thanks to Julie Weiskopf for sharing this citation.

2. Laurie to Trant, July 10, 1950, Dr. Trant Files, NIMR, Mwanza.

3. Hope Trant, *Not Merrion Square: Anecdotes of a Woman's Medical Career in Africa* (Toronto: Thornhill Press, 1970), 99.

4. Mwami Teresa Ntare was an unusual woman, and what little we know of her biography was gathered by Susan Geiger in an interview in 1992. Geiger writes: "Born in 1928 into the ruling Tutsi family and educated at Tabora Girls' Secondary School, Teresa Ntare succeeded her father as chief (*mwami*) of Kasulu (Buha) in 1946. A progressive farmer whose husband led the Kasulu Coffee Cooperative Society in the 1950s, Mwami Ntare was one of the most powerful chiefs in the territory. . . . Both she and her sister, Anna Gwassa, were active TANU supporters from the late 1950s. Mwami Ntare became legislative councilor for Kasulu in 1960. Junior minister for youth and culture from 1962 to 1965, she was the first member of Parliament for Kasulu constituency in Kigoma Region." Susan Geiger, *TANU Women: Gender and Culture in the Making of Tanganyikan Nationalism, 1955–1965* (Portsmouth, NH: Heinemann, 1997), 167. Additional information about Teresa Ntare in Johan Herman Scherer, *Marriage and Bride-Wealth in the Highlands of Buha (Tanganyika)* (Groningen: V. R. B. Kleine der A, 1965), 91; Margot Lovett, "Elders,

Migrants and Wives: Labor Migration and the Renegotiation of Intergenerational, Patronage and Gender Relations in Highland Buha, Western Tanzania" (PhD diss., Columbia University, 1995), 383, citing Kasulu District Report, Rural Local Government, 1961, 967.823, TNA.

5. Trant, *Not Merrion Square*, 137.

6. Ibid., 137–38.

7. Ibid., 137–39.

8. There is some confusion about the ultimate fate of the EAMS in Kasulu. The project was stopped in August 1954, without plans to continue. Yet the annual report for 1954–55 reported that the Buha survey (consisting of Kibondo and Kasulu) was completed in March 1955, noting, "there having been interruption of the work that at one stage threatened failure." If the survey did continue, there are no records in Trant's papers of her participating. There are also no final reports available summarizing the work in Buha. East African Medical Survey and Research Institute Annual Report, 1954–55.

Modern Narrative: A Male Circumcision Trial Canceled in Rakai, Uganda, 2005

1. In the early 1990s, it was estimated that 15 percent of the children in Rakai district had lost one or both parents. Fred Nalugoda et al., "HIV Infections in Rural Households, Rakai District, Uganda," in "Evidence of the Socio-Demographic Impact of AIDS in Africa," supplement, *Health Transition Review* 7 (1997): 127, 129, citing Serwadda et al., "HIV Risk Factors in Three Geographic Strata of Rural Rakai District, Uganda," *AIDS* 6, no. 9 (September 1992): 983–89.

2. Nalugoda et al., "HIV Infections in Rural Households," 127; Women and Girls Empowerment Project, "Rakai District Uganda: Fact Sheet," 2012, citing National HIV sero survey, ANCE sentinel sites, and RHSP cohort.

3. United Nations Department of Public Information, "Global Summary of the HIV/AIDS Epidemic," DPI/2199, background information sheet distributed at the African Summit on HIV/AIDS, Tuberculosis and Other Related Infectious Diseases, Abuja, Nigeria, April 2001, http://www.un.org/ga/aids/pdf/stats.pdf; and World Health Organization, "Overview: HIV/AIDS," accessed January 26, 2014, http://www.afro.who.int/en/clusters-a-programmes/dpc/acquired-immune-deficiency-syndrome/overview.html.

4. B. Auvert et al., "Randomized, Controlled Intervention Trial of Male Circumcision for Reduction of HIV Infection Risk: The ANRS 1265 Trial," *PLoS Medicine* 2, no. 11 (2005): e298; Robert C. Bailey et al., "Male Circumcision for HIV Prevention in Young Men in Kisumu, Kenya: A Randomized, Controlled Trial," *Lancet* 369, no. 9562 (2009): 643–56.

5. H. A. Weiss, M. Quigley, and R. Hayes, "Male Circumcision and Risk of HIV Infection in Sub-Saharan Africa: A Systemic Review and Meta-Analysis," *AIDS* 14, no. 15 (2000): 2361–70; B. Auvert et al., "Ecological and Individual Level Analysis of Risk Factors for HIV Infection in Four Urban Populations in Sub-Saharan Africa with Different Levels of HIV Infection," in "The Multicentre Study of Factors

Determining the Different Prevalences of HIV in Sub-Saharan Africa," supplement 4, *AIDS* 15 (August 2001): S15–S30.

6. Rakai Health Sciences Program, "History of the Rakai Health Sciences Program," November 23, 2010, http://www.rhsp.org/content/history-rakai-health-sciences-program.

7. Ronald H. Gray, "Circumcision: HIV/STIS and Behaviors in a RCT and Post-RCT Surveillance, Rakai, Uganda. U01-AI075115-01A1 Protocol with Appendixes." Report of NIH-funded trial. Accessed November 15, 2013.

8. United States Department of Health and Human Services, "Guidance for Clinical Trial Sponsors: Establishment and Operation of Clinical Trial Data Monitoring Committees" (March 2006), http://www.fda.gov/downloads/RegulatoryInformation/Guidances/ucm127073.pdf.

9. There remains a lot of murkiness about DSMBs. In many cases the names of those on the board are not made public. There is no effective requirement to report monitoring board recommendations to the public. There has also been speculation that the boards are not as independent as they should be, since they are often funded by the sponsors of the trials they monitor. Trudie Lang et al., "Data Safety and Monitoring Boards for African Clinical Trials," *Transactions of the Royal Society of Tropical Medicine and Hygiene* 102, no. 12 (2008): 1190.

10. Ron Gray et al., "Male Circumcision for HIV Prevention in Men in Rakai, Uganda: a Randomised Trial," *Lancet* 369, no. 9562 (February 20, 2007): 657–66.

11. DSMBs have also stopped other trials in Africa, and globally, where an intervention was clearly ineffective or exposed subjects to unacceptable levels of risk. In 2011, an experiment testing the oral and vaginal use of the drug tenofovir in South Africa was stopped by a DSMB associated with the US National Institutes of Health after data from roughly two years indicated the drug was no better than a placebo. "The 'VOICE' Trial (Vaginal and Oral Interventions to Control the Epidemic)," accessed January 23, 2014, http://www.mtnstopshiv.org/news/studies/mtn003. Another multisite microbicide trial of PRO2000 was also cut short. Sheena McCormack et al., "PRO2000 Vaginal Gel for Prevention of HIV-1 Infection (Microbicides Development Programme 301): A Phase 3, Randomised, Double-Blind, Parallel-Group Trial," *Lancet* 376, no. 9749 (2010): 1329–37; Donald G. McNeil, "Anti-H.I.V. Trial in Africa Canceled over Failure to Prevent Infection," *New York Times*, November 25, 2011. In 2013, trials of the drug perifosine were cancelled when the DSMB reviewed the data and found it unlikely that the study would find measurable effects. "Aeterna Zentaris Cancels Phase III Trial of Blood Cancer Drug Perifosine," accessed January 23, 2014, http://www.pharmaceutical-technology.com/news/newsaeterna-zentaris-cancels-phase-iii-trial-of-blood-cancer-drug-perifosine.

12. Male circumcision removes special cells (Langerhans cells) that are targets for HIV, results in the toughening of the remaining foreskin (a process called keratinization), and decreases the risk of abrasion, inflammation, or ulcer on the foreskin by removing it. Ron Gray et al., "The Effectiveness of Male Circumcision for HIV Prevention and Effects on Risk Behaviors in a Posttrial Follow-Up Study,"

Aids 26, no. 5 (2012): 609–15; Maria J. Wawer et al., "Circumcision in HIV-Infected Men and Its Effect on HIV Transmission to Female Partners in Rakai, Uganda: A Randomised Controlled Trial," *Lancet* 374, no. 9685 (2009).

13. There is not unanimous agreement that male circumcision is actually a good strategy for global HIV prevention. The most thorough review of the data and a very reasonable criticism is presented in Michel Garenne, Alain Giami, and Christophe Perrey, "Male Circumcision and HIV Control in Africa: Questioning Scientific Evidence and the Decision-Making Process," in *Global Health in Africa: Historical Perspectives on Disease Control*, ed. Tamara Giles-Vernick and James L. A. Webb, Jr. (Ohio: Ohio University Press, 2013), 185–210.

14. In its own materials, the RHSP calls itself "one of the largest and oldest community-based research endeavors . . . in Africa." Rakai Health Sciences Program, "History of the Rakai Health Sciences Program." There are other long-term cohort studies in developing countries. The Wellcome Trust (UK) supports twelve different longitudinal studies in Eastern Europe, Southern Africa, South America, Thailand, and the Middle East, which focus on diseases such as malaria, HIV/AIDs, and non-infectious chronic diseases. The "Birth to Twenty" project, run by the South African Medical Research Council, was a longitudinal cohort study tracking roughly three thousand children from 1990 to 2010. Linda Richter et al., "Cohort Profile: Mandela's Children: The 1990 Birth to Twenty Study in South Africa," *International Journal of Epidemiology* 36, no. 3 (2007): 504–11.

15. The 2002 data was reported in Thiessen et al., "Personal and Community Benefits and Harms of Research: Views from Rakai, Uganda," *AIDS* 21, no. 18 (2007): 2493–501.

16. The "Rakai Project" began with the first community cohort in 1988, and included 21 villages and 1,280 participants. It was later enlarged to include 31 villages and roughly 4,000 people. In 1994 the cohort was enlarged again to roughly 14,000 people. The interviewing and sampling takes roughly an hour and participants are paid approximately $1.50. Rakai Health Sciences Program, "History of the Rakai Health Sciences Program."

17. These benefits cannot be directly attributed to the presence of medical research, since arguably these gains could have been achieved through purely public health interventions, or by strengthening existing medical structures. J. Matovu et al., "The Rakai Project Counselling Programme Experience," *Tropical Medicine & International Health* 7, no. 12 (2002): 1064–67; T. Lutalo et al., "Trends and Determinants of Contraceptive Use in Rakai District, Uganda, 1995–1998," *Studies in Family Planning* 31, no. 3 (2000): 217–27.

18. Maria J. Wawer and David Serwadda, "Randomized Trial of Male Circumcision: STD, HIV and Behavioral Effects in Men, Women and the Community Protocol," *Circ Gates Protocol Stage 2* (October 21, 2004): 49.

19. Although it is remarkable that more than 2,500 people in Rakai District have been provided with free antiretroviral therapy, it is not ideal that patients must wait to begin ART until they have a CD4 count of 250 or less,

or be considered WHO stage IV—the most severe stage of HIV/AIDS. Official policy from the US Department of Health and Human Services recommends starting treatment as soon as the HIV diagnosis is made, although that does not match with Medicaid protocol. Currently, Medicaid only provides free ART when a person has a CD4 cell count below 200, a diagnosable opportunistic infection, or meets other criteria beyond just a positive HIV test. AVERT, "HIV Treatment in the U.S.," last revised July 23, 2014, http://www.avert.org/hiv-treatment-us.htm.

Chapter 5: Balancing Risks and Benefits

1. Warwick Anderson, "Immunities of Empire: Race, Disease, and the New Tropical Medicine, 1900–1920," *Bulletin of the History of Medicine* 70, no. 1 (1996): 94–118; Warwick Anderson, "Postcolonial Technoscience," *Social Studies of Science* 32, no. 5–6 (2002): 643–58; Warwick Anderson, "Natural Histories of Infectious Disease: Ecological Vision in Twentieth-Century Biomedical Science," *OSIRIS* 19 (2004): 39–61; David Arnold, *Colonizing the Body: State Medicine and Epidemic Disease in Nineteenth-Century India* (Berkeley: University of California Press, 1993); Wolfgang Eckart, "The Colony as Laboratory: German Sleeping Sickness Campaigns in German East Africa and in Togo, 1900–1914," *History and Philosophy of the Life Sciences* 24, no. 1 (2002): 69–89; Helen Tilley, "Global Histories, Vernacular Science, and African Genealogies; or, Is the History of Science Ready for the World?" *Isis* 101, no. 1 (2010): 110–19; Shula Marks, "What Is Colonial about Colonial Medicine? And What Has Happened to Imperialism and Health?" *Social History of Medicine* 10, no. 2 (1997): 205–19; Helen Tilley, *Africa as a Living Laboratory: Empire, Development, and the Problem of Scientific Knowledge, 1870–1950* (Chicago: University of Chicago Press, 2011); Michael Worboys, "The Discovery of Colonial Malnutrition between the Wars," in *Imperial Medicine and Indigenous Societies,* ed. David Arnold (New York: Manchester University Press, 1988), 208–25.

2. Alan Irwin and Brian Wynne, eds. *Misunderstanding Science? The Public Reconstruction of Science and Technology* (Cambridge: Cambridge University Press, 1996), 42.

3. M. Hayes, "On the Epistemology of Risk: Language, Logic and Social Sciences," *Social Science & Medicine* 35, no. 4 (1992): 401–7; P. Slovic, *The Perception of Risk* (London: Earthscan Publications, 2000).

4. Rachel R. Chapman, *Family Secrets: Risking Reproduction in Central Mozambique* (Nashville: Vanderbilt University Press, 2010), 21, quoting Dorothy Nelkin, "Foreword: The Social Meaning of Risk," in *Risk, Culture, and Health Inequality: Shifting Perceptions of Danger and Blame,* ed. Barbara Herr Harthorn and Laury Oaks (Westport, CT: Praeger, 2003), vii-xiii; Patrick Caplan, ed., *Risk Revisited* (London: Pluto Press, 2000); Denise Roth Allen, *Managing Motherhood, Managing Risk: Fertility and Danger in West Central Tanzania* (Ann Arbor: University of Michigan Press, 2002).

5. C. Taylor, "Condoms and Cosmology: The 'Fractal' Person and Sexual Risk in Rwanda," *Social Science & Medicine* 31, no. 9 (1990): 1023–28.

6. Sassy Molyneux et al., "Benefits and Payments for Research Participants: Experiences and Views from a Research Centre on the Kenyan Coast," *BMC Medical Ethics* 13, no. 13 (2012): 6.

7. Nicola Barsdorf et al., "Access to Treatment in HIV Prevention Trials: Perspectives from a South African Community," *Developing World Bioethics* 10, no. 2 (2010): 78.

8. Christine Grady et al., "Research Benefits for Hypothetical HIV Vaccine Trials: The Views of Ugandans in the Rakai District," *IRB: Ethics & Human Research* 30, no. 2 (2008): 1–7; Barsdorf et al., "Access to Treatment," 82.

9. It is unclear whether the Swahili term *mumiani* (variously translated as "vampires," "medicine made from blood," or in reference to the people who procured the blood) was used in Heru Juu, although Trant was well aware of the concept. As the entomologist Alec Smith described it in his memoir, Trant was particularly cautious when participating in births on Ukara Island, making sure to handle the placenta carefully because of mumiani fears. Alec Smith, *Insect Man: A Fight against Malaria in Africa* (New York: Palgrave Macmillan, 1993), 30–31.

10. Trant to Senior Medical Officer, Tabora, August 17, 1954, Dr. Trant Files, NIMR, Mwanza.

11. "Medical Procedures: Hemoglobin Estimation," accessed January 27, 2014, http://www.medprocedures.com/2012/11/hemoglobin-estimation.html.

12. Rebecca Kreston, "Suck It: The Ins and Outs of Mouth Pipetting," *Body Horrors* (blog), *Discover*, March 20, 2013, http://blogs.discovermagazine.com/bodyhorrors/2013/03/20/mouth_pipetting. A movie clip of the practice of mouth pipetting is available online (accessed January 27, 2014): http://www.musee-afrappier.qc.ca/en/index.php?pageid=3134ba&image=3134ba_bouche.

13. James Brennan, "Blood Enemies: Exploitation and Urban Citizenship in the Nationalist Political Thought of Tanzania, 1958–1975," *Journal of African History* 47, no. 3 (2006): 395–96.

14. Ibid., 398.

15. Hope Trant, *Not Merrion Square: Anecdotes of a Woman's Medical Career in Africa* (Toronto: Thornhill Press, 1970), 128.

16. East African Medical Survey Annual Report, 1953.

17. For a general and thorough discussion of the changes in Buha, with a special focus on gender, Margot Lovett's dissertation provides a strong sense of the region. Julie Weiskopf's recent work has also traced the logistics of the mass movement of people during the sleeping sickness resettlements and the effect it had on local communities and social practices. Margot Lovett, "Elders, Migrants and Wives: Labor Migration and the Renegotiation of Intergenerational, Patronage and Gender Relations in Highland Buha, Western Tanzania" (PhD diss., Columbia University, 1995); Julie Weiskopf, "Rooting New Communities in Tanzania's Sleeping Sickness Concentrations" (April 16, 2013), accessed February 9, 2014, http://ssrn.com/abstract=2251606.

18. *Trypanosoma brucei rhodesiense* is the form of sleeping sickness found in East Africa. There are two stages to the disease. In the early stage, symptoms include

swollen lymph glands and aching muscles and joints. In the advanced stage, the central nervous system is affected and sufferers often become lethargic and fall into comas. The British used both a medical and environmental approach to controlling sleeping sickness, although they put greater emphasis on bush clearance and forced village resettlement outside of fly-infested areas. The German colonial government had relied upon policies of forced medical treatment. Sleeping sickness has been treated since the 1920s, although the drugs used had extremely limited efficacy and bad side effects. One of the first drugs used was atoxyl, an arsenic-based drug that was extremely toxic and often caused blindness. The sleeping sickness camps were only one part of a multipronged attack on sleeping sickness. Kirk Arden Hoppe, *Lords of the Fly: Sleeping Sickness Control in East Africa, 1900–1960* (Westport, CT: Preager, 2003); Eckart, "Colony as Laboratory"; Mari Webel, "Ziba Politics and the German Sleeping Sickness Camp at Kigarama, Tanzania, 1907–1914," *International Journal of African Historical Studies* 47, no. 3 (2104): 399–424; Deborah Neill, *Networks in Tropical Medicine: Internationalism, Colonialism, and the Rise of a Medical Specialty, 1890–1930* (Palo Alto: Stanford University Press, 2012).

19. Annual Report 1950 Western Province, Tanganyika, 450/1501/4, TNA. Thanks to Julie Weiskopf for sharing this citation.

20. "Sleeping Sickness and Reclamation Kigoma and Kasulu Districts," 523A/M.11/2, TNA. Thanks to Julie Weiskopf for sharing this citation.

21. Johan Herman Scherer, *Marriage and Bride-Wealth in the Highlands of Buha (Tanganyika)* (Groningen: V. R. B. Kleine der A, 1965), 8.

22. EAMS Director to Trant, February 7, 1951, Dr. Trant Files, NIMR, Mwanza.

23. EAMS Director to Trant, April 28, 1952, Dr. Trant Files, NIMR, Mwanza.

24. EAMS Director to Trant, November 1, 1950, Dr. Trant Files, NIMR, Mwanza.

25. Administrative Officer to Trant, November 13, 1950, Dr. Trant Files, NIMR, Mwanza.

26. EAMS Director to Trant, February 27, 1951, Dr. Trant Files, NIMR, Mwanza.

27. Trant to EAMS Director, November 14, 1950, Dr. Trant Files, NIMR, Mwanza.

28. Bozman to Trant, May 9, 1955, Dr. Trant Files, NIMR, Mwanza.

29. Trant to Laurie, November 8, 1950, Dr. Trant Files, NIMR, Mwanza.

30. Hope Trant was unusual partly because she was a professional woman (hired as a medical officer) at a time and in a place with virtually no other female professionals. Because some of her other attributes are so unusual, it's difficult to say how important gender was in creating these alternative viewpoints. I hope to pursue a project in the future that more systematically considers female scientists in the region, many of whom were the wives of formally hired medical officers.

31. Trant to Laurie, November 21, 1953, Dr. Trant Files, NIMR, Mwanza.

32. She continued to push for treatment for local residents on Ukara Island and also in later work with the Pare-Taveta Malaria Scheme. Notably, she ran into the same problems there, as the director of the project, Dr. Donald Bagster Wilson, was

unwilling to pay for drugs for participants, and Trant was forced to beg for supplies from the district commissioner and drug companies. Trant to Bozman and Holmes, May 22, 1955, Dr. Trant Files, NIMR, Mwanza.

33. Laurie to Trant, November 1, 1950, Dr. Trant Files, NIMR, Mwanza.

34. Laurie to Trant, November 21, 1950, Dr. Trant Files, NIMR, Mwanza.

35. Trant to Laurie, December 11, 1950, Dr. Trant Files, NIMR, Mwanza.

36. Trant to Laurie, November 8, 1950, Dr. Trant Files, NIMR, Mwanza.

37. Administrative Officer to Trant, November 13, 1950, Dr. Trant Files, NIMR, Mwanza.

38. EAMS Director to Meek, cc Trant, August 3, 1951, Dr. Trant Files, NIMR, Mwanza.

39. H. Cullumbine, "The Taveta-Pare Malaria Control Scheme," n.d. [1954], Box 23, NIMR, Amani.

40. Bagster Wilson to Cullumbine, December 30, 1954, Box 23, NIMR, Amani.

41. "British Red Cross Society: Upare Clinic Scheme: 2nd Annual Progress Report," n.d. [1956], Box 23, NIMR, Amani.

42. Acting Provincial Medical Officer, Nyanza, to Director of Medical Services, Kenya, "Medical Survey," May 5, 1952, MOH/22/15, KNA.

43. Division of Insect-Borne Diseases, Nyanza Province, Annual Report, 1951.

44. The self-experimentation involved testing drug-resistant strains of malaria, and he had plans to inject himself with plague, although he relented after his supervisor, Farnworth Anderson, expressed his displeasure. S. Avery Jones, "Experiment to Determine if a Strain of P. Falciparum Resistant to Proguanil Would Respond to Large Doses of Daraprim," n.d., Box 23, NIMR, Amani; Anderson to Avery Jones, January 30, 1952, BY/66/3315, Confidential, KNA. More broadly, the issue of self-experimentation in medicine has been addressed by Lawrence Altman, *Who Goes First? The Story of Self-Experimentation in Medicine* (Berkeley: University of California Press, 1998).

45. Lewthwaite to Anderson, December 5, 1953, BY/66/3315, KNA.

46. Anna Crozier, *Practising Colonial Medicine: The Colonial Medical Service in British East Africa* (London: I. B. Tauris, 2007), 73.

47. There is an intriguing note found in the Kenya National Archives that raises the possibility that Avery Jones acted as—or threatened to be—a whistleblower about medical activities in Kenya. In May 1952, Avery Jones sent a confidential note to Anderson, the director of medical services. He wrote, "I wish to bring certain aspects of my conditions of service to the notice of the Sec[retary] of State for the Colonies. I should be grateful for information as to the correct procedure to be followed. As these matters are likely to be of considerable concern to the British Medical Association and, possibly, to the Medical Defence Society I have to enquire whether it is in order to bring them to their attention at the same time." However, there are no other citations. Folder: Dr. S. Avery Jones, BY/66/3315, KNA.

48. Avery Jones, "Progress Report: Makueni Experiment," n.d., Box 23, NIMR, Amani.

49. Ibid.

50. Ibid.

51. Avery Jones to Bagster Wilson, "Confidential," April 14, 1953, Box 23, NIMR, Amani. Emphasis in original.

52. Avery Jones to Director of Medical Services, Kenya, cc Bagster Wilson, "Makueni Experiment, Confidential," April 28, 1953, Box 23, NIMR, Amani.

53. Anderson to Bagster Wilson, April 17, 1953, Box 23, NIMR, Amani. Avery Jones, "Progress Report," n.d., Box 23, NIMR, Amani.

54. Avery Jones to Bagster Wilson, "Confidential," April 14, 1953, Box 23, NIMR, Amani.

55. Ibid.

56. Bagster Wilson to Avery Jones, cc Director of Medical Services, Nairobi, Senior Parasitologist Nairobi, Professor Garnham, April 24, 1953, Box 23, NIMR, Amani.

57. Ibid.

58. Bagster Wilson is a complicated figure. His involvement with the Nandi and Makueni Malaria Schemes indicates that, when given a choice, he typically leaned toward continuing the experiment even when risks increased. Yet, in broader discussions about the future of malaria control in Africa, he led a group of "conservative" scientists who argued that eradication attempts shouldn't be made in tropical Africa until more research had been done and scientists could be sure that their attempts were not going to inadvertently destroy local peoples' acquired immunity. M. J. Dobson, M. Malowany, and R. W. Snow, "Malaria Control in East Africa: The Kampala Conference and the Pare-Taveta Scheme: A Meeting of Common and High Ground," *Parissitologia* 42, no. 1–2 (2000): 149–66; Melissa Graboyes, "'The Malaria Imbroglio': Ethics, Eradication, and Endings in Pare-Taveta, East Africa, 1959–1960," *International Journal of African Historical Studies* 47, no. 3 (2014): 445–72.

59. Bagster Wilson to Director of Medical Services, Kenya, "Daraprim Scheme—Nandi," February 6, 1954, Box 23, NIMR, Amani.

60. Avery Jones to Director of Medical Services, Kenya, cc Bagster Wilson, "Makueni Experiment, Confidential," April 28, 1953, Box 23, NIMR, Amani.

61. Ibid.

62. Division of Insect Borne Diseases, Nyanza Province, Annual Report, 1953.

63. Avery Jones to Bagster Wilson, December 11, 1953, Box 23, NIMR, Amani; Division of Insect Borne Diseases, Nyanza Province, Annual Report, 1953.

64. Avery Jones to Bagster Wilson, December 11, 1953, Box 23, NIMR, Amani.

65. Avery Jones to Director of Medical Services, January 1, 1954, Box 23, NIMR, Amani.

66. Ibid.

67. Bagster Wilson to Director of Medical Services, Kenya, "Daraprim Scheme—Nandi," February 6, 1954, Box 23, NIMR, Amani.

68. Crozier, *Practising Colonial Medicine,* 115.

69. Ibid., 118.

70. Avery Jones to Bagster Wilson, September 6, 1965, Box 23, NIMR, Amani.

71. Leonard Jan Bruce-Chwatt. "Chemotherapy in Relation to Possibilities of Malaria Eradication in Tropical Africa," *Bulletin of the World Health Organization* 15, nos. 3–5 (1956): 852–62.

72. Based on knowledge gleaned over the past fifty years, it is now clear that resistance can exist naturally and grow through selective pressure (as in Makueni) *or* appear when a drug is given in sub-therapeutic doses. Bruce-Chwatt, "Chemotherapy in Relation to Possibilities," 860; U. D'Alessandro and H. Buttiëns, "History and Importance of Antimalarial Drug Resistance," *Tropical Medicine & International Health* 6, no. 11 (2001): 845–46; Michelle Gatton, Laura Martin, and Qin Cheng, "Evolution of Resistance to Sulfadoxine-Pyrimethamine in *Plasmodium falciparum*," *Antimicrobial Agents and Chemotherapy* 48, no. 6 (2004): 2116–23.

73. Principal Secretary, Ministry of Health Tanganyika, to World Health Organization Representative, "WHO Malarial Consultant Dr. S. Avery Jones," October 12, 1965, Box 11, NIMR, Amani.

74. Clyde to Bruce-Chwatt, June 19, 1964, Box 2, NIMR, Amani.

75. Smith, *Insect Man*, 91.

76. John Iliffe, *East African Doctors: A History of the Modern Profession* (Cambridge: Cambridge University Press, 1998), 229, citing Magdalene Ngaiza, "Women's Bargaining Power in Sexual Relations: Practical Issues and Problems in Relation to HIV Transmission and Control in Tanzania," *Tanzanian Medical Journal* 6, no. 2 (1991): 59; G. G. Mbungua et al., "Epidemiology of HIV Infection among Long-Distance Truck Drivers in Kenya," *East African Medical Journal* 72, no. 8 (1995): 515.

77. P. Wenzel Geissler et al., "'He Is Now Like a Brother, I Can Even Give Him Some Blood'—Relational Ethics and Material Exchanges in a Malaria Vaccine 'Trial Community' in The Gambia," *Social Science & Medicine* 67, no. 5 (2008): 700.

78. D. N. Shaffer et al., "Equitable Treatment for HIV/AIDS Clinical Trial Participants: A Focus Group Study of Patients, Clinician Researchers, and Administrators in Western Kenya," *Journal of Medical Ethics* 32, no. 1 (2006): 58.

79. Ibid., 55.

80. In the older English-Swahili dictionaries, "cash" is not listed, but "money" was defined as *fetha* or *fedha*; there is no mention of *hela*. Only in the 1939 Johnson dictionary is *hela* listed, but as an interjection which he translates as, "Well then! Come then! Make way!" Frederick Johnson, *A Standard Swahili-English Dictionary* (Oxford: Oxford University Press, 1939), 131–32.

81. Vinay Kamat, "Negotiating Illness and Misfortune in Post-Socialist Tanzania: An Ethnographic Study in Temeke District, Dar es Salaam" (PhD diss., Emory University, 2004), 64.

82. Lovett, *Elders, Migrants and Wives*, 241–42.

83. P. Wenzel Geissler, "'Transport to Where?': Reflections on the Problem of Value and Time à Propos an Awkward Practice in Medical Research," *Journal of Cultural Economy* 4, no. 1 (2011): 45–64.

84. Carl Elliot and Roberto Abadie, "Exploiting a Research Underclass in Phase 1 Clinical Trials," *New England Journal of Medicine* 358, no. 22 (2008): 2316–17; Roberto Abadie, *The Professional Guinea Pig: Big Pharma and the Risky World of Human Subjects* (Durham: Duke University Press, 2010).

85. Molyneux et al. "Benefits and Payments," 2, citing J. Koen et al., "Payment of Trial Participants Can Be Ethically Sound: Moving Past a Flat Rate," *South African Medical Journal* 98, no. 12 (2008): 926–29.

86. Council of Europe, *Texts of the Council of Europe on Bioethical Matters: Volume I* (Strasbourg: Council of Europe, April 2014), 27.

87. Molyneux, "Benefits and Payments, 2."

88. Angela Ballantyne, "Benefits to Research Subjects in International Trials: Do They Reduce Exploitation or Increase Undue Inducement?" *Developing World Bioethics* 8, no. 3 (2008): 190.

89. International codes and reports do speak specifically to undue inducement but rarely define it clearly. Council for International Organizations of Medical Sciences, *International Ethical Guidelines for Biomedical Research Involving Human Subjects* (Geneva: CIOMS, 2002), guideline 7, "Inducement to Participate in Research"; Nuffield Council on Bioethics, *The Ethics of Research Related to Healthcare in Developing Countries* (London: Nuffield Council on Bioethics, 2002), paragraph 6.29; National Bioethics Advisory Commission, Ethical and Policy Issues *in International Research: Clinical Trials in Developing Countries* (Bethesda, MD: NBAC, 2001), 60; Joint United Nations Programme on HIV/AIDS (UNAIDS), *Ethical Considerations in HIV Preventive Vaccine Research* (Geneva: UNAIDS, 2000), 42.

90. Ballantyne, "Benefits to Research Subjects," 178.

91. Christine Grady, "Money for Research Participation: Does It Jeopardize Informed Consent?" *American Journal of Bioethics* 1, no. 2 (2001): 40–44.

92. Ballantyne, "Benefits to Research Subjects," 190.

93. Ibid.

EXITS AND LONGER-TERM OBLIGATIONS

Historical Narrative: "Almost Completely Eradicated": The Pare-Taveta Malaria Scheme, 1955

1. "Five Year Experiment Shows the Way to Defeat of Malaria," *East African Standard,* July 10, 1959, Box 24, NIMR, Amani.

2. Dieldrin is a chlorinated hydrocarbon developed in the 1940s and related to DDT, but slightly more irritating and dangerous to humans. It is an extremely effective insecticide, but because of its toxicity to humans it is no longer used. It was likely used in Pare-Taveta because it was cheaper than other insecticides.

3. M. J. Dobson, M. Malowany, and R. W. Snow, "Malaria Control in East Africa: The Kampala Conference and the Pare-Taveta Scheme: A Meeting of Common and High Ground," *Parissitologia* 42, no. 1–2 (2000): 161.

4. The Global Malaria Eradication Program, and specifically how it played out in Africa, has been discussed by James Webb, *Humanity's Burden: A Global History of*

Malaria (Cambridge: Cambridge University Press, 2009), 160–87; Randall Packard, *The Making of a Tropical Disease* (Baltimore: Johns Hopkins University Press, 2007), 177–246; and more generally by Nancy Leys Stepan, *Eradication: Ridding the World of Disease Forever?* (Ithaca: Cornell University Press, 2011).

5. Bagster Wilson, "Application for Research Funds in Respect of Proposals for an Investigation of the Effects of Malaria and Malaria Control in Hyperendemic African Areas," n.d. [early 1951], Box 23, NIMR, Amani. Annual Report of the East African Institute of Malaria and Vector-Borne Diseases, July 1959–June 1960.

6. C. C. Draper to District Officer, Taveta, April 16, 1956, Box 23, NIMR, Amani; "Five Year Experiment"; "5th Meeting, East African Medical Research Scientific Advisory Committee Dar es Salaam," January 11–12, 1960, Box 22, NIMR, Amani; Dobson, Malowany, and Snow, "Malaria Control in East Africa," 161.

7. Bagster Wilson to Director of Medical Services Tanganyika, "Study of Malaria in Hyperendemic Areas, Confidential," May 1, 1951, Box 23, NIMR, Amani.

8. "Five Year Experiment."

9. Ibid., my emphasis. Unfortunately, the organizations involved in this project used the terms "elimination" and "eradication" interchangeably. I will use the terms as they are understood today: "elimination," to completely remove a disease from an area; "eradication," to remove it from the entire world.

10. Some articles make a distinction between "resurgent" malaria, where case rates return to pre-intervention levels, versus "rebound" malaria, where the number of cases may temporarily rise higher than pre-intervention levels. I am using these terms interchangeably.

11. Malariologists had already observed the large toll it took on other African communities when malaria control measures had ended, notably after stopping control measures in Monrovia, Liberia, in the late 1940s and again in 1957. James Webb, "The First Large-Scale Use of Synthetic Insecticides for Malaria Control in Tropical Africa: Lessons from Liberia, 1945–1962," *Journal of the History of Medicine and Allied Sciences* 66, no. 3 (2011): 347–76.

12. G. Davidson, "Results of Recent Experiments on the Use of DDT and BHC against Adult Mosquitos at Taveta, Kenya," *Bulletin of the World Health Organization* 4, no. 3 (1951): 329; C. C. Draper and A. Smith, "Malaria in the Pare Area of Tanganyika: Part II: Effects of Three Years' Spraying of Huts with Dieldrin," *Transactions of the Royal Society of Tropical Medicine* 54, no. 4 (1960): 342–57; A. Smith and C. C. Draper, "Malaria in the Taveta Area of Kenya and Tanganyika: I: Epidemiology," *East African Medical Journal* 36, no. 2 (1959): 99; A. Smith, "Malaria in the Taveta Area of Kenya and Tanzania. IV. Entomological Findings Six Years after the Spraying Period," *East African Medical Journal* 43, no. 1 (1966): 7–18; G. Pringle, C. C. Draper, and D. F. Clyde, "A New Approach to the Measurement of Residual Transmission in a Malaria Control Scheme in East Africa," *Transactions of the Royal Society of Tropical Medicine* 54, no. 5 (1960): 434–38.

13. "6th Meeting, East African Council for Medical Research, Entebbe," February 5, 1960, Box 22, NIMR, Amani.

14. "A Report of the East African Institute of Malaria and Vector-Borne Diseases, April 1963–March 1964," Box 24, NIMR, Amani. "Summary of Work Carried out by the East African Institute of Malaria and Vector-Borne Diseases during April 1959 and March 1960," Annual Report of the East African Institute of Malaria and Vector-Borne Diseases, 1966, Box 24, NIMR, Amani.

15. Infant mortality remained low. Conacher to Provincial Commissioner Tanga, "Pare-Taveta Malaria Scheme," August 10, 1960, Box 21, NIMR, Amani.

16. C. C. Draper et al., "Malaria in the Pare Area of Tanzania: IV: Malaria in the Human Population 11 Years after the Suspension of Residual Insecticide Spraying, with Special Reference to the Serological Findings," *Transactions of the Royal Society of Tropical Medicine* 66, no. 6 (1972): 905–12; A. Smith, "Malaria in the Taveta area of Kenya and Tanganyika: III: Entomological Findings Three Years after the Spraying Period," *East African Medical Journal* 39 (1962): 553–64; A. Smith and G. Pringle, "Malaria in the Taveta area of Kenya and Tanzania: V: Transmission Eight Years after the Spraying Period," *East African Medical Journal* 44 (1967): 469–74; G. Pringle, "Malaria in the Pare Area of Tanzania: III: The Course of Malaria Transmission since the Suspension of an Experimental Programme of Residual Insecticide Spraying," *Transactions of the Royal Society of Tropical Medicine and Hygiene* 61, no. 1 (1967): 69–79; G. Pringle and S. Avery Jones, "Observations on the Early Course of Untreated Falciparum Malaria in Semi-Immune African Children following a Short Period of Protection," *Bulletin of the World Health Organization* 34, no. 2 (1966): 269–72; D. J. Bradley, "Morbidity and Mortality at Pare-Taveta, Kenya and Tanzania, 1954–66: The Effects of a Period of Malaria Control," in *Disease and Mortality in Sub-Saharan Africa,* ed. D. T. Jamison et al. (Washington, DC: World Bank, 1991), 248–63.

17. Webb, "First Large-Scale Use of Synthetic Insecticides," 352, 366; Dobson, Malowany, and Snow, "Malaria Control in East Africa," 152; G. Corbellini, "Acquired Immunity against Malaria as a Tool for the Control of the Disease: The Strategy Proposed by the Malaria Commission of the League of Nations in 1933," *Parassitologia* 40, no. 1–2 (1998): 109–15.

18. Dobson, Malowany, and Snow, "Malaria Control in East Africa," 152, quoting "The Therapeutics of Malaria; Third General Report of the Malaria Commission," *Quarterly Bulletin of the Health Organisation of the League of Nations* 2, no. 2 (1933): 202–3.

19. Dobson, Malowany, and Snow, "Malaria Control in East Africa."

20. District Commissioner Tanga to unknown recipient, "Pare-Taveta Malaria Scheme," September 1, 1960, Box 21, NIMR, Amani.

21. Pringle to Clyde, July 27, 1960, Box 2, NIMR, Amani.

22. It is unclear if this was an effective policy, as it had been roundly criticized by the WHO when used in Liberia in 1957: "Such a rather ill-planned and unmethodical drugging may have some propagandistic value and may be prompted by humanitarian consideration, but their [*sic*] value as an effective means of malaria control is more than doubtful." "Report on the Joint UNICEF/WHO Malaria Project, Kpain (Liberia)," 2, SJ 2, JKT III, Liberia, WH07.0022, WHO Archives. Cited in Webb, "First Large-Scale Use of Synthetic Insecticides," 375.

23. The original grant was awarded in June 1960 and was expected to provide a total of £29,000 over five years. It appears that in 1964 the Vital Statistics Survey's annual budget was £58,000, with the following contributions: £4,000 from the Nuffield Foundation, £20,000 from the East African Common Services Organization, £24,000 from a British Government Matching Grant. It is unclear who provided the additional £10,000. Hyder to Pringle, November 1964, Box 24, NIMR, Amani.

24. When the Vital Statistics Survey ended is a bit unclear. While Dobson, Malowany, and Snow list the date as 1966, documents from Amani indicate that the researcher Thomas Fletcher was still in the Taveta region collecting "vital data" in 1968. This may have been data related to his work on sickle cell, but it may also have been a reference to the conclusion of the VSS. Dobson, Malowany, and Snow, "Malaria Control in East Africa." "Taveta Pare Survey, Dr. Trant," n.d., Box 2, NIMR, Amani. Pringle to District Commissioner Same, February 3, 1961, Box 2, NIMR, Amani.

25. The Malaria Institute delayed accepting the grant until it was absolutely clear local communities would *not* continue with indoor residual spraying. The grant had been awarded in September 1960 but was not formally accepted until March 1961. Unknown author [Pringle?] to Mac, "Personal," October 6, 1960, Box 2, NIMR, Amani.

26. Pringle to Officer in Charge, Pare, "Malaria Cases," July 18, 1960, Box 21, NIMR, Amani.

Modern Narrative: A New Malaria Vaccine? Testing the RTS, S Vaccine across Africa, 2010

1. After three decades of research and roughly $350 million spent, GlaxoSmithKline was looking for partners. It's estimated that an additional $260 million will be spent before the vaccine is complete. David Busse, "Another Addition to the Arsenal: GlaxoSmithKline's Malaria Vaccine," *LuSci* (blog), *Leeds University Union Science Magazine*, November 2, 2013, http://www.lusci.co.uk/another-addition-to-the-arsenal-glaxosmithklines-malaria-vaccine; W. R. Ballou, "The Development of the RTS,S Malaria Vaccine Candidate: Challenges and Lessons," *Parasite Immunology* 31, no. 9 (September 2009): 498.

2. The European Medicines Agency will review the product under Article 58 and issue a scientific opinion intended to help guide African national regulatory authorities. The World Health Organization will review all data, including that of the EMA, and deliver a formal recommendation. WHO, "Questions and Answers on Malaria Vaccines," October 2013, http://www.who.int/immunization/topics/malaria/vaccine_roadmap/WHO_malaria_vaccine_q_a_Oct2013.pdf. The most up-to-date information on all of the experimental malaria vaccines are kept on the WHO's "Rainbow Tables," accessible at http://www.who.int/immunization/research/development/Rainbow_tables/en/index.html. Lauren Schwartz et al., "A Review of Malaria Vaccine Clinical Projects Based on the WHO Rainbow Table," *Malaria Journal* 11, no. 1 (2012): 1–2.

3. In Kenya, the sites were in Kilifi (KEMRI-Wellcome Trust Research Program), Kisumu (KEMRI-Walter Reed Project), and Kisumu (KEMRI-CDC Research and Public Health Collaboration). In Tanzania, the sites were in Bagamoyo (Ifakara Health Institute and Swiss Tropical and Public Health Institute) and Korogwe (National Institute for Medical Research and Kilimanjaro Christian Medical Centre). The sites represented the different malaria transmission settings in Africa, ranging from intense seasonal epidemics to year-round endemic forms.

4. Those 6–12 weeks of age were randomized into the experimental group given the malaria vaccine and a set of standard vaccinations given to all children *or* a control vaccine (meningococcal C conjugate vaccine) and the standard EPI. The older children, aged 5–7 months, were given either the experimental malaria vaccine or the control vaccine (rabies vaccine).

5. There is some variation in the number of visits, since the first 200 children enrolled were visited more frequently to collect information on side effects and adverse events. The RTS,S Clinical Trials Partnership, "Protocol for: The RTS,S Clinical Trials Partnership: A phase 3 trial of RTS,S/AS01 malaria vaccine in African infants: Supplementary Documents," November 8, 2012, http://www.nejm.org/doi/suppl/10.1056/NEJMoa1208394/suppl_file/nejmoa1208394_protocol.pdf.

6. RTS,S Clinical Trials Partnership, "A Phase 3 Trial of RTS,S/AS01 Malaria Vaccine in African Infants," *New England Journal of Medicine* 367, no. 24 (2012): 2284–95.

7. Ally Olotu et al., "Four-Year Efficacy of RTS,S/AS01E and Its Interaction with Malaria Exposure," *New England Journal of Medicine* 368, no. 12 (2013): 1117.

8. Ibid., 1120.

9. This already proved to be a stumbling block during clinical testing when one of the international-grade laboratories failed to store a batch at the proper temperature.

10. World Health Organization, "Questions and Answers on Malaria Vaccines"

11. Debates concerning the vaccination calendar have been raised with the new meningitis A vaccine, and concerns about stable partnerships and long-term funding are also relevant to the Expanded Program on Immunization, which is a WHO program intended to provide universal immunization for all children.

12. The history of the RTS,S vaccine development has been well covered. It is worth mentioning that the vaccine was originally developed for use by the military and short-term foreign visitors, hence the involvement of the US Walter Reed Army Institute of Research. The decision to change the focus to children was a pragmatic one since children typically respond better to vaccines than adults. Joe Cohen et al., "From the Circumsporozoite Protein to the RTS,S/AS Candidate Vaccine," *Human Vaccines* 6, no. 1 (2010): 90; Ballou, "Development of the RTS,S Malaria Vaccine," 495.

13. Olusegun G. Ademowo and Joseph O. Fadare, "Ethical Issues in Malaria Vaccine Clinical Trials: A Principle-Based Approach," *Annals of Tropical Medicine & Public Health* 3, no. 1 (2010): 35.

14. Malaria Vaccine Funders Group, *Malaria Vaccine Technology Roadmap*, November 2013, 3, http://www.who.int/immunization/topics/malaria/vaccine_roadmap/TRM_update_nov13.pdf, citing Ilona Carneiro et al., "Age-Patterns of Malaria Vary with Severity, Transmission Intensity and Seasonality in Sub-Saharan Africa: A Systematic Review and Pooled Analysis," *PLoS ONE* 5, no. 2 (2010): e8988; W. P. O'Meara et al., "Effect of a Fall in Malaria Transmission on Morbidity and Mortality in Kilifi, Kenya," *Lancet* 372, no. 9649 (2008): 1555–62. S. J. Ceesay et al., "Changes in Malaria Indices between 1999 and 2007 in The Gambia: A Retrospective Analysis," *Lancet* 372, no. 9649 (2008): 1545–54.

15. Ballou, "Development of the RTS,S Malaria Vaccine," 498.

16. Schwartz et al., "A Review of Malaria Vaccine Clinical Projects," 11.

17. RTS,S Clinical Trials Partnership, "Protocol for: The RTS,S Clinical Trials Partnership," 72.

Chapter 6: Endings, Exits, and Longer-Term Obligations

1. D. N. Shaffer et al., "Equitable Treatment for HIV/AIDS Clinical Trial Participants: A Focus Group Study of Patients, Clinician Researchers, and Administrators in Western Kenya," *Journal of Medical Ethics* 32, no. 1 (2006): 55.

2. Timms to Pringle, September 6, 1960, Box 2, NIMR, Amani.

3. I will often refer to the Institute of Malaria and Vector-Borne Diseases as the "Malaria Institute," following the organization's own habit. The organization had a variety of names over the years. In 1949 it was referred to as the East African Malaria Unit. In 1951 it became the East African Malaria Institute, and in 1954 it became the East African Institute of Malaria and Vector-Borne Diseases. Pringle to Timms, July 8, 1960, Box 2, NIMR, Amani.

4. In South Cameroon, malaria transmission was interrupted, and it was "virtually" halted in Uganda. Other WHO pilot projects occurred in Garki, Nigeria; Kisumu, Kenya; South Cameroon; southwestern Uganda; Liberia; Upper Volta; Senegal; and Ghana. L. Molineaux and G. Gramiccia, *The Garki Project: Research on the Epidemiology and Control of Malaria in the Sudan Savanna of West Africa* (Geneva: World Health Organization, 1980); James Webb, "The First Large-Scale Use of Synthetic Insecticides for Malaria Control in Tropical Africa: Lessons from Liberia, 1945–1962" *Journal of the History of Medicine and Allied Sciences* 66, no. 3 (2011): 347–76.

5. Malaria transmission only occurs between 16°C and 33°C (60–91°F) and at less than 2,000 meters (1.2 miles) elevation. Gordon Cook and Alimuddin Zumla, eds., *Manson's Tropical Diseases*, 22nd ed. (Philadelphia: Saunders, 2009), 1202.

6. Yvonne Geissbuher et al., "Interdependence of Domestic Malaria Prevention Measures and Mosquito-Human Interactions in Urban Dar es Salaam, Tanzania," *Malaria Journal* 6 (September 19, 2007): 126; Tanya Russell et al., "Increased Proportions of Outdoor Feeding among Residual Malaria Vector Populations following Increased Use of Insecticide-Treated Nets in Rural Tanzania," *Malaria Journal* 10 (April 9, 2011): 80.

7. Cook and Zumla, *Manson's Tropical Diseases*, 1202.

8. The other forms of malaria are *P. vivax*, *P. ovale*, and *P. malariae.*

9. The early history of malaria is well covered by James Webb in chapters 1–3 of *Humanity's Burden: A Global History of Malaria* (Cambridge: Cambridge University Press, 2009).

10. "Duffy negativity" actually refers to Red Blood Cell Duffy antigen negativity, a genetic mutation that provides nearly complete protection against vivax malaria; it is widespread across sub-Saharan Africa, although the highest levels are found in West Africa. Sickle cell hemoglobin (also known as hemoglobin S, or the sickle cell gene) provides partial protection and reduces the severity of falciparum malaria. However, as this mutation is inherited from one's parents, if a child inherits two mutated genes it will develop sickle cell anemia, which is often fatal. Webb, *Humanity's Burden*, 21–29; Randall Packard, *The Making of a Tropical Disease* (Baltimore: Johns Hopkins University Press, 2007), 29–31.

11. Cook and Zumla, *Manson's Tropical Diseases*, 1205.

12. Ibid.

13. This premise of low adult mortality rates has been questioned by Murray et al. in their 2012 paper. They argue that in 2010 the WHO underestimated adult malaria deaths by more than 400,000. Their findings, if validated, imply that our understanding of acquired immunity and the morbidity/mortality toll on adults may still need to be revised. Christopher J. L. Murray et al., "Global Malaria Mortality between 1980 and 2010: A Systematic Analysis," *Lancet* 379, no. 9814 (2012): 413–31.

14. D. L. Doolan, C. Dobano, and J. K. Baird, "Acquired Immunity to Malaria," *Clinical Microbiology Reviews* 22, no. 1 (January 8, 2009): 13–36; Jean Langhorne et al., "Immunity to Malaria: More Questions Than Answers," *Nature Immunology* 9, no. 7 (2008): 725–32.

15. Some articles make a distinction between "resurgent" malaria, where case rates return to pre-intervention levels, versus "rebound" malaria, where the number of cases may temporarily rise higher than pre-intervention levels. I am using these terms interchangeably.

16. Abdisalan M. Noor et al., "The Changing Risk of *Plasmodium falciparum* Malaria Infection in Africa, 2000–10: A Spatial and Temporal Analysis of Transmission Intensity," *Lancet* 383, no. 9930 (2014): 1739–47; Justin Cohen et al., "Malaria Resurgence: A Systematic Review and Assessment of Its Causes," *Malaria Journal* 11, no. 1 (2012): 9–10.

17. Cohen et al., "Malaria Resurgence," 1, 13.

18. J. M. Roberts, "The Control of Epidemic Malaria in the Highlands of Western Kenya: 3: After the Campaign," *Journal of Tropical Medicine and Hygiene* 67 (September 1964): 230–37; V. K. Barbiero et al., *Project Completion Report: Zanzibar Malaria Control Project* (Arlington, VA: Vector Biology and Control Project, Medical Service Corporation International, 1990).

19. Webb, "First Large-Scale Use of Synthetic Insecticides," 366

20. Ibid., 364, quoting Dr. W. Mullhausen, "The Activities of the Malaria Team for the Quarter January-March 1957 (Quarterly Report)," 2, SJ 1, JKT III, Liberia, WHO 7.0022, WHO Archives.

21. New evidence from Senegal presented by Jean-François Trape raises the possibility that rebound malaria could occur even if specific interventions had not stopped, by showing that rebound epidemics may occur following the introduction of insecticide-treated bednets. These conclusions are far from well accepted, but present challenging new interpretations that await further investigation. Jean-François Trape et al., "The Rise and Fall of Malaria in a West African Rural Community, Dielmo, Senegal, from 1990 to 2012: A 22 Year Longitudinal Study," *Lancet Infectious Diseases* 14, no. 6 (2014): 476–88; Jean-François Trape et al., "Malaria Morbidity and Pyrethroid Resistance after the Introduction of Insecticide-Treated Bednets and Artemisinin-Based Combination Therapies: A Longitudinal Study," *Lancet Infectious Diseases* 11, no. 12 (2011): 925–32.

22. Pringle to Clyde, July 27, 1960, Box 2, NIMR, Amani.

23. Timms to Pringle, "Nuffield Foundation Grant," July 5, 1960, Box 2, NIMR, Amani.

24. Leask to Draper, April 24, 1956, Box 23, NIMR, Amani.

25. Bagster Wilson, "The Future of Control in the Taveta/Pare Area," December 1956, Box 22, NIMR, Amani.

26. Thompson to Provincial Commissioner Coast Province, cc Bagster Wilson, "Taveta/Pare Malaria Scheme," May 28, 1959, Box 23, NIMR, Amani. Emphasis in original.

27. Ibid.

28. Ibid.

29. Ibid.

30. Ibid.

31. Evans to Medical Research Secretary Nairobi, cc Provincial Commissioner, Tanga, Provincial Medical Officer, Tanga, Director, Amani, August 8, 1960, Box 2, NIMR, Amani.

32. Pringle to Timms, July 8, 1960, Box 2, NIMR, Amani.

33. Unknown author [Pringle?] to Clyde, July 27, 1960, Box 2, NIMR, Amani.

34. "5th Meeting, East African Medical Research Scientific Advisory Committee, Dar es Salaam," January 11–12, 1960, Box 22, NIMR, Amani.

35. Pringle to Marcus, June 25, 1962, Box 24, NIMR, Amani.

36. Notes from a 1957 meeting of the Committee indicate that the following people participated: the director of medical services, Uganda, Zanzibar, and Tanganyika; the deputy director of medical services for Kenya; the directors of the Malaria Institute, the East African Virus Institute, and the East Africa Leprosy Research Center; members of the Colonial Medical Research Committee; the dean and members of the faculty of medicine at Makerere University; members of the East Africa Trypanosomiasis Research Organization and the Veterinarian Research Organization; and a representative from the Wellcome Institute.

37. "East African Council for Medical Research, Second Meeting," n.d. [1956], Box 22, NIMR, Amani.

38. "East African Medical Research Scientific Advisory Committee, Nairobi," January 20–21, 1958, Box 22, NIMR, Amani.

39. Ibid.

40. Ibid.

41. "5th Meeting, East African Medical Research Scientific Advisory Committee, Dar es Salaam," January 11–12, 1960, Box 22, NIMR, Amani.

42. "6th Meeting, East African Council for Medical Research, Entebbe, February 5, 1960, Box 22, NIMR, Amani.

43. Other potential routes of transmission, according to residents, could include changes in the wind, hot sun, and hard work. The association between mosquitoes and malaria wasn't entirely new. Susan Beckerleg, "Maintaining Order, Creating Chaos: Swahili Medicine in Kenya" (PhD diss., University of London, 1989); Susanna Hausmann Muela et al., "Medical Syncretism with Reference to Malaria in a Tanzanian Community," *Social Science & Medicine* 55, no. 3 (2002): 403–13.

44. Modern anthropological research in East Africa has shown that children affected by the high fevers and seizures characterizing falciparum malaria are often described as having an entirely different disease.

45. Mfumwa wa Gonja to Malaria Officer, Gonja Juu, "Dawa ya Mmbu (Dieldrin)," n.d. [February 1957], Box 24, NIMR, Amani; "To the Editor, Habari Za Upare, Same, for publication," February 11, 1957, Box 24, NIMR, Amani.

46. Msangi to the Editor, Habari Za Upare (Same), "Baraka za Kupungukiwa Na Malaria Katika Wilaya Za Upare na Taveta," February 23, 1957, Box 24, NIMR, Amani.

47. Ibid.

48. Timms to Farrer-Brown, September 29, 1960, Box 2, NIMR, Amani.

49. The actual language used by the council has not been recorded. District Commissioner Tanga to unknown recipient, "Pare-Taveta Malaria Scheme," September 1, 1960, Box 21, NIMR, Amani.

50. Field Officer to Pringle, "Taveta Council," December 19, 1960, NIMR, Amani.

51. Ibid.

52. Msangi, "Public Relations Work," n.d. [1960], Box 23, NIMR, Amani.

53. A recently published article in *Lancet* showed that it was "least feasible" to consider elimination in sub-Saharan Africa, and that it may not even be economically prudent. Pam Das and Richard Horton, "Malaria Elimination: Worthy, Challenging, and Just Possible," *Lancet* 376, no. 9752 (2010): 1515. Marcel Tanner and Don de Savigny, "Malaria Eradication Back on the Table," *Bulletin of the World Health Organization* 86, no. 2 (2008): 81.

54. Estimates for 2010 were 216 million malaria cases globally and 655,000 malaria deaths. Christabel Ligami, "African Countries on Track to Meeting Four MDGs as 2015 Deadline Comes Closer," *East African*, June 22, 2013, http://www

.theeastafrican.co.ke/news/Africa-on-course-for-meeting-four-MDGs-as-2015-deadline-looms-/-/2558/1891226/-/lw739k/-/index.html.

55. In 2010, it was estimated that malaria was responsible for 22 percent of all under-five deaths in Uganda, 16 percent in Tanzania, and 11 percent in Kenya. Statistics from UNICEF, accessed July 18, 2013, http://www.childinfo.org.

56. That there are negative "effects" of malaria is not in doubt; however, quantifying these effects is more fraught than one would expect. In terms of morbidity and mortality, research by Murray et al. in 2012 indicated that global rates of malaria were actually far higher than those reported by the WHO. This claim was hotly disputed. On the economic front, a paper by Sachs and Malaney, which first put a price tag on the "cost" of malaria on the state level, has been subsequently criticized and debated by scholars from all disciplines. Malaria morbidity is also notoriously tricky to pin down, as this varies depending on the type of transmission and the level of acquired immunity within a population. Christopher J. L. Murray et al., "Global Malaria Mortality between 1980 and 2010: A Systematic Analysis," *Lancet* 379, no. 9814 (2012): 413–31; Jeffrey Sachs and Pia Malaney, "The Economic and Social Burden of Malaria," *Nature* 415, no. 6872 (2002): 680–85; Randall Packard, "'Roll Back Malaria, Roll in Development'? Reassessing the Economic Burden of Malaria," *Population and Development Review* 35, no. 1 (2009): 53–87.

57. "Goal 6: Combat HIV/AIDS, Malaria, and Other Diseases," accessed July 18, 2013, http://www.un.org/millenniumgoals/aids.shtml.

58. Geoffrey Mbaabu Lairumbi et al., "Promoting the Social Value of Research in Kenya: Examining the Practical Aspects of Collaborative Partnerships Using an Ethical Framework," *Social Science & Medicine* 67, no. 5 (2008): 735.

59. Das and Horton, "Malaria Elimination," 1516.

60. W. R. Ballou, "The Development of the RTS,S Malaria Vaccine Candidate: Challenges and Lessons," *Parasite Immunology* 31, no. 9 (September 2009): 499.

61. Nuffield Council on Bioethics, *The Ethics of Research Related to Healthcare in Developing Countries* (London: Nuffield Council on Bioethics, 2002), 38.

62. Jon Cohen, "A Good Enough Malaria Vaccine?" *Latest News* (blog), *Science*, October 8, 2013, http://news.sciencemag.org/health/2013/10/good-enough-malaria-vaccine.

63. Ballou, "Development of the RTS,S. Malaria Vaccine," 498.

64. Global Alliance for Vaccines and Immunization, "GAVI Statement on Latest Trial Data on Malaria Vaccine Candidate RTS,S," October 8, 2013, http://www.gavialliance.org/library/news/statements/2013/gavi-statement-on-latest-trial-data-on-malaria-vaccine-candidate-rts-s.

65. These codes and laws include World Medical Association, "World Medical Association Declaration of Helsinki: Ethical Principles for Medical Research Involving Human Subjects," *Journal of the American Medical Association* 310, no. 20 (2013): 2191–94, paragraph 30 and Special Note; Council for International Organizations of Medical Sciences, *International Ethical Guidelines for Biomedical Research Involving Human Subjects* (Geneva: CIOMS, 2002), guideline 10; European Group on Ethics in Science and New Technologies, *Opinion Nr 17 on Ethical Aspects of*

Clinical Research in Developing Countries (Brussels: EGE, 2003); and European Council and European Parliament, "Directive 2001/201 EC." A good overview of these issues can be found in Nuffield Council on Bioethics, "Ethics of Research Related to Healthcare," and Nuffield Council on Bioethics, *The Ethics of Clinical Research in Developing Countries* (London: Nuffield Council on Bioethics, 1999).

66. Council for International Organizations of Medical Sciences, *International Ethical Guidelines,*, guideline 10, also discussed in guideline 21.

67. Nuffield Council on Bioethics, *Ethics of Research Related to Healthcare*, paragraph 9.36.

68. European Group on Ethics in Science and New Technologies, *Opinion Nr 17*, paragraph 2.13.

69. My emphasis. J. D. Seema Shah, Stacey Elmer, and Christine Grady, "Planning for Posttrial Access to Antiretroviral Treatment for Research Participants in Developing Countries," *American Journal of Public Health* 99, no. 9 (2009): 1556–62; J. D. Seema Shah and Christine Grady, "Shah and Grady Respond," *American Journal of Public Health* 100, no. 6 (June 2010): 967; Onyekachi Sunny Onyeabor, "Ethical Imperative of Posttrial Access to Antiretroviral Treatment," *American Journal of Public Health* 100, no. 6 (June 2010): 966–67.

70. GlaxoSmithKline, "Clinical Trials in the Developing World," Position paper, November 1, 2011, http://www.glaxosmithkline.ch/Attachments/GSK-on-clinical-trials-in-the-developing-world.pdf.

71. Ibid.

72. D. N. Shaffer et al., "Equitable Treatment for HIV/AIDS Clinical Trial Participants," 57.

73. Christine Grady et al., "Research Benefits for Hypothetical HIV Vaccine Trials: The Views of Ugandans in the Rakai District," *IRB: Ethics & Human Research* 30, no. 2 (2008): 5.

74. Ibid.

75. Jintanat Ananworanich et al., "Creation of a Drug Fund for Post-Clinical Trial Access to Antiretrovirals," *Lancet* 364, no. 9428 (2004): 101–2.

76. Das and Horton, "Malaria Elimination," 1515.

77. Shaffer et al., "Equitable Treatment for HIV/AIDS Clinical Trial Participants," 57.

THE EXPERIMENT ENDS?

Chapter 7: Modern Medical Research and Historical Residue

1. "What Is the Purpose of Medical Research?" *Lancet* 381, no. 9864 (2013): 347.

2. East African Medical Survey and Research Institute Annual Report, 1954–55.

3. Avery Jones was just one of the researchers working on the Nandi Scheme. African staff members ran nearly all the field operations, and included Messrs. Omondi, Kandie, Olindo, Otieno, Mboga, and Chesa. "Nandi Dieldrin 1957: Operation Report," April 10, 1957, Box 23, NIMR, Amani.

4. Anderson to Bagster Wilson, November 13, 1952, Box 15, NIMR, Mwanza.

5. Avery Jones to Director of Medical Services, Kenya, "Confidential," April 26, 1953, Box 15, NIMR, Mwanza.

6. Sandra Crouse Quinn, "Protecting Human Subjects: the Role of Community Advisory Boards," *American Journal of Public Health* 94, no. 6 (2004): 921.

7. Fletcher to Marijani, November 7, 1963, Box 23, NIMR, Amani.

8. Ibid.

9. Ibid.

10. The archives in Amani hold a few different drafts of the document, and it is unclear which one was circulated. The information in each is largely the same. Thomas Fletcher, "An Account of the Work at Present Being Undertaken by the Malaria Institute in the Taveta Area," November, 1963, Box 23, NIMR, Amani.

11. Ibid.

12. Ibid.

13. Fletcher to Yohana [Matola?], June 24, 1965, Box 21, NIMR, Amani.

14. Ibid.

15. Marcia Angell, "The Body Hunters," *New York Review of Books* 52, no. 15 (October 6, 2005): 52–58. A vast majority of trials still occur in the United States. These estimates come from interviews with pharmaceutical company executives and date to 2005. More recent information from Petryna does not indicate this trend is waning. Abrahm Lustgarten, "Drug Testing Goes Offshore," *Fortune* 152, no. 3 (2005): 66. Adriana Petryna, "The Competitive Logic of Global Clinical Trials," *Social Research: An International Quarterly* 78, no. 3 (2011): 949–74.

16. In the United States costs of drug trials are approximately $30,000 per patient, while trials in poorer countries amount to a tenth of that amount. The cost of testing drugs on humans is the most expensive phase of the research and development process, and recruitment alone accounts for up to 40 percent of the average $900 million needed to bring a drug to market. Lustgarten, "Drug Testing Goes Offshore," 66.

17. Petryna has also noted that a 1980 US ban on using prisoners during some stages of research also pushed trials overseas. Petryna, "Competitive Logic," 959.

18. Sonia Shah, *The Body Hunters: Testing New Drugs on the World's Poorest Patients*, (New York: New Press, 2006), 9. Shah's work is a bit dated now, but still seems to accurately describe aspects of the global pharmaceutical industry.

19. The FDA's Clinical Trials database records 758 trials completed or occurring in East Africa since 2008 (265 in Kenya, 205 in Tanzania, 288 in Uganda); search conducted on May 26, 2014, http://clinicaltrials.gov. For drugs that were registered within the EU between 2005 and 2010, the most trials in East Africa are based in Kenya (4 sites, 446 subjects), then Uganda (1 site, 176 subjects), then Tanzania (1 site, 97 subjects). Within Africa as a whole, South Africa is the site of most trials (928 sites involving 18,712 subjects). European Medicines Agency, "Clinical Trials Submitted in Marketing-Authorisation Applications to the European Medicines Agency: Overview of Patient Recruitment and the Geographical Location of Investigator Sites: Containing data from 2005 to 2011," December 11, 2013, 36, http://www.ema.europa.eu/docs/en_GB/document_library/Other/2009/12/WC500016819.pdf.

20. Department of Health and Human Services Office of the Inspector General. "The Globalization of Clinical Trials: A Growing Challenge in Protecting Human Subjects: Executive Summary," *Journal International de Bioéthique* 14, no. 1–2 (2003): 165–69.

21. Randomized controlled trials were conducted in eighteen different nations, yet East Africa accounted for 37 percent of all studies—significantly higher than any other region. While East Africa is heavily used, South Africa appears to be the place where the most medical research involving human subjects has been, and is, being conducted. This isn't surprising since South Africa has a far more developed health care system and a more straightforward system for medical research approval (governed by one board rather than three different national ones). Nandi Siegfried, Mike Clarke, and Jimmy Volmink, "Randomised Controlled Trials in Africa of HIV and AIDS: Descriptive Study and Spatial Distribution," *British Medical Journal* 331, no. 7519 (2005): 742.

22. The RTS,S Clinical Trials Partnership, "A Phase 3 Trial of RTS,S/AS01 Malaria Vaccine in African Infants," *New England Journal of Medicine* 367, no. 24 (2012): 2284–95; Johanna P. Daily, "Malaria Vaccine Trials—Beyond Efficacy End Points," *New England Journal of Medicine* 367, no. 24 (2012): 2349–51.

23. B. Pedrique et al., "The Drug and Vaccine Landscape for Neglected Diseases (2000–11): a Systematic Assessment," *Lancet Global Health* 1, no. 6 (2013): 371–79; Doctors Without Borders, "Deadly Gaps Persist in New Drug Development for Neglected Diseases," press release, October 24, 2013, http://www.doctorswithoutborders.org/news-stories/press-release/deadly-gaps-persist-new-drug-development-neglected-diseases.

24. Other public-private partnerships that have begun to show signs of success include GAVI: the Vaccine Alliance, which involves the WHO, UNICEF, the World Bank, and the Gates Foundation, in addition to national governments and the pharmaceutical industry. Another example is the Meningitis Vaccine Project, a partnership between PATH and the WHO.

25. One of the partnerships focused on research and development for new treatments on the most neglected diseases is the Drugs for Neglected Diseases Initiative (DNDi). The group was established in 2003 by Médecins Sans Frontières, the Indian Council of Medical Research, Brazil's Oswaldo Cruz Foundation, the Kenya Medical Research Institute, the Ministry of Health of Malaysia, and the Institut Pasteur in France, with the Special Programme for Research and Training in Tropical Diseases (WHO-TDR) as a permanent observer.

26. Adriana Petryna, *When Experiments Travel: Clinical Trials and the Global Search for Human Subjects* (Princeton: Princeton University Press, 2009), 12–13.

27. MU-JHU Research Collaboration, accessed July 18, 2013, http://www.mujhu.org/history.html.

28. Geissler states the annual budget is roughly $25 million, and that there are approximately 1,200 staff members and hundreds of thousands of trial participants. CDC/KEMRI programs are the largest single employer in the area outside of the civil service. P. Wenzel Geissler, "'Transport to Where?' Reflections on the Problem

of Value and Time à Propos an Awkward Practice in Medical Research," *Journal of Cultural Economy* 4, no. 1 (2011): 45–64; P. Wenzel Geissler, "Stuck in Ruins, or Up and Coming? The Shifting Geography of Urban Public Health Research in Kisumu, Kenya," *Africa* 83, no. 4 (2013): 547.

29. Kenya/CDC partnership information can be found at http://www.cdc.gov/globalhealth/countries/kenya/partners/ and http://www.kemri.org/index.php/about-kemri/typography-mainmenu-33.

30. Johanna Crane, *Scrambling for Africa: AIDS, Expertise, and the Rise of American Global Health Science* (Ithaca: Cornell University Press, 2013), 149.

31. Ibid., 11.

32. The Human Heredity and Health in Africa (H3Africa) Initiative is a new example of a partnership meant to build human and physical capacity. The group focuses on the study of genomics and environmental determinants of common diseases by developing the necessary expertise among African scientists, ensuring access to relevant genomic technologies, facilitating training at all levels, and establishing necessary research infrastructure. The work is supported by the NIH (US) and Wellcome Trust (UK). http://h3africa.org.

33. Other recent accounts of global medical research include Shah, *The Body Hunters*; Karen DeYoung and Deborah Nelson, "The Body Hunters: Testing's El Dorado: Latin America is Ripe for Trials, and Fraud, Frantic Pace Could Overwhelm Controls," *Washington Post*, December 22, 2000.

34. In 2009 the US Government Accountability Office (GAO) investigated the oversight of Institutional Review Boards in the United States, which are the first step for receiving regulatory approval for research. The GAO uncovered many disturbing problems. http://gao.gov/assets/130/122142.pdf. The research was duplicated by *Dateline NBC* in India, highlighting the role of contract research organizations. Tim Sandler, "In India, Oversight Lacking in Outsourced Drug Trials," NBCNews.com, March 4, 2012, http://investigations.nbcnews.com/_news/2012/03/04/10562883-in-india-oversight-lacking-in-outsourced-drug-trials.

35. David Rothman, "The Shame of Medical Research," *New York Review of Books* 47, no. 19 (November 30, 2000): 60–64.

36. Peter Lurie and Sidney Wolfe, "Unethical Trials of Interventions to Reduce Perinatal Transmission of the Human Immunodeficiency Virus in Developing Countries," *New England Journal of Medicine* 337, no. 12 (1997): 853–56.

37. Marcia Angell, "The Ethics of Clinical Research in the Third World," *New England Journal of Medicine* 337, no. 12 (1997): 847–49.

38. Harold Varmus and David Satcher, "Ethical Complexities of Conducting Research in Developing Countries," *New England Journal of Medicine* 337, no. 14 (1996): 1003–5.

39. Angell, "Body Hunters."

40. Sharon LaFraniere, Mary Pat Flaherty, and Joe Stevens, "The Body Hunters: The Dilemma: Submit or Suffer: 'Uninformed Consent' is Rising Ethic of the Drug Test Boom," *Washington Post*, December 19, 2000; Mary Pat Flaherty and

Doug Struck, "The Body Hunters: Perils of Placebos: Life by Luck of the Draw: In Third World Drug Tests, Some Subjects Go Untreated," *Washington Post*, December 22, 2000; DeYoung and Nelson, "The "Body Hunters: Testing's El Dorado"; John Pomfret and Deborah Nelson, "The Body Hunters: An Isolated Region's Genetic Mother Lode," *Washington Post*, December 20, 2000; Joe Stephens, "The Body Hunters: Where Profits and Lives Hang in the Balance," *Washington Post*, December 17, 2000.

41. These cases included *Abdullahi v. Pfizer*, *Zango v. Pfizer*, and *Adamu v. Pfizer*. Amy Wollensack, "Closing the Constant Garden: The Regulation and Responsibility of U.S. Pharmaceutical Companies Doing Research on Human Subjects in Developing Nations," *Washington University Global Studies Law Review* 6, no. 3 (2007): 747–71.

42. Donald McNeil, "Nigerians Receive First Payments for Children Who Died in 1996 Meningitis Drug Trial," *New York Times*, August 11, 2011; Bill Berkrot, "Pfizer Settles Remaining Nigeria, U.S. Trovan Suits," *Reuters*, February 23, 2011; Joe Stephens, "Pfizer to Pay $75 Million to Settle Nigerian Trovan Drug-Testing Suit," *Washington Post*, July 31, 2009.

43. Petryna, "Competitive Logic," 969.

44. Ibid., quoting Arthur Caplan, "Commentary: Report Paints Grim Picture of Drug Trial Safety," NBCNews.com, September 28, 2007, http://www.nbcnews.com/id/21029879.

45. Udo Schuklenk, "Protecting the Vulnerable: Testing Times for Clinical Research Ethics," *Social Science & Medicine* 51, no. 6 (2000): 976, citing *US News and World Report*, "Dying for a Cure" (October 11, 1999).

46. Schuklenk, "Protecting the Vulnerable," 976.

47. A 2012 "reflection paper" on the ethical aspects of clinical trials occurring outside of the EU area stated that, between 1997 and 2011, of a total 70,291 investigator sites, only 357 had been inspected, "giving an idea of the very small sample of sites that are, or can be, inspected" European Medicines Agency, "Reflection Paper," 24.

48. Harriet Washington, *Medical Apartheid: The Dark History of Medical Experimentation on Black Americans from Colonial Times to the Present* (New York: Anchor, 2006), 388.

49. James Ferguson has made clear the potentially damaging role played by non-state actors taking on state responsibilities. However, without sounding too cynical, I would argue that those non-state actors are already present in East Africa, already have a great deal of power, but rarely are forced to deliver tangible benefits that accrue to a majority, or even a large number, of citizens. James Ferguson, *The Anti-Politics Machine: Development, Depoliticization, and Bureaucratic Power in Lesotho* (Minneapolis: University of Minnesota Press, 1994).

50. Paul Farmer, "Rich World, Poor World: Medical Ethics and Global Inequality," in *Partner to the Poor: A Paul Farmer Reader*, ed. Haun Saussy (Berkeley: University of California Press, 2010), 542.

51. District Commissioner, Pare to Bagster Wilson, "Medical—General—Malaria Research," June 7, 1952, Box 23, NIMR, Amani.

52. East African Medical Survey and Research Institute Annual Report, July 1955–June 1956.

53. East African Institute of Malaria and Vector-Borne Diseases Annual Report, July 1955–June 1956.

54. Malaria Field Officer to Teacher, Gonja Primary School, November 7, 1962, Box 21, NIMR, Amani.

55. Pringle to Woodruff, November 14, 1962, Box 23, NIMR, Amani; Matola to Pringle, "Community Development Seminars," June 17, 1965, Box 2, NIMR, Amani.

56. East African Institute of Malaria and Vector-Borne Diseases Annual Report, 1975; East African Institute of Malaria and Vector-Borne Diseases Annual Report, January 1974–December 1975.

57. Trant to Bozman and Holmes, May 22 1955, Dr. Trant Files, NIMR, Mwanza.

58. "Five Year Experiment Shows the Way to Defeat of Malaria," *East African Standard*, July 10, 1959.

BIBLIOGRAPHY

Archival Sources

National Institute of Medical Research, Amani Files, Tanzania
National Institute of Medical Research, Mwanza Files, Tanzania
Kenya National Archives
Public Records Office, United Kingdom
Tanzania National Archives
Wellcome Library, Papers of Stanley George Browne
Zanzibar National Archives

Reports

Bukumbi Hospital (TZ) Annual Reports. 1970–79
Colonial Medical Research Committee Tenth Annual Report, 1954–55
Division of Insect-Borne Diseases Annual Reports, Nyanza. 1948, 1951, 1953, 1956–60
East African Institute of Malaria and Vector-Borne Diseases Annual Reports. 1951, 1954–66, 1968–76
East African Medical Survey, Monograph No. 1, 1954
East African Medical Survey/East African Institute for Medical Research Annual Reports. 1950–74
East African Trypanosomiasis Research Organization Annual Reports. 1955–64, 1966–71
Filariasis Research Unit Annual Reports. 1950–53
Medical Research Council (UK) Annual Reports. 1950–56, 1960–61, 1966–71
Ministry of Health Annual Reports, Mwanza Region, Tanzania. 1965, 1967–72
Mwanza Hospital Annual Reports. 1972–73
Tanganyika Medical Department Annual Reports. 1929, 1941–45, 1947–62
Tanganyika Medical Laboratory and Autopsies Annual Reports. 1957–63, 1968–69
Tanganyika Provincial Medical Office Western Province, Tabora, Annual Report Western Region. 1950
Tropical Pesticide Research Institute Annual Reports. 1957–62
Uganda Medical Department Annual Reports. 1950–53
World Health Organization/ Tanzania Schistosomiasis Scheme Quarterly Reports. 1969–72

Interviews

Dr. Yahya Athyumani, May 6, 2008. Tanga, Tanzania.
Mzee Joseph Myamba, May 6–7, 2008. Tanga, Tanzania.
Dr. Leonard Mboera, May 7, 2008. Tanga, Tanzania.
Dr. Erling Pedersen, May 8, 2008. Tanga, Tanzania.

Dr. Steven Magesa, May 9, 2008. Tanga, Tanzania.
Mzee Joseph Myamba, June 4–5, 2008. Muheza, Tanzania.
Mzee Kettle, June 6, 2008. Muheza, Tanzania.
Mzee Selestin, July 31, 2008. Mwanza, Tanzania.
Mzee Gaspor, July 31, 2008. Mwanza, Tanzania.
Dr. Malenganisho, August 1, 2008. Mwanza, Tanzania.
Mama Christine Nzito, August 3 and 6, 2008. Bukumbi, Tanzania.
Dr. Shija, August 4, 2008. Mwanza, Tanzania.
Mzee Dominic Gelazi, August 6, 2008. Bukumbi, Tanzania.
Mzee John Lyang'ombe, August 6, 2008. Bukumbi, Tanzania.
Bwana Julius Nyangaki, August 6, 2008. Bukumbi Health Center, Tanzania.
Mzee Elias Manoni Izengo, August 7, 2008. Bukumbi, Tanzania.
Mzee Oswald Kahurananga, August 7, 2008. Kigongo, Tanzania.
Mzee Thomas Inyassi, August 7, 2008. Kigongo, Tanzania.
Father John Somers, August 7, 2008. Bukumbi Parish, Tanzania.
Bibi Agnes Kagera, August 8, 2008. Bukumbi, Tanzania.
Bwana Allen Luzozakwenda, August 8, 2008. Bukumbi, Tanzania.
Bibi Maria Petro, August 8, 2008. Bukumbi, Tanzania.
Mzee Frances Kalengi, August 8, 2008. Bukumbi, Tanzania.
Bibi Teresa Wilburt, August 8–9, 2008. Bukumbi, Tanzania.
Mzee Edward Bunduki, August 8–9, 2008. Bukumbi, Tanzania.
Father Mwalimu, August 11, 2008. Mwanza Parish, Tanzania.
Mzee Salvatory Nambiza, August 25, 2008. Nansio, Ukerewe Island, Tanzania.
Mzee Mkanzabi, August 25, 2008. Nansio, Ukerewe Island, Tanzania.
Mzee Masatu Ndege, August 26, 2008. Bwisya, Ukara Island, Tanzania.
Mzee Manyama Mtaki, August 26, 2008. Bwisya, Ukara Island, Tanzania.
Bibi Scholastica Kuchagga, August 26, 2008, Bwisya, Ukara Island, Tanzania.
Mzee Deogratias Mwanika, August 26, 2008. Namanga, Ukara Island, Tanzania.
Mzee Chiloli, August 27, 2008. Nyang'ombe, Ukara Island, Tanzania.
Mzee Donat, August 27, 2008. Namanga, Ukara Island, Tanzania.
Mzee Damian Mumwi, August 27, 2008. Bwisya, Ukara Island, Tanzania.
Mzee Mwendadi, October 17, 2008. Amani, Tanzania.
Mama S____, 2008.
Mzee D____, 2008.
Mzee J_____, 2008.
Mzee P ____, 2008.
Mzee N____, 2008.
Mzee R____, 2008.
Mzee M____, 2008.

Secondary Sources

Abadie, Roberto. *The Professional Guinea Pig: Big Pharma and the Risky World of Human Subjects.* Durham: Duke University Press, 2010.

Adas, Michael. *Machines as the Measure of Men: Science, Technology, and Ideologies of Western Dominance.* London: Cornell University Press, 1989.

Ademowo, Olusegun G. and Joseph O. Fadare. "Ethical Issues in Malaria Vaccine Clinical Trials: A Principle-Based Approach." *Annals of Tropical Medicine & Public Health* 3, no. 1 (2010): 35–38.

"Aeterna Zentaris Cancels Phase III Trial of Blood Cancer Drug Perifosine." Accessed January 23, 2014. http://www.pharmaceutical-technology.com/news/newsaeterna-zentaris-cancels-phase-iii-trial-of-blood-cancer-drug-perifosine.

Allen, Denise Roth. *Managing Motherhood, Managing Risk: Fertility and Danger in West Central Tanzania.* Ann Arbor: University of Michigan Press, 2002.

Allen, J. W. T., ed. and trans. *The Customs of the Swahili People: The Desturi Za Waswahili of Mtoro Bin Mwinyi Bakari and Other Swahili Persons.* Berkeley: University of California Press, 1981.

Altman, Lawrence K. *Who Goes First? The Story of Self-Experimentation in Medicine.* Berkeley: University of California Press, 1998.

Ananworanich, Jintanat, Theshinee Cheunyam, Somsong Teeratakulpisarn, Mark A. Boyd, Kiat Ruxrungtham, Joep Lange, David Cooper, and Praphan Phanuphak. "Creation of a Drug Fund for Post-Clinical Trial Access to Antiretrovirals." *Lancet* 364, no. 9428 (2004): 101–2.

Anderson, David. *Histories of the Hanged: The Dirty War in Kenya and the End of Empire.* New York: WW Norton, 2005.

Anderson, Warwick. *The Collectors of Lost Souls: Turning Kuru Scientists into Whitemen.* Baltimore: Johns Hopkins University Press, 2008.

———. *Colonial Pathologies: American Tropical Medicine, Race, and Hygiene in the Philippines.* Durham, NC: Duke University Press, 2006.

———. "Immunities of Empire: Race, Disease, and the New Tropical Medicine, 1900–1920." *Bulletin of the History of Medicine* 70, no. 1 (1996): 94–118.

———. "Natural Histories of Infectious Disease: Ecological Vision in Twentieth-Century Biomedical Science." *OSIRIS* 19 (2004): 39–61.

———. "Postcolonial Technoscience." *Social Studies of Science* 32, no. 5–6 (2002): 643–58.

Angell, Marcia. "The Body Hunters." *New York Review of Books* 52, no. 15 (October 6, 2005): 52–58.

———. "The Ethics of Clinical Research in the Third World." *New England Journal of Medicine* 337, no. 12 (1997): 847–49.

———. "Investigators' Responsibilities for Human Subjects in Developing Countries." *New England Journal of Medicine* 342, no. 13 (2000): 967–69.

———. *The Truth about Drug Companies: How They Deceive Us and What to Do about It.* New York: Random House, 2004.

Angwenyi, Vibian, Dorcas Kamuya, Dorothy Mwachiro, Vicki Marsh, Patricia Njuguna and Sassy Molyneux. "Working with Community Health Workers as 'Volunteers' in a Vaccine Trial: Practical and Ethical Experiences and Implications." *Developing World Bioethics* 13, no. 1 (2013): 38–47.

Angwenyi, Vibian, Dorcas Kamuya, Dorothy Mwachiro, Betty Kalama, Vicki Marsh, Patricia Njuguna, and Sassy Molyneux. "Complex Realities: Community Engagement for a Paediatric Randomized Controlled Malaria Vaccine Trial in Kilifi, Kenya." *Trials* 15, no. 1 (2014): 1–16.

Annas, George, and Michael Grodin. *The Nazi Doctors and the Nuremberg Code: Human Rights in Human Experimentation*. New York: Oxford University Press, 1992.

Appelbaum, Paul S., Loren H. Roth, Charles W. Lidz, Paul Benson, and William Winslade. "False Hopes and Best Data: Consent to Research and the Therapeutic Misconception." *Hastings Center Report* 17, no. 21 (1987): 20–24.

Arnold, David. *Colonizing the Body: State Medicine and Epidemic Disease in Nineteenth-Century India*. Berkeley: University of California Press, 1993.

———, ed. *Imperial Medicine and Indigenous Societies.* Manchester: Manchester University Press, 1988.

Auvert, Bertran, A. Buvé, B. Ferry, M. Caraël, L. Morison, E. Lagarde, N. J. Robinson, M. Kahindo, J. Chege, and N. Rutenberg. "Ecological and Individual Level Analysis of Risk Factors for HIV Infection in Four Urban Populations in Sub-Saharan Africa with Different Levels of HIV Infection." In "The Multicentre Study of Factors Determining the Different Prevalences of HIV in Sub-Saharan Africa," supplement 4, *AIDS* 15 (August 2001): S15–S30.

Auvert, Bertran, Dirk Taljaard, Emmanuel Lagarde, Joëlle Sobngwi-Tambekou, Rémi Sitta, and Adrian Puren. "Randomized, Controlled Intervention Trial of Male Circumcision for Reduction of HIV Infection Risk: The ANRS 1265 Trial." *PLoS Medicine* 2, no. 11 (2005): e298.

AVERT, "HIV Treatment in the U.S." Last revised July 23, 2014. http://www.avert.org/hiv-treatment-us.htm.

Baader, Gerhard, Susan E. Lederer, Morris Low, Florian Schmaltz, and Alexander V. Schwerin. "Pathways to Human Experimentation, 1933–1945: Germany, Japan, and the United States." *Osiris*, 2nd ser., 20 (2005): 205–31.

Baalbaki, Munir, and Rohi Baalbaki. *Al-Mawrid Al-Quareeb Pocket Dictionary, English-Arabic Arabic-English*. 20th ed. Beirut: Dar El-Ilm Lilmalayin, 2007.

Bado, Jean-Paul. *Médecine Coloniale et Grandes Endémies en Afrique 1900–1960: Lèpre, Trypanosomiase Humaine et Onchocercose.* Paris: Editions Karthala, 1996.

Bailey, Robert C., Stephen Moses, Corette B. Parker, Kawango Agot, Ian Maclean, John N. Krieger, Carolyn F. M. Williams, Richard T. Campbell, and Jeckoniah O. Ndinya-Achola. "Male Circumcision for HIV Prevention in Young Men in Kisumu, Kenya: A Randomized, Controlled Trial." *Lancet* 369, no. 9562 (2007): 643–56.

Baker, R. B., ed. *The Codification of Medical Morality: Historical and Philosophical Studies of the Formalization of Western Medical Morality in the Eighteenth and Nineteenth Centuries.* Vol. 2, *Anglo-American Medical Ethics and Medical Jurisprudence in the Nineteenth Century.* Dordrecht, The Netherlands: Kluwer Academic Publishers, 1995.

Ballantyne, Angela. "Benefits to Research Subjects in International Trials: Do They Reduce Exploitation or Increase Undue Inducement?" *Developing World Bioethics* 8, no. 3 (2008): 178–91.

Ballou, W. R. "The Development of the RTS,S Malaria Vaccine Candidate: Challenges and Lessons." *Parasite Immunology* 31, no. 9 (September 2009): 492–500.

Barbiero, V. K., A. DeGorges, J. Minjas, and R. J. Tonn. *Project Completion Report: Zanzibar Malaria Control Project*. Arlington, VA: Vector Biology and Control Project, Medical Service Corporation International, 1990.

Barry, M. "Ethical Considerations of Human Investigation in Developing Countries: The AIDS Dilemma." *New England Journal of Medicine* 319, no. 16 (1988): 1083–86.

Barsdorf, Nicola, Suzanne Maman, Nancy Kass, and Catherine Slack. "Access to Treatment in HIV Prevention Trials: Perspectives from a South African Community." *Developing World Bioethics* 10, no. 2 (2009): 78–87.

Beck, Ann. "The East African Community and Regional Research in Science and Medicine." *African Affairs* 72, no. 288 (1973): 300–308.

Beckerleg, Susan. "Maintaining Order, Creating Chaos: Swahili Medicine in Kenya." PhD diss., University of London, 1989.

———. "Medical Pluralism and Islam in Swahili Communities in Kenya." *Medical Anthropology Quarterly* 8, no. 3 (1994): 299–313.

Berkrot, Bill. "Pfizer Settles Remaining Nigeria, U.S. Trovan Suits." *Reuters*, February 23, 2011.

Beidelman, T. O. "Witchcraft in Ukaguru." In *Witchcraft and Sorcery in East Africa*, edited by J. Middleton and E. H. Winter, 57–98. London: Routledge & Kegan Paul, 1963.

Benitez, Oscar, D. Devaux, and J. Dausset. "Audiovisual Documentation of Oral Consent: A New Method of Informed Consent for Illiterate Populations." *Lancet* 359, no. 9315 (April 20, 2002): 1406–7.

Berman, Bruce. *Control and Crisis in Colonial Kenya: The Dialectic of Domination*. London: James Currey, 1980.

Berry, Sara. "Hegemony on a Shoestring: Indirect Rule and Access to Agricultural Land." *Africa* 62, no. 3 (1992): 327–55.

Boga, Mwanamvua, Alun Davies, Dorcas Kamuya, Samson Kinyanjuli, Ester Kivaya, Francis Kombe, and Trudie Lang, et al. "Strengthening the Informed Consent Process in International Health Research through Community Engagement: The KEMRI-Wellcome Trust Research Programme Experience." *PLoS Medicine* 8, no. 9 (2011): 1–4.

Bonah, Christian. "'You Should Not Use Our Senegalese Infantry as Guinea Pigs': Human Vaccination Experiments in the French Army, 1916–1933." In *Man, Medicine and the State: The Human Body as an Object of Government Sponsored Medical Research in the 20th Century*, edited by Wolfgang Eckart, 15–34. Stuttgart: Franz Steiner Verlag, 2006.

Bonah, Christian. *Histoire de l'experimentation humaine en France: Discours et pratiques 1900–1940*. Paris: Les Belles Lettres, 2007.

Bradley, D. J. "Morbidity and Mortality at Pare-Taveta, Kenya and Tanzania, 1954–66: The Effects of a Period of Malaria Control." In *Disease and Mortality in sub-Saharan Africa,* edited by D. T. Jamison, R. G. Feachem, M. W. Makgoba, E. R. Bos, F. K. Baingana, K. J. Hofman, and K. O. Rogo, 248–63. Washington, DC: World Bank, 1991.

Brandt, Allan. "Racism and Research: The Case of the Tuskegee Syphilis Study." *Hastings Center Report* 8, no. 6 (1978): 21–29.

Brennan, James. "Blood Enemies: Exploitation and Urban Citizenship in the Nationalist Political Thought of Tanzania, 1958–1975." *Journal of African History* 47, no. 3 (2006): 389–413.

———. "Destroying *Mumiani:* Cause, Context and Violence in Late Colonial Dar es Salaam." *Journal of Eastern African Studies* 2, no. 1 (2008): 95–111.

Browne, Stanley George. "The Indigenous Medical Evangelist in Congo." *International Review of Mission* 35, no. 1 (1946): 59–67.

Bruce-Chwatt, Leonard Jan. "Chemotherapy in Relation to Possibilities of Malaria Eradication in Tropical Africa." *Bulletin of the World Health Organization* 15, no. 3-5 (1956):852-62.

Buckley, Thomas C. T., and Alma Gottlieb, eds. *Blood Magic: The Anthropology of Menstruation*. Berkeley: University of California Press, 1988.

Busse, David. "Another Addition to the Arsenal: GlaxoSmithKline's Malaria Vaccine." *LuSci* (blog), *Leeds University Union Science Magazine*, November 2, 2013, http://www.lusci.co.uk/another-addition-to-the-arsenal-glaxosmithklines-malaria-vaccine.

Caplan, Arthur. "Commentary: Report Paints Grim Picture of Drug Trial Safety." NBCNews.com, September 28, 2007. http://www.nbcnews.com/id/21029879.

Caplan, Patrick, ed. *Risk Revisited*. London: Pluto Press, 2000.

Carneiro, Ilona, A. Roca-Feltrer, J. T. Griffin, L. Smith, M. Tanner, J. Armstrong Schellenberg, B. Greenwood, and D. Schellenberg. "Age-Patterns of Malaria Vary with Severity, Transmission Intensity and Seasonality in Sub-Saharan Africa: A Systematic Review and Pooled Analysis." *PLoS ONE* 5, no. 2 (2010): e8988.

Carothers, J. C. *The Psychology of Mau Mau*. Nairobi: Government Printer, 1954.

Ceesay, S. J., C. Casals-Pascual, J. Erskine, S. E. Anya, N. O. Duah, A. J. Fulford, S. S. Sesay, I. Abubakar, S. Dunyo, O. Sey, A. Palmer, M. Fofana, T. Corrah, K. A. Bojang, H. C. Whittle, B. M. Greenwood, D. J. Conway. "Changes in Malaria Indices between 1990 and 2007 in The Gambia: A Retrospective Analysis." *Lancet* 372, no. 9649 (2008): 1545–54.

Cham, Mamady. "Maternal Mortality in the Gambia: Contributing Factors and What Can be Done to Reduce Them." PhD diss., University of Oslo, 2003.

Chantler, Tracey, Faith Otewa, Peter Onyango, Ben Okoth, Frank Odhiambo, Michael Parker, and P. Wenzel Geissler. "Ethical Challenges That Arise at the

Community Interface of Health Research: Village Reporters' Experiences in Western Kenya." *Developing World Bioethics* 13, no. 1 (2013): 30–37.

Chapman, Rachel R. *Family Secrets: Risking Reproduction in Central Mozambique.* Nashville: Vanderbilt University Press, 2010.

Christakis, N. A. "The Ethical Design of an AIDS Vaccine Trial in Africa." *Hastings Centers Report* 18, no. 3 (1988): 31–37.

———. "Ethics Are Local: Engaging Cross-Cultural Variation in the Ethics for Clinical Research." *Social Science & Medicine* 35, no. 9 (1982): 1079–91.

Clarke, Sabine. "The Research Council System and the Politics of Medical and Agricultural Research for the British Colonial Empire, 1940–52." *Medical History* 57, no. 3 (2013): 338–58.

———. "A Technocratic Imperial State? The Colonial Office and Scientific Research, 1940–1960." *Twentieth Century British History* 18, no. 4 (2007): 453–80.

Clyde, David. *Malaria in Tanzania.* London: Oxford University Press, 1967.

Cohen, Joe, Victor Nussenzweig, Johan Vekemans, and Amanda Leach. "From the Circumsporozoite Protein to the RTS,S/AS Candidate Vaccine." *Human Vaccines* 6, no. 1 (2010): 90–96.

Cohen, Jon. "A Good Enough Malaria Vaccine?" *Latest News* (blog), *Science*, October 8, 2013. http://news.sciencemag.org/health/2013/10/good-enough-malaria-vaccine.

Cohen, Justin M., David L. Smith, Chris Cotter, Abigail Ward, Gavin Yamey, Oliver J. Sabot, and Bruno Moonen. "Malaria Resurgence: A Systematic Review and Assessment of Its Causes." *Malaria Journal* 11, no. 1 (2012): 1–17.

Comaroff, Jean. *Body of Power, Spirit of Resistance: The Culture and History of a South African People.* Chicago: University of Chicago Press, 1985.

Comaroff, Jean, and John Comaroff. "Occult Economies and the Violence of Abstraction: Notes from the South African Postcolony." *American Ethnologist* 26, no. 2 (1999): 279–303.

Conte, Christopher. "Imperial Science, Tropical Ecology and Indigenous History: Tropical Research Stations in Northwestern German East Africa, 1896 to the Present." In *Colonialism and the Modern World,* edited by Gregory Blue, Martin Bunton, and Ralph Crozier, 246–61. Armonk, NY: ME Sharpe, 2002.

Cook, Gordon, and Alimuddin Zumla, eds. *Manson's Tropical Diseases.* 22nd ed. (Philadelphia: Saunders, 2009)

Cook, A. R. "Syphilis in Uganda." *Lancet* (December 12, 1908): 1771.

———. *Uganda Memories, 1897–1940.* Kampala: Uganda Society, 1945.

Cooper, Frederick, and Anne Stoler. "Introduction: Tensions of Empire: Colonial Control and Visions of Rule." *American Ethnologist* 16, no. 4 (1989): 609–21.

Corbellini, G. "Acquired Immunity against Malaria as a Tool for the Control of the Disease: The Strategy Proposed by the Malaria Commission of the League of Nations in 1933." *Parassitologia* 40, no. 1–2 (1998): 109–15.

Costello, Matthew. "Administration Triumphs Over Politics: The Transformation of the Tanzanian State." *African Studies review* 39, no. 1 (April 1996): 123–48.

Council for International Organizations of Medical Sciences. *International Ethical Guidelines for Biomedical Research Involving Human Subjects.* Geneva: CIOMS, 2002.

Council of Europe. *Texts of the Council of Europe on Bioethical Matters: Volume I.* Strasbourg: Council of Europe, March 2012.

Crais, Clifton. *The Politics of Evil: Magic, State Power and the Political Imagination in South Africa.* Cambridge: Cambridge University Press, 2002.

Crane, Johanna. *Scrambling for Africa: AIDS, Expertise, and the Rise of American Global Health Science.* Ithaca: Cornell University Press, 2013.

———. "Unequal 'Partners': AIDS, Academia, and the Rise of Global Health." *Behemoth: A Journal on Civilisation* 3, no. 3 (2010): 78–97.

Crozier, Anna. *Practising Colonial Medicine: The Colonial Medical Service in British East Africa.* London: I. B. Tauris, 2007.

D'Alessandro, U., and H. Buttiëns. "History and Importance of Antimalarial Drug Resistance." *Tropical Medicine & International Health* 6, no. 11 (2001): 845–48.

Daily, Johanna P. "Malaria Vaccine Trials—Beyond Efficacy End Points. *New England Journal of Medicine* 367, no. 24 (2012): 2349–51.

Das, Pam, and Richard Horton. "Malaria Elimination: Worthy, Challenging, and Just Possible." *Lancet* 376, no. 9752 (2010): 1515–17.

Davidson, G. "Results of Recent Experiments on the Use of DDT and BHC against Adult Mosquitos at Taveta, Kenya." *Bulletin of the World Health Organization* 4, no. 3 (1951): 329–32.

Department of Health and Human Services Office of the Inspector General. "The Globalization of Clinical Trials: A Growing Challenge in Protecting Human Subjects: Executive Summary." *Journal International de Bioéthique* 14, no. 1–2 (2003): 165–169.

de Sardan, J. P. Olivier. "A Moral Economy of Corruption in Africa?" *Journal of Modern African Studies* 37, no. 1 (1995): 25–52.

Dewhurst, K. S. "Observations on East African Blood Donors." *East African Medical Journal* 22 (1945): 276–78.

DeYoung, Karen, and Deborah Nelson. "The Body Hunters: Testing's El Dorado: Latin America is Ripe for Trials, and Fraud, Frantic Pace Could Overwhelm Controls." *Washington Post*, December 22, 2000.

Dobson, M. J., M. Malowany, and R. W. Snow. "Malaria Control in East Africa: The Kampala Conference and the Pare-Taveta Scheme: A Meeting of Common and High Ground." *Parissitologia* 42, no. 1–2 (2000): 149–66.

Doctors Without Borders. "Deadly Gaps Persist in New Drug Development for Neglected Diseases." Press release, October 24, 2013. http://www.doctorswithoutborders.org/news-stories/press-release/deadly-gaps-persist-new-drug-development-neglected-diseases.

Doolan, D. L., C. Dobano, and J. K. Baird. "Acquired Immunity to Malaria." *Clinical Microbiology Reviews* 22, no. 1 (2009): 13–36.

Draper, C. C., J. L. M. Lelijveld, Y. G. Matola, and G. B. White. "Malaria in the Pare Area of Tanzania: IV: Malaria in the Human Population 11 Years after the Suspension of

Residual Insecticide Spraying, With Special Reference to the Serological Findings." *Transactions of the Royal Society of Tropical Medicine* 66, no. 6 (1972): 905–12.
Draper, C. C., and A. Smith. "Malaria in the Pare Area of Tanganyika: Part II: Effects of Three Years' Spraying of Huts with Dieldrin." *Transactions of the Royal Society of Tropical Medicine* 54, no. 4 (1960): 342–57.
Dunderdale, G. "Notes on the Incidence of Filarial Infection in the Neighbourhood of Lamu, British East Africa." *Transactions of the Royal Society of Tropical Medicine and Hygiene* 15 (1921): 190–97.
Eckart, Wolfgang. "The Colony as Laboratory: German Sleeping Sickness Campaigns in German East Africa and in Togo, 1900–1914." *History and Philosophy of the Life Sciences* 24, no. 1 (2002): 69–89.
Eckart, Wolfgang, ed. *Man, Medicine and the State: The Human Body as an Object of Government Sponsored Medical Research in the 20th Century.* Stuttgart: Franz Steiner Verlag, 2006.
Eckart, Wolfgang, and H. Vondra. "Malaria and World War II: German Malaria Experiments 1939–45." *Parassitologia* 42 (2000): 53–58.
Eichstaedt, Peter. *If You Poison Us: Uranium and Native Americans.* Santa Fe: Red Crane Books, 1994.
Elkins, Caroline. *Imperial Reckoning: The Untold Story of Britain's Gulag in Kenya.* New York: Owl Books, 2005.
Elliot, Carl, and Roberto Abadie. "Exploiting a Research Underclass in Phase 1 Clinical Trials." *New England Journal of Medicine* 358, no. 22 (2008): 2316–17.
Emanuel, Ezekiel, Christine Grady, Robert Crouch, Reidar Lie, Franklin Miller, and David Wendler, eds. *The Oxford Textbook of Clinical Research Ethics.* Oxford: Oxford University Press, 2008.
European Group on Ethics in Science and New Technologies. *Opinion Nr 17 on Ethical Aspects of Clinical Research in Developing Countries.* Brussels: EGE, 2003.
European Medicines Agency. "Clinical Trials Submitted in Marketing-Authorisation Applications to the European Medicines Agency: Overview of Patient Recruitment and the Geographical Location of Investigator Sites: Containing Data from 2005 to 2011." December 11, 2013. http://www.ema.europa.eu/docs/en_GB/document_library/Other/2009/12/WC500016819.pdf.
European Medicines Agency. "Reflection Paper on Ethical and GCP Aspects of Clinical Trials of Medicinal Products for Human Use Conducted Outside of the EU/EEA and Submitted in Marketing Authorisation Applications to the EU Regulatory Authorities." April 16, 2012. http://www.ema.europa.eu/docs/en_GB/document_library/Regulatory_and_procedural_guideline/2012/04/WC500125437.pdf
Fairhead, James, Melissa Leach, and Mary Small. "Where Techno-Science Meets Poverty: Medical Research and the Economy of Blood in The Gambia, West Africa." *Social Science & Medicine* 63, no. 4 (2006): 1109–20.
Faden, Ruth, and Tom Beauchamp. *A History and Theory of Informed Consent.* New York: Oxford University Press, 1986.

Farmer, Paul. "Rich World, Poor World: Medical Ethics and Global Inequality." In *Partner to the Poor: A Paul Farmer Reader*, edited by Haun Saussy, 528–44. Berkeley: University of California Press, 2010.

———. *Peasant Intellectuals: Anthropology and History in Tanzania*. Madison: University of Wisconsin Press, 1990.

———. "Struggles for Control: The Social Roots of Health and Healing in Modern Africa." *African Studies Review* 28, no. 2 (1985): 73–147.

Feierman, Steven, and John Janzen, eds. *The Social Basis of Health and Healing in Africa*. Berkeley: University of California Press, 1992.

Feierman, Steven. "Explanation and Uncertainty in the Medical World of Ghaambo." *Bulletin of the History of Medicine* 74, no. 2 (2000): 317–44.

Ferguson, Douglas Edwin. "The Political Economy of Health and Medicine in Colonial Tanganyika." In *Tanzania under Colonial Rule,* edited by Martin H.Y. Kaniki, 307–43. London: Longman, 1980.

Ferguson, James. *The Anti-Politics Machine: Development, Depoliticization, and Bureaucratic Power in Lesotho.* Minneapolis: University of Minnesota Press, 1994.

Fisher, Jill. *Medical Research for Hire: The Political Economy of Pharmaceutical Clinical Trials.* New Brunswick: Rutgers University Press, 2008.

Fitzgerald, Daniel W., Cecile Marotte, Rose Irene Verdier, Warren D. Johnson, and Jean William Pape. "Comprehension during Informed Consent in a Less-Developed Country. *Lancet* 360, no. 9342 (October 26, 2002): 1301–2.

"Five Year Experiment Shows the Way to Defeat of Malaria." *East African Standard* (July 10, 1959).

Flaherty, Mary Pat, and Doug Struck. "The Body Hunters: Perils of Placebos: Life by Luck of the Draw: In Third World Drug Tests, Some Subjects Go Untreated." *Washington Post,* December 22, 2000.

Garenne, Michel, Alain Giami, and Christophe Perrey. "Male Circumcision and HIV Control in Africa: Questioning Scientific Evidence and the Decision-Making Process." In *Global Health in Africa: Historical Perspectives on Disease Control*, edited by Tamara Giles-Vernick and James L. A. Webb, Jr., 185–210. Ohio: Ohio University Press, 2013.

Gatton, Michelle L., Laura B. Martin, and Qin Cheng. "Evolution of Resistance to Sulfadoxine-Pyrimethamine in *Plasmodium falciparum*." *Antimicrobial Agents and Chemotherapy* 48, no. 6 (2004): 2116–23.

Global Alliance for Vaccines and Immunization. "GAVI Statement on Latest Trial Data on Malaria Vaccine Candidate RTS,S." October 8, 2013. http://www.gavialliance.org/library/news/statements/2013/gavi-statement-on-latest-trial-data-on-malaria-vaccine-candidate-rts-s.

Gearhart, Rebecca, and Munib Said Abdulrehman. "Concepts of Illness among the Swahili of Lamu, Kenya." *Journal of Transcultural Nursing* (December 31, 2013): 1–5. doi:10.1177/1043659613515713.

———. "Purity, Balance, and Wellness among the Swahili of Lamu, Kenya." *Journal of Global Health* (Columbia University) 2, no. 1 (Spring 2012): 42–44.

Geiger, Susan. *TANU Women: Gender and Culture in the Making of Tanganyikan Nationalism, 1955–1965*. Portsmouth, NH: Heinemann, 1997.

Geissbuher, Yvonne, P. Chaki, B. Emidi, N. J., Govella, R. Shirima, V. Mayagaya, D. Mtasiwa, et al. "Interdependence of Domestic Malaria Prevention Measures and Mosquito-Human Interactions in Urban Dar es Salaam, Tanzania." *Malaria Journal* 6 (September 19, 2007): 126.

Geissler, P. Wenzel. "'*Kachinja* Are Coming!': Encounters Around Medical Research Work in a Kenyan Village." *Africa* 75, no. 2 (2005): 173–202.

———. "Parasite Lost: Remembering Modern Times with Kenyan Government Medical Scientists." In *Evidence, Ethos and Experiment: The Anthropology and History of Medical Research in Africa,* edited by P. Wenzel Geissler and Catherine Molyneux, 297–331. New York: Berghahn Books, 2011.

———. "Public Secrets in Public Health: Knowing Not to Know While Making Scientific Knowledge." *American Ethnologist* 40, no. 1 (2013): 13–34.

———. "Stuck in Ruins, or Up and Coming? The Shifting Geography of Urban Public Health Research in Kisumu, Kenya." *Africa* 83, no. 4 (2013): 539–60.

———. "'Transport to Where?': Reflections on the Problem of Value and Time à Propos an Awkward Practice in Medical Research." *Journal of Cultural Economy* 4, no. 1 (2011): 45–64.

Geissler, P. Wenzel, Ann Kelly, Babatunde Imoukhede, and Robert Pool. "'He Is Now Like a Brother, I Can Even Give Him Some Blood'—Relational Ethics and Material Exchanges in a Malaria Vaccine 'Trial Community' in the Gambia." *Social Science & Medicine* 67, no. 5 (2008): 696–707.

Geissler, P. Wenzel, and Catherine Molyneux, eds. *Evidence, Ethos and Experiment: The Anthropology and History of Medical Research in Africa.* New York: Berghahn Books, 2011.

Geissler, P. Wenzel, and Robert Pool. "Editorial: Popular Concerns about Medical Research Projects in Sub-Saharan Africa—a Critical Voice in Debates about Medical Research Ethics." *Tropical Medicine & International Health* 11, no. 7 (2006): 975–82.

Gelfand, M. *Witch Doctor: Traditional Medicine Man of Rhodesia.* London: Harvill Press, 1964.

Giblin, James. *The Politics of Environmental Control in Northeastern Tanzania, 1840–1940.* Philadelphia: University of Pennsylvania Press, 1992.

Gikonyo, C., P. Bejon, V. Marsh, and S. Molyneux. "Taking Social Relationships Seriously: Lessons Learned from the Informed Consent Practices of a Vaccine Trial on the Kenyan Coast." *Social Science & Medicine* 67, no. 5 (2008): 708–20.

Gikonyo, Caroline, Dorcas Kamuya, Bibi Mbete, Patricia Njuguna, Ally Olotu, Philip Bejon, Vicki Marsh, and Sassy Molyneux. "Feedback of Research Findings for Vaccine Trials: Experiences from Two Malaria Vaccine Trials Involving Healthy Children on the Kenyan Coast." *Developing World Bioethics* 13, no. 1 (2013): 48–56.

Giles-Vernick, Tamara, and James L. A. Webb, Jr., eds. *Global Health in Africa: Historical Perspectives on Disease Control*. Athens: Ohio University Press, 2013.

Glantz, Leonard, George Annas, Michael Grodin, and Wendy Mariner. "Research in Developing Countries: Taking 'Benefit' Seriously." *Hastings Center Report* 28, no. 6 (1998): 38–42.

GlaxoSmithKline. "Clinical Trials in the Developing World." Position paper, November 1, 2011. http://www.glaxosmithkline.ch/Attachments/GSK-on-clinical-trials-in-the-developing-world.pdf.

Global Alliance for Vaccines and Immunization. "GAVI Statement on Latest Trial Data on Malaria Vaccine Candidate RTS,S." October 8, 2013. http://www.gavialliance.org/library/news/statements/2013/gavi-statement-on-latest-trial-data-on-malaria-vaccine-candidate-rts-s.

Graboyes, Melissa. "Fines, Orders, Fear . . . and Consent? Medical Research in East Africa, c. 1950s." *Developing World Bioethics* 10, no. 1 (2010): 34–41.

———. "From *Mumiani* to Transfusions: Economies of Blood in East Africa." Paper presented at University of Oregon, May 2014.

———. "Introduction to the Special Issue: Incorporating Medical Research into the History of Medicine in East Africa." *International Journal of African Historical Studies* 47, no. 3 (2014): 379–98.

———. "'The Malaria Imbroglio': Ethics, Eradication, and Endings in Pare Taveta, East Africa, 1959–1960." *International Journal of African Historical Studies* 47, no. 3 (2014): 445–72.

———. "Surveying the 'Pathological Museum': A History of Medical Research and Ethics in East Africa, 1940–1965." PhD diss., Boston University, 2010.

Grady, Christine. "Money for Research Participation: Does it Jeopardize Informed Consent?" *American Journal of Bioethics* 1, no. 2 (2001): 40–44.

Grady, Christine, Jennifer Wagman, Robert Ssekubugu, Maria J. Wawer, David Serwadda, Mohammed Kiddugavu, Fred Nalugoda, Ronald H. Gray, David Wendler, and Qian Dong. "Research Benefits for Hypothetical HIV Vaccine Trials: The Views of Ugandans in the Rakai District." *IRB: Ethics & Human Research* 30, no. 2 (2008): 1–7.

Gray, Ron, Godfrey Kigozi, Xiangrong Kong, Victor Ssempiija, Frederick Makumbi, Stephen Wattya, David Serwadda, Fred Nalugoda, Nelson K. Sewenkambo, and Maria J. Wawer. "The Effectiveness of Male Circumcision for HIV Prevention and Effects on Risk Behaviors in a Posttrial Follow-Up Study." *Aids* 26, no. 5 (2012): 609–15.

Gray, Ronald H. "Circumcision: HIV/STIS and Behaviors in a RCT and Post-RCT Surveliance, Rakai, Uganda. U01-AI075115-01A1 Protocol with Appendixes." Accessed November 15, 2013. https://www.google.com/url?sa=t&rct=j&q=&esrc=s&source=web&cd=1&cad=rja&uact=8&ved=0CB4QFjAAahUKEwiMwI6Kp-bGAhUVLYgKHQpmBww&url=http%3A%2F%2Fwww.plosone.org%2Farticle%2FfetchSingleRepresentation.action%3Furi%3Dinfo%3Adoi%2F10.1371%2Fjournal.pone.0008422.s01&ei

=iiCrVYzTB5XaoASKzJ1g&us=AFQjCNEfCoqUvow9PmzEz86L581R-WVIWRQ&sig2=zdnIN2LTxRZBSE6cXRP6kw.

Green, Maia. "After Ujamaa? Cultures of Governance and the Representation of Power in Tanzania." *Social Analysis* 54, no. 1 (2010): 15–34.

———. "Medicines and the Embodiment of Substances among Pogoro Catholics, Southern Tanzania." *Journal of the Royal Anthropological Institute* 2, no. 3 (1996): 485–98.

Grodin, Michael, and Leonard Glantz, eds. *Children as Research Subjects: Science, Ethics, and Law.* New York: Oxford University Press, 1994.

Harkness, Jon. "Research behind Bars: A History of Nontherapeutic Research on American Prisoners." PhD diss., University of Wisconsin-Madison, 1996.

Harris, Sheldon. *Factories of Death: Japanese Biological Warfare 1932–45 and the American Cover-up.* London: Routledge, 2001.

Haugerud, Angelique. *The Culture of Politics in Modern Kenya.* Cambridge: Cambridge University Press, 1993.

Hayes, M. "On the Epistemology of Risk: Language, Logic and Social Sciences." *Social Science & Medicine* 35, no. 4 (1992): 401–7.

Headrick, Daniel. *The Tools of Empire: Technology and European Imperialism in the Nineteenth Century.* New York: Oxford University Press, 1981.

Heisch, R. B., G. S. Nelson, and M. Furlong. "Studies in Filariasis in East Africa: 1. Filariasis on the Island of Pate, Kenya." *Transactions of the Royal Society of Tropical Medicine and Hygiene* 53, no. 1 (1959): 41–53.

Hobsbawm, Eric, and Terence Ranger, eds. *The Invention of Tradition.* Cambridge: Cambridge University Press, 1992.

Hoppe, Kirk Arden. *Lords of the Fly: Sleeping Sickness Control in East Africa, 1900–1960.* Westport, CT: Preager, 2003.

Hornblum, Allen. *Acres of Skin: Human Experiments at Holmesburg Prison.* New York: Routledge, 1999.

Hornblum, Allen, Judith Newman, and Gregory Dober. *Against Their Will: The Secret History of Medical Experimentation on Children in Cold War America.* New York: Palgrave Macmillan, 2013.

Howard-Jones, N. "Human Experimentation in Historical and Ethical Perspectives." *Social Science & Medicine* 16, no. 15 (1982): 1429–48.

Iliffe, John. *East African Doctors: A History of the Modern Profession.* Cambridge: Cambridge University Press, 1998.

Ijsselmuiden, C., and R. Faden. "Research and Informed Consent in Africa—Another Look." In *Health and Human Rights: A Reader*, edited by Jonathan M. Mann, Sofia Gruskin, Michael A. Grodin, and George J. Annas: 363–72. New York: Routledge, 1999.

"Investigating Parents' Understanding and Decision Making Process for Children's Participation in a Malaria Vaccine Trial, in Kenya: Qualitative Data Analysis." MSc thesis, London School of Hygiene and Tropical Medicine, 2012.

Irwin, Alan, and Brian Wynne, eds. *Misunderstanding Science? The Public Reconstruction of Science and Technology.* Cambridge: Cambridge University Press, 1996.
Jacobs, B., and Z. A. Berege. "Attitudes and Beliefs about Blood Donation among Adults in Mwanza Region, Tanzania." *East African Medical Journal* 72, no. 6 (1995): 345–48.
Johnson, Frederick. *A Standard Swahili-English Dictionary.* Oxford: Oxford University Press, 1939.
Joint United Nations Programme on HIV/AIDS (UNAIDS). *Ethical Considerations in HIV Preventive Vaccine Research.* Geneva: UNAIDS, 2000.
Jones, James. *Bad Blood: The Tuskegee Syphilis Experiment.* New York: Free Press, 1993.
Kamat, Vinay. "Dying Under the Bird's Shadow: Narrative Representations of *Degedege* and Child Survival among the Zaramo of Tanzania." *Medical Anthropology Quarterly* 22, no. 1 (2008): 67–93.
———. "Negotiating Illness and Misfortune in Post-Socialist Tanzania: An Ethnographic Study in Temeke District, Dar es Salaam." PhD diss., Emory University, 2004.
Kamusi Ya Kiswahili Sanifu. Dar es Salaam: Oxford University Press, 1981.
Kamuya, Dorcas M., Vicki Marsh, Francis K. Kombe, P. Wenzel Geissler, and Sassy C. Molyneux. "Engaging Communities to Strengthen Research Ethics in Low-Income Settings: Selection and Perceptions of Members of a Network of Representatives in Coastal Kenya." *Developing World Bioethics* 13, no. 1 (2013): 10–20.
Kamuya, Dorcas M., Sally J. Theobald, Patrick K. Munywoki, Dorothy Koech, P. Wenzel Geissler, and Sassy C. Molyneux. "Evolving Friendships and Shifting Ethical Dilemmas: Fieldworkers' Experiences in a Short Term Community Based Study in Kenya." *Developing World Bioethics* 13, no. 1 (2013): 1–9.
Kerasidou, Angeliki. "Therapeutic Misconception in Research in Developing Countries." *Ethox Blog.* Accessed May 5, 2014. http://www.ethox.org.uk/ethox-blog/therapeutic-misconception-in-research-in-developing-countries.
Killingray, David. "The Maintenance of Law and Order in British Colonial Africa." *African Affairs* 85, no. 340 (1986): 411–37.
Kirk-Greene, A. H. M. "The Thin White Line: The Size of the British Colonial Service in Africa." *African Affairs* 79, no. 314 (1980): 25–44.
Kitereza, Aniceti. *Mr. Myombekere and His Wife Bugonoka, Their Son Ntulanolwo and Daughter Bulihwali.* Dar es Salaam: Mkuki na Nyota Publishers, 2002.
Kithinji, Caroline, and Nancy Kass. "Assessing the Readability of Non-English-Language Consent Forms: The Case of Kiswahili for Research Conducted in Kenya." *IRB: Ethics & Human Research* 32, no. 4 (2010): 10–15.
Kjekshus, Helge. *Ecology Control and Economic Development in East African History: The Case of Tanganyika, 1850–1950.* Berkeley: University of California Press, 1977.
Kleinman, Arthur. *Writing at the Margin: Discourse between Anthropology and Medicine.* Berkeley: University of California Press, 1995.
Koen, J., C. Slack, N. Barsdorf, and Z. Essack. "Payment of Trial Participants Can Be Ethically Sound: Moving Past a Flat Rate." *South African Medical Journal* 98, no. 12 (December 1, 2008): 926–29.

Krapf, Ludwig, ed. *A Dictionary of the Suahili Language: With Introduction Containing an Outline of Suahili Grammar.* London: Tauber, 1882.

Kreston, Rebecca. "Suck It: The Ins and Outs of Mouth Pipetting." *Body Horrors* (blog), *Discover*, March 20, 2013. http://blogs.discovermagazine.com/bodyhorrors/2013/03/20/mouth_pipetting.

Lachenal, Guillaume. "The Doctor Who Would be King." *Lancet* 376, no. 9748 (9 October 2010): 1216–17.

———. "Franco-African Familiarities: A History of the Pasteur Institute of Cameroun, 1945–2000." In *Hospitals beyond the West: from Western Medicine to Global Medicine,* edited by Mark Harrison and Belinda White, 411–44. New Delhi: Orient-Longman, 2009.

LaFraniere, Sharon, Mary Pat Flaherty, and Joe Stevens. "The Body Hunters: The Dilemma: Submit or Suffer: 'Uniformed Consent' is Rising Ethic of the Drug Test Boom." *Washington Post*, December 19, 2000.

Lairumbi, Geoffrey Mbaabu, Sassy Molyneux, Robert W. Snow, Kevin Marsh, Norbert Peshu, and Mike English. "Promoting the Social Value of Research in Kenya: Examining the Practical Aspects of Collaborative Partnerships Using an Ethical Framework." *Social Science & Medicine* 67, no. 5 (2008): 734–47.

Lang, Trudie, Chilengi Roma, Ramadhani A. Noor, Bernhards Ogutu, James E. Todd, Wen L. Kilama, and Geoffrey A. Targett. "Data Safety and Monitoring Boards for African Clinical Trials." *Transactions of the Royal Society of Tropical Medicine and Hygiene* 102, no. 12 (2008): 1189–94.

Langhorne, Jean, Francis M. Ndungu, Anne-Marit Sponaas, and Kevin Marsh. "Immunity to Malaria: More Questions Than Answers." *Nature Immunology* 9, no. 7 (2008): 725–32.

Langwick, Stacey. "Articulate(d) Bodies: Traditional Medicine in a Tanzanian Hospital." *American Ethnologist* 35, no. 3 (2008): 428–39.

———. *Bodies, Politics and African Healing: The Matter of Maladies in Tanzania.* Bloomington: Indiana University Press, 2011.

———. "Devils, Parasites, and Fierce Needles: Healing and the Politics of Translation in Southern Tanzania." *Science, Technology & Human Values* 32, no. 1 (2007): 88–117.

Lawrance, Benjamin, Emily Osborn, and Richard Roberts, eds. *Intermediaries, Interpreters, and Clerks: African Employees in the Making of Colonial Africa.* Madison: University of Wisconsin Press, 2006.

Leach, Melissa, and James Fairhead. *Vaccine Anxieties: Global Science, Child Health and Society.* London: Earthscan, 2007.

Lederer, Susan. *Subjected to Science: Human Experimentation in America before the Second World War.* Baltimore: Johns Hopkins University Press, 1995.

Lema, V. M. "Therapeutic Misconception and Clinical Trials in Sub-Saharan Africa: A Review." *East African Medical Journal* 86, no. 6 (June 2009): 291–99.

Lidz, Charles, Paul Appelbaum, Thomas Grisso, and Michelle Renaud. "Therapeutic Misconception and the Appreciation of Risks in Clinical Trials." *Social Science & Medicine* 58, no. 9 (May 2004): 1689–97.

Lienhardt, Peter. Introduction to *The Medicine Man: Swifa Ya Nguvumali*, by Hasani Bin Ismail, edited and translated by Peter Lienhardt. Oxford: Clarendon Press, 1968.

———. "The Mosque College of Lamu and Its Social Background." *Tanganyika Notes and Records* 53 (1959): 228–42.

Ligami, Christabel. "African Countries on Track to Meeting Four MDGs as 2015 Deadline Comes Closer." *East African*, June 22, 2013. http://www.theeastafrican.co.ke/news/Africa-on-course-for-meeting-four-MDGs-as-2015-deadline-looms-/-/2558/1891226/-/lw739k/-/index.html.

Liu, Margaret, and Kate Davis. *A Clinical Trials Manual from the Duke Clinical Research Institute: Lessons from a Horse Named Jim*. 2nd ed. Durham: Duke Clinical Research Institute, 2010.

Livingston, Julie. *Debility and the Moral Imagination in Botswana*. Bloomington: Indiana University Press, 2005.

———. "Productive Misunderstandings and the Dynamism of Plural Medicine in Mid-Century Bechuanaland." *Journal of Southern African Studies* 33, no. 4 (December 2007): 801–10.

Lonsdale, John. "The Moral Economy of Mau Mau: Wealth, Poverty and Civic Virtue in Kikuyu Political Thought." In *Unhappy Valley: Conflict in Kenya and Africa*, Book 2, *Violence and Ethnicity*, edited by Bruce Berman and John Lonsdale, 315–468. London: James Currey, 1992.

Lovett, Margot. "Elders, Migrants and Wives: Labor Migration and the Renegotiation of Intergenerational, Patronage and Gender Relations in Highland Buha, Western Tanzania." PhD diss., Columbia University, 1996.

Lurie, Peter, and Sidney Wolfe. "Unethical Trials of Interventions to Reduce Perinatal Transmission of the Human Immunodeficiency Virus in Developing Countries." *New England Journal of Medicine* 337, no. 12 (1997): 853–56.

Lustgarten, Abrahm. "Drug Testing Goes Offshore." *Fortune* 152, no. 3 (2005): 66–72.

Lutalo, T., M. Kidugavu, M. J. Wawer, D. Serwadda, L. Zabin, and R. H. Gray. "Trends and Determinants of Contraceptive Use in Raka District, Uganda, 1995–1998." *Studies in Family Planning* 31, no. 3 (2000): 217–27.

Lyons, Maryinez. *The Colonial Disease: A Social History of Sleeping Sickness in Northern Zaire, 1900–1940*. Cambridge: Cambridge University Press, 1992.

MacLeod, Roy, and Milton Lewis, eds. *Disease, Medicine, and Empire: Perspectives on Western Medicine and the Experience of European Expansion*. New York: Routledge, 1988.

Macklin, Ruth. *Against Relativism: Cultural Diversity and the Search for Ethical Universals in Medicine*. New York: Oxford University Press, 1999.

———. *Double Standards in Medical Research in Developing Countries*. New York: Cambridge University Press, 2004.

Madan, A. C. *English-Swahili Vocabulary*. London: Society for Promoting Christian Knowledge, 1917.

———. *Swahili-English Dictionary*. Oxford: Clarendon Press, 1903.

Malaria Vaccine Funders Group. *Malaria Vaccine Technology Roadmap*. November 2013. http://www.who.int/immunization/topics/malaria/vaccine_roadmap/TRM_update_nov13.pdf.

Malloy, Patrick Thomas. "'Holding [Tanganyika] by the Sindano': Networks of Medicine in Colonial Tanganyika." PhD diss., University of California, Los Angeles, 2003.

———. "Research Material and Necromancy: Imagining the Political-Economy of Biomedicine in Colonial Tanganyika." *International Journal of African Historical Studies* 47, no. 3 (2014): 425–44.

Malowany, Maureen. "Medical Pluralism: Disease, Health and Healing on the Coast of Kenya, 1840–1940." PhD diss., McGill University, 1997.

———. "Unfinished Agendas: Writing the History of Medicine of Sub-Saharan Africa." *African Affairs* 99, no. 395 (2000): 325–49.

Mann, Gregory. "An Africanist's Apostasy: On Luise White's Speaking with Vampires." *International Journal of African Historical Studies* 41 (2008): 117–22.

Marks, Shula. "What Is Colonial about Colonial Medicine? And What Has Happened to Imperialism and Health?" *Social History of Medicine* 10, no. 2 (1997): 205–19.

Marsh, Vicki, Dorcas Kamuya, Albert Mlamba, Thomas Williams, and Sassy Molyneux. "Experiences with Community Engagement and Informed Consent in a Genetic Cohort Study of Severe Childhood Diseases in Kenya." *BMC Medical Ethics* 11, no. 13 (2010).

Marsh, Vicki, Dorcas Kamuya, M. J. Parker, and C. S. Molyneux. "Working with Concepts: The Role of Community in International Collaborative Biomedical Research." *Public Health Ethics* 4, no. 1 (2011): 26–39.

Marsh, Vicki, Dorcas Kamuya, Yvonne Rowa, Caroline Gikonyo, and Sassy Molyneux. "Beginning Community Engagement at a Busy Biomedical Research Programme: Experiences from the KEMRI CGMRC–Wellcome Trust Research Programme, Kilifi, Kenya." *Social Science & Medicine* 67, no. 5 (September 2008): 721–33.

Marx, Karl. *Capital: Volume 1: A Critique of Political Economy.* New York: Vintage Books, 1976.

Matovu, J., G. Kigozi, F. Nalugoda, F. Wabwire-Mangen, and R. H. Gray. "The Rakai Project Counselling Programme Experience." *Tropical Medicine & International Health* 7, no. 12 (2002): 1064–67.

Mbungua, G. G., L. N. Muthami, C. W. Mutura, S. A. Oogo, P. G. Waiyaka, and N. Hearst. "Epidemiology of HIV Infection among Long-Distance Truck Drivers in Kenya." *East African Medical Journal* 72, no. 8 (1995): 515–18.

McCormack, Sheena, Gita Ramjee, Anatoli Kamali, Helen Rees, Angela M Crook, Mitzy Gafos, Ute Jentsch, et al. "PRO2000 Vaginal Gel for Prevention of HIV-1 Infection (Microbicides Development Programme 301): A Phase 3, Randomised, Double-Blind, Parallel-Group Trial." *Lancet* 376, no. 9749 (2010): 1329–37.

McNeil, Donald G. "Anti-H.I.V. Trial in Africa Canceled over Failure to Prevent Infection." *New York Times,* November 25, 2011.

———. "Nigerians Receive First Payments for Children Who Died in 1996 Meningitis Drug Trial." *New York Times,* August 11, 2011.

"Medical Procedures: Hemoglobin Estimation," accessed January 27, 2014. http://www.medprocedures.com/2012/11/hemoglobin-estimation.html.

Medical Research Council. "Responsibility in Investigations on Human Subjects." *British Medical Journal* 2 (July 18, 1964): 178–80.

Melrose, Wayne. *Lymphatic Filariasis: A Review 1862–2002.* Killarney, Australia: Warwick Educational Publishing, 2004.

Mesaki, Simeon. "Witchcraft and Witch-Killings in Tanzania: Paradox and Dilemma." PhD diss., University of Minnesota, 1993.

Mohammed, Khalfan A., David H. Molyneux, Marco Albonico, and Francesco Rio. "Progress towards Eliminating Lymphatic Filariasis in Zanzibar: A Model Programme." *Trends in Parasitology* 22, no. 7 (July 2006): 340–44.

Molineaux, L., and G. Gramiccia. *The Garki Project: Research on the Epidemiology and Control of Malaria in the Sudan Savanna of West Africa.* Geneva: World Health Organization, 1980.

Molyneux, C. S., N. Peshu, and K. Marsh. "Trust and Informed Consent: Insights from Community Members on the Kenyan Coast." *Social Science & Medicine* 61, no. 7 (October 2005): 1463–73.

———. "Understanding of Informed Consent in a Low-Income Setting: Three Case Studies from the Kenyan Coast." *Social Science & Medicine* 59, no. 12 (2004): 2547–59.

Molyneux, C. S., D. R. Wassenaar, N. Peshu, and K. Marsh. "'Even If They Ask You to Stand by a Tree All Day, You Will Have to Do It (Laughter)...!': Community Voices on the Notion and Practice of Informed Consent for Biomedical Research in Developing Countries." *Social Science & Medicine* 61, no. 2 (2005): 443–54.

Molyneux, Sassy, Stephen Mulupi, Lairumbi Mbaabu, and Vicki Marsh. "Benefits and Payments for Research Participants: Experiences and Views from a Research Centre on the Kenyan Coast." *BMC Medical Ethics* 13, no. 13 (2012): 1–15

Molyneux, Sassy, Dorcas Kamuya, Philister Adhiambo Madiega, Tracey Chantler, Vibian Angwenyi, and P. Wenzel Geissler. "Editorial: Field Workers at the Interface." *Developing World Bioethics* 13, no. 1 (2013): ii–iv.

Moore, Roland, and A. Roberts. "An Investigation of the Pattern of Disease Prevalent in Parts of the Rufiji District." East African Medical Survey and Research Institute, n.d.

Moreno, Jonathan. *Undue Risk: Secret State Experiments on Humans.* New York: Routledge, 2001.

Morrice, Andrew A. G. "'Honour and Interests': Medical Ethics and the British Medical Association." In *Historical and Philosophical Perspectives on Biomedical*

Ethics: From Paternalism to Autonomy? edited by Maehle and Geyer-Kordesch, 11–35. Burlington: Ashgate Publishing, 2002.

Moulin, Anne Marie. "Defenseless Bodies and Violent Afflictions in the Global World: Blood, Iatrogenesis, and Hepatitis C Transmission in Egypt." In *Global Health in Africa: Historical Perspectives on Disease Control,* edited by Tamara Giles-Vernick and James L. A. Webb, Jr., 138–58. Athens: Ohio University Press, 2013.

———. "Patriarchal Science: The Network of the Overseas Pasteur Institutes." In *Sciences and Empires: Historical Studies about Scientific Development and European Expansion,* edited by P. Petitjean, Catherine Jami, and A. M. Moulin, 307–22. Dordrecht: Springer Science & Business Media, 1992.

Muela, Susanna Hausmann, Joan Muela Ribera, Adiel Mushi, and Marcel Tanner. "Medical Syncretism with Reference to Malaria in a Tanzanian Community." *Social Science & Medicine* 55, no. 3 (2002): 403–13.

Muela, Susanna Hausmann, Joan Muela Ribera, and Marcel Tanner. "Fake Malaria and Hidden Parasites—the Ambiguity of Malaria." *Anthropology & Medicine* 5, no. 1 (1998): 43–61.

Muriuki, Godfrey. *History of the Kikuyu, 1500–1900.* Oxford: Oxford University Press, 1975.

Murray, Christopher J. L., Lisa C. Rosenfeld, Stephen S. Lim, Kathryn G. Andrews, Kyle J. Foreman, Diana Haring, Nancy Fullman, Mohsen Naghavi, Rafael Lozano, and Alan D. Lopez. "Global Malaria Mortality between 1980 and 2010: A Systematic Analysis." *Lancet* 379, no. 9814 (2012): 413–31.

Murray-Lyon, R. M. "Important Diseases Affecting West African Native Troops." *Transactions of the Royal Society of Tropical Medicine and Hygiene* 37, no. 5 (1944): 287–96.

Mwandawiro, C. S. "Studies on Filarial Infection in Lamu and Tana River Districts." MS thesis, University of Nairobi, 1990.

Nalugoda, Fred, Maria J. Wawer, Joseph K. Konde-Lule, Rekha Menon, Ronald H. Gray, David Serwadda, Nelson K. Sewankambo, and Chuanjin Li. "HIV Infections in Rural Households, Rakai District, Uganda." In "Evidence of the Socio-Demographic Impact of AIDS in Africa," supplement, *Health Transition Review* 7 (1997): 127–40.

National Bioethics Advisory Commission. *Ethical and Policy Issues in International Research: Clinical Trials in Developing Countries.* Bethesda, MD: NBAC, 2001.

National Council for Law Reporting. *Laws of Kenya: Chiefs' Act: Chapter 128.* Rev. ed. 2012 [1998]. http://www.icnl.org/research/library/files/Kenya/chief.pdf.

Neill, Deborah. *Networks in Tropical Medicine: Internationalism, Colonialism, and the Rise of a Medical Specialty, 1890–1930.* Palo Alto: Stanford University Press, 2012.

Nelkin, Dorothy. "Foreword: The Social Meaning of Risk." In *Risk, Culture, and Health Inequality: Shifting Perceptions of Danger and Blame,* edited by Barbara Herr Harthorn and Laury Oaks, vii–xiii. Westport, CT: Praeger, 2003.

Nelson, George S. "A Preliminary Report on the Out-Patient Treatment of Onchocerciasis with Antrypol in the West Nile District of Uganda." *East African Medical Journal* 32, no. 11 (1955): 413–29.

Ngaiza, Madgalene. "Women's Bargaining Power in Sexual Relations: Practical Issues and Problems in Relation to HIV Transmission and Control in Tanzania." *Tanzanian Medical Journal* 6, no. 2 (1991): 55–59.

Ngubane, Harriet. *Body and Mind in Zulu Medicine: An Ethnography of Health and Disease in Nyuswa-Zulu Thought and Practice.* London: Academic Press, 1977.

Njouom, Richard, Eric Nerrienet, Martine Dubois, Guillaume Lachenal, Dominique Rousset, Aurelia Vessière, Ahidjo Ayouba, Christophe Pasquier, and Régis Pouillot. "The Hepatitis C Virus Epidemic in Cameroon: Genetic Evidence for Rapid Transmission between 1920 and 1960." *Infection, Genetics and Evolution* 7, no. 3 (June 2007): 361–67.

Noor, Abdisalan M., Damaris K. Kinyoki, Clara W. Mundia, Caroline W. Kabaria, Jonesmus W. Mutua, Victor A. Alegana, Ibrahima Soce Fall, and Robert W. Snow. "The Changing Risk of *Plasmodium falciparum* Malaria Infection in Africa, 2000–10: A Spatial and Temporal Analysis of Transmission Intensity." *Lancet* 383, no. 9930 (2014): 1739–47.

Nsimba, Stephen E. D. "How Sulfadoxine-Pyrimethamine (SP) Was Perceived in Some Rural Communities After Phasing Out Chloroquine (CQ) as a First-Line Drug for Uncomplicated Malaria in Tanzania: Lessons to Learn towards Moving from Monotherapy to Fixed Combination Therapy." *Journal of Ethnobiology and Ethnomedicine* 2, no. 5 (2006): 1–8.

Nuffield Council on Bioethics. *The Ethics of Clinical Research in Developing Countries.* London: Nuffield Council on Bioethics, 1999.

———. *The Ethics of Research Related to Healthcare in Developing Countries.* London: Nuffield Council on Bioethics, 2002.

O'Meara, W. P., P. Bejon, T. W. Mwangi, E. A. Okiro, N. Peshu, R. W. Snow, C. R. Newton, and K. Marsh. "Effect of a Fall in Malaria Transmission on Morbidity and Mortality in Kilifi, Kenya." *Lancet* 372, no. 9649 (2008): 1555–62.

Olotu, Ally, Gregory Fegan, Juliana Wambua, George Nyangweso, Ken O. Awuondo, Amanda Leach, Marc Lievens, et al. "Four-Year Efficacy of RTS,S/AS01E and Its Interaction with Malaria Exposure." *New England Journal of Medicine* 368, no. 12 (2013): 1111–20.

Onyeabor, Onyekachi Sunny. "Ethical Imperative of Posttrial Access to Antiretroviral Treatment." *American Journal of Public Health* 100, no. 6 (2010): 966–67.

Ortner, Sherry. "Resistance and the Problem of Ethnographic Refusal." *Comparative Studies in Society and History* 37, no. 1 (1995): 173–93.

Ottesen, Eric. "Lymphatic Filariasis: Treatment, Control and Elimination." *Advances in Parasitology* 61 (2006): 395–441.

Paasche, Hans. *The Journey of Lukanga Mukara into the Innermost of Germany.* Last revised September 3, 1998. http://wp.cs.ucl.ac.uk/anthonysteed/lukanga.

Packard, Randall. *The Making of a Tropical Disease.* Baltimore: Johns Hopkins University Press, 2007.

———. "'No Other Logical Choice': Global Malaria Eradication and the Politics of International Health in the Post-War Era." *Parassitologia* 40, no. 1–2 (1998): 217–29.

———. "'Roll Back Malaria, Roll in Development'? Reassessing the Economic Burden of Malaria." *Population and Development Review* 35, no. 1 (2009): 53–87.

———. *White Plague, Black Labor: Tuberculosis and the Political Economy of Health and Disease in South Africa.* Berkeley: University of California Press, 1989.

Pappworth, Maurice. *Human Guinea Pigs: Experimentation on Man.* Boston: Beacon Press, 1967.

Pedrique, Belen, Nathalie Strub-Wourgraft, Claudette Some, Piero Olliaro, Patrice Trouiller, Nathan Ford, Bernard Pécoul, and Jean-Hervé Bradol. "The Drug and Vaccine Landscape for Neglected Diseases (2000–11): a Systematic Assessment." *Lancet Global Health* 1, no. 6 (2013): 371–79.

Pels, Peter. "Mumiani: The White Vampire: A Neo-Diffusionist Analysis of Rumour." *Etnofoor* 5, nos. 1–2 (1992): 165–87.

Pépin, J., and A. C. Labbé. "Noble Goals, Unforeseen Consequences: Control of Tropical Diseases in Colonial Central Africa and the Iatrogenic Transmission of Blood-Borne Viruses." *Tropical Medicine & International Health* 13, no. 6 (June 2008): 744–53.

Pépin, J., M. Lavoie, O. G. Pybus, R. Pouillot, Y. Foupouapouognigni, D. Rousset, A. C. Labbé, and R. Njouom. "Risk Factors for Hepatitis C Virus Transmission in Colonial Cameroon." *Clinical Infectious Diseases* 51, no. 7 (October 2010): 768–76.

Petryna, Adriana. "The Competitive Logic of Global Clinical Trials." *Social Research: An International Quarterly* 78, no. 3 (2011): 949–74.

———. *When Experiments Travel: Clinical Trials and the Global Search for Human Subjects.* Princeton: Princeton University Press, 2009.

"Poliomyelitis: A New Approach." *Lancet* 1 (1952): 552.

Pomfret, John, and Deborah Nelson. "The Body Hunters: An Isolated Region's Genetic Mother Lode." *Washington Post*, December 20, 2000.

Power, Helen. "'For Their Own Good': Drug Testing in Liverpool, West and East Africa, 1917–1938." In *Coping with Sickness: Medicine, Law, and Human Rights—Historical Perspectives,* edited by John Woodward and Robert Jütte, 107–26. Sheffield: European Association for the History of Medicine, 2000.

Prince, Ruth. "HIV and the Moral Economy of Survival in an East African City." *Medical Anthropology Quarterly* 26, no. 4 (2012): 534–56.

Pringle, G. "Malaria in the Pare Area of Tanzania: III: The Course of Malaria Transmission since the Suspension of an Experimental Programme of Residual Insecticide Spraying." *Transactions of the Royal Society of Tropical Medicine and Hygiene* 61, no. 1 (1967): 69–79.

Pringle, G., C. C. Draper, and D. F. Clyde. "A New Approach to the Measurement of Residual Transmission in a Malaria Control Scheme in East Africa." *Transactions of the Royal Society of Tropical Medicine* 54, no. 5 (1960): 434–38.

Pringle, G., and S. Avery Jones. "Observations on the Early Course of Untreated Falciparum Malaria in Semi-Immune African Children following a Short Period of Protection." *Bulletin of the World Health Organization* 34, no. 2 (1966): 269–72.

Proctor, Robert. *Racial Hygiene: Medicine under the Nazis.* Cambridge: Harvard University Press, 1998.

Quinn, Sandra Crouse. "Protecting Human Subjects: The Role of Community Advisory Boards." *American Journal of Public Health* 94, no. 6 (2004): 918–22.

Rakai Health Sciences Program. "History of the Rakai Health Sciences Program." November 23, 2010. http://www.rhsp.org/content/history-rakai-health-sciences-program.

Rekdal, Ole Bjorn. "Cross-Cultural Healing in East African Ethnography." *Medical Anthropology Quarterly* 13, no. 4 (1999): 458–82.

Reverby, Susan M. "Ethical Failures and History Lessons: The U.S. Public Health Service Research Studies in Tuskegee and Guatemala." *Public Health Reviews* 34, no. 1 (2012): 189–206.

———. *Examining Tuskegee: The Infamous Syphilis Study and Its Legacy.* Chapel Hill: University of North Carolina Press, 2009.

———, ed. *Tuskegee's Truths: Rethinking the Tuskegee Syphilis Study.* Chapel Hill: University of North Carolina Press, 2000.

Richards, Audrey. *Chisungu: A Girl's Initiation Ceremony among the Bemba of Zambia.* Edison, NJ: Tavistock, 1982.

Richter, Linda, Shane Norris, John Pettifor, Derek Yach, and Noel Cameron. "Cohort Profile: Mandela's Children: The 1990 Birth to Twenty Study in South Africa." *International Journal of Epidemiology* 36, no. 3 (2007): 504–11.

Roberts, J. M. "The Control of Epidemic Malaria in the Highlands of Western Kenya: 3: After the Campaign." *Journal of Tropical Medicine and Hygiene* 67 (September 1964): 230–37.

Roelsgaard, E., and J. Nyboe. "A Tuberculosis Survey in Kenya." *Bulletin of the World Health Organization* 25 (1961): 851–70.

Romero, Patricia. *Lamu: History, Society and Family in an East African Port City.* Princeton: Markus Wiener, 1997.

Rosenberg, Clifford. "The International Politics of Vaccine Testing in Interwar Algiers." *American Historical Review* 117, no. 3 (2012): 671–97.

Rothman, David. "The Shame of Medical Research." *New York Review of Books* 47, no. 19 (November 30, 2000): 60–64.

———. *Strangers at the Bedside: A History of How Law and Bioethics Transformed Medical Decision Making.* New York: Basic Books, 1991.

Rothman, David, and Sheila Rothman. *Trust Is Not Enough: Bringing Human Rights to Medicine.* New York: New York Review of Books, 2006.

———. *The Willowbrook Wars.* New York: Harper & Row, 1984.

The RTS,S Clinical Trials Partnership. "A Phase 3 Trial of RTS,S/AS01 Malaria Vaccine in African Infants." *New England Journal of Medicine* 367, no. 24 (2012): 2284–95.

———. "First Results of Phase 3 Trial of RTS,S/AS01 Malaria Vaccine in African Children." *New England Journal of Medicine* 365, no. 20 (2011): 1863–75.

———. "Protocol for: The RTS,S Clinical Trials Partnership: A phase 3 trial of RTS,S/AS01 malaria vaccine in African infants: Supplementary Documents" November 8, 2012. http://www.nejm.org/doi/suppl/10.1056/NEJMoa1208394/suppl_file/nejmoa1208394_protocol.pdf.

Russell, Tanya, Nicodem J. Govella, Salum Azizi, Christopher J. Drakeley, S. Patrick Kachur, and Gerry F. Killeen. "Increased Proportions of Outdoor Feeding among Residual Malaria Vector Populations following Increased Use of Insecticide-Treated Nets in Rural Tanzania." *Malaria Journal* 10 (April 9, 2011): 80.

Sachs, Jeffrey, and Pia Malaney. "The Economic and Social Burden of Malaria." *Nature* 415, no. 6872 (2002): 680–85.

Sacleux, Charles. *Dictionnaire Swahili-Français.* Paris: Institut d'Ethnologie, 1939.

Salim, Ahmed. *The Swahili-Speaking Peoples of Kenya's Coast, 1895–1965.* Nairobi: East African Pub. House, 1973.

Sandler, Tim. "In India, Oversight Lacking in Outsourced Drug Trials." *NBC News,* March 4, 2012, http://investigations.nbcnews.com/_news/2012/03/04/10562883-in-india-oversight-lacking-in-outsourced-drug-trials.

Samoff, Joel. "The Bureaucracy and the Bourgeoisie: Decentralization and Class Structure in Tanzania." *Comparative Studies in Society and History* 21, no. 1 (1979): 30–62.

Savitt, Todd. "The Use of Blacks for Medical Experimentation and Demonstration in the Old South." *Journal of Southern History* 43, no. 3 (1982): 331–48.

Scherer, Johan Herman. *Marriage and Bride-Wealth in the Highlands of Buha (Tanganyika).* Groningen: V. R. B. Kleine der A, 1965.

Schiebinger, Londa. "Medical Experimentation and Race in the Eighteenth-Century Atlantic World." *Social History of Medicine* 26, no. 3 (2013): 377–82.

Schneider, William. *The History of Blood Transfusion in Sub-Saharan Africa.* Athens: Ohio University Press, 2013.

Schneider, William, and Ernest Drucker. "Blood Transfusions in the Early Years of AIDS in Sub-Saharan Africa." *American Journal of Public Health* 96, no. 6 (2006): 984–94.

Schoenbrun, David L. "Conjuring the Modern in Africa: Durability and Rupture in Histories of Public Healing between the Great Lakes of East Africa." *American Historical Review* 111, no. 5 (2006): 1403–39.

Schroeder, D., and E. Gefenas. "Vulnerability: Too Vague and Too Broad?" *Cambridge Quarterly of Healthcare Ethics* 18, no. 2 (2009): 113–21.

Schuklenk, Udo. "Protecting the Vulnerable: Testing Times for Clinical Research Ethics." *Social Science & Medicine* 51, no. 6 (2000): 969–77.

Schwartz, Lauren, G.V. Brown, B. Genton, and V. S. Moorthy. "A Review of Malaria Vaccine Clinical Projects Based on the WHO Rainbow Table." *Malaria Journal* 11, no. 1 (2012): 1–2.

Seema Shah, J. D., Stacey Elmer, and Christine Grady. "Planning for Posttrial Access to Antiretroviral Treatment for Research Participants in Developing Countries." *American Journal of Public Health* 99, no. 9 (2009): 1556–62.

Seema Shah, J. D., and Christine Grady. "Shah and Grady Respond." *American Journal of Public Health* 100, no. 6 (2010): 967.

Serwadda, D., M. J. Wawer, S. D. Wawer, S. D. Musgrave, N. K. Sewankambo, J. E. Kaplan, and R. H. Gray. "HIV Risk Factors in Three Geographic Strata of Rural Rakai District, Uganda." *AIDS* 6, no. 9 (September 1992): 983–89.

Shaffer, D. N., V. N. Yebei, J. B. Ballidawa, J. E. Sidle, J. Y. Greene, E. M. Meslin, S. J. N. Kimaiyo, and W. M. Tierney. "Equitable Treatment for HIV/AIDS Clinical Trial Participants: A Focus Group Study of Patients, Clinician Researchers, and Administrators in Western Kenya." *Journal of Medical Ethics* 32, no. 1 (2006): 55–60.

Shah, Sonia. *The Body Hunters: Testing New Drugs on the World's Poorest Patients.* New York: New Press, 2006.

Shipton, Parker. *The Nature of Entrustment: Intimacy, Exchange, and the Sacred in Africa.* New Haven: Yale University Press, 2007.

Siegfried, Nandi, Mike Clark, and Jimmy Volmink. "Randomised Controlled Trials in Africa of HIV and AIDS: Descriptive Study and Spatial Distribution." *British Medical Journal* 331, no. 7519 (2005): 742.

Silla, Eric. *People Are Not the Same: Leprosy and Identity in Twentieth-Century Mali.* Portsmouth, NH: Heinemann, 1998.

Simonsen, Paul E., Mwele N. Malecela, Edwin Michael, and Charles D. Mackenzie. *Lymphatic Filariasis: Research and Control in Eastern and Southern Africa.* Frederiksberg, Denmark: DBL – Centre for Health Research and Development, 2008.

Skloot, Rebecca. *The Immortal Life of Henrietta Lacks.* New York: Broadway, 2010.

Slovic, P. *The Perception of Risk.* London: Earthscan Publications, 2000.

Smith, A. "Malaria in the Taveta Area of Kenya and Tanganyika: III: Entomological Findings Three Years after the Spraying Period." *East African Medical Journal* 39 (1962): 553–64.

———. "Malaria in the Taveta Area of Kenya and Tanzania: IV: Entomological Findings Six Years after the Spraying Period." *East African Medical Journal* 43, no. 1 (1966): 7–18.

Smith, A., and C. C. Draper. "Malaria in the Taveta Area of Kenya and Tanganyika: I: Epidemiology." *East African Medical Journal* 36, no. 2 (1959): 99–113.

Smith, A., and G. Pringle. "Malaria in the Taveta Area of Kenya and Tanzania: V: Transmission Eight Years after the Spraying Period." *East African Medical Journal* 44 (1967): 469–74.

Smith, Alec. *Insect Man: A Fight against Malaria in Africa.* New York: Palgrave Macmillan, 1993.

Smith, David, and Alison Mitchell. "Sacrifices for the Miracle: The Polio Vaccine Research and Children with Mental Retardation." *Mental Retardation* 39, no. 5 (2001): 405–9.

Spear, Thomas. "Neo-Traditionalism and the Limits of Invention in British Colonial Africa." *Journal of African History* 44, no. 1 (2003): 3–27.

Stadler, Jonathan, and Eirik Saethre. "Rumours about Blood and Reimbursements in a Microbicide Gel Trial." *African Journal of AIDS Research* 9, no. 4 (2010): 345–53.

Stepan, Nancy Leys. *Eradication: Ridding the World of Disease Forever?* Ithaca: Cornell University Press, 2011.

Stephens, Joe. "The Body Hunters: Where Profits and Lives Hang in the Balance." *Washington Post*, December 17, 2000.

———. "Pfizer to Pay $75 Million to Settle Nigerian Trovan Drug-Testing Suit." *Washington Post*, July 31, 2009.

Strickland, G. T. "An Epidemic of Hepatitis C Virus Infection While Treating Endemic Infectious Diseases in Equatorial Africa More Than a Half Century Ago: Did It Also Jump-Start the AIDS Pandemic?" *Clinical Infectious Diseases* 51, no. 7 (October 2010): 785–87.

———. "Liver Disease in Egypt: Hepatitis C Superseded Schistosomiasis as a Result of Iatrogenic and Biological Factors." *Hepatology* 43, no. 5 (May 2006): 915–22.

Swantz, Lloyd. *The Medicine Man: Among the Zaramo of Dar es Salaam*. Uppsala, Sweden: Nordic Africa Institute, 1990.

Swantz, Marja-Liisa. *Blood, Milk and Death: Blood Symbols and the Power of Regeneration among the Zaramo of Tanzania*. London: Bergin & Garvey, 1995.

Sweet, James. *Recreating Africa: Culture, Kinship, and Religion in the African-Portuguese World, 1441–1770*. Chapel Hill: University of North Carolina Press, 2003.

Tanner, Marcel, and Don de Savigny. "Malaria Eradication Back on the Table." *Bulletin of the World Health Organization* 86, no. 2 (2008): 81.

Tappan, Jennifer. "Blood Work and 'Rumors' of Blood: Nutritional Research and Insurrection in Buganda, 1935–1970." *International Journal of African Historical Studies* 47, no. 3 (2014): 473–94.

Taylor, C. E. "Clinical Trials and International Health Research." *American Journal of Public Health* 69, no. 10 (1979): 981–83.

———. "Condoms and Cosmology: The 'Fractal' Person and Sexual Risk in Rwanda." *Social Science & Medicine* 31, no. 9 (1990): 1023–28.

"The Therapeutics of Malaria; Third General Report of the Malaria Commission." *Quarterly Bulletin of the Health Organisation of the League of Nations* 2, no. 2 (1933): 202–3.

Thiessen, Carrie, Robert Ssekubugu, Jennifer Wagman, Mohammed Kiddugavu, Maria Wawer, Ezekiel Emanuel, Robert Gray, David Serwadda, and Christine Grady. "Personal and Community Benefits and Harms of Research: Views from Rakai, Uganda." *AIDS* 21, no. 18 (2007): 2493–501.

Thornton, John. "On the Trail of Voodoo: African Christianity in Africa and the Americas." *Americas* 44, no. 3 (1988): 261–78.

———. "Perspectives on African Christianity." In *Race, Discourse, and the Making of the Americas,* edited by Vera Hyatt and Rex Nettleford, 169–98. Washington, DC: Smithsonian Institution, 1994.

Tilley, Helen. *Africa as a Living Laboratory: Empire, Development, and the Problem of Scientific Knowledge, 1870–1950*. Chicago: University of Chicago Press, 2011.

———. "Conclusion: Experimentation in Colonial East Africa and Beyond." *International Journal of African Historical Studies* 47, no. 3 (2014): 495–505.

———. "Global Histories, Vernacular Science, and African Genealogies; or, Is the History of Science Ready for the World?" *Isis* 101, no. 1 (2010): 110–19.

Titmuss, Richard. *The Gift Relationship: From Human Blood to Social Policy.* New York: Vintage Books, 1973.

Tosh, John. *Clan Leaders and Colonial Chiefs in Lango: The Political History of an East African Stateless Society c.1800–1939.* Oxford: Oxford University Press, 1978.

Trant, Hope. *Not Merrion Square: Anecdotes of a Woman's Medical Career in Africa.* Toronto: Thornhill Press, 1970.

Trape, Jean-François, Adama Tall, Cheikh Sokhna, Alioune Badara Ly, Nafissatou Diagne, Ousmane Ndiath, Catherin Mazenot, et al. "The Rise and Fall of Malaria in a West African Rural Community, Dielmo, Senegal, from 1990 to 2012: A 22 Year Longitudinal Study." *Lancet Infectious Diseases* 14, no. 6 (2014): 476–88.

Trape, J. F., A. Tall, N. Diagne, O. Ndaith, A. B. Ly, J. Faye, F. Dieye-Ba, et al. "Malaria Morbidity and Pyrethroid Resistance after the Introduction of Insecticide-Treated Bednets and Artemisinin-Based Combination Therapies: A Longitudinal Study." *Lancet Infectious Diseases* 11, no. 12 (2011): 925–32.

Turshen, Meredeth. *The Political Ecology of Disease in Tanzania.* Piscataway: Rutgers University Press, 1984.

Umeora, O. U. J., S. O. Onuh, and M. C. Umeora. "Socio-Cultural Barriers to Voluntary Blood Donation for Obstetric Use in a Rural Nigerian Village." *African Journal of Reproductive Health* 9, no. 6 (2005): 72–76.

United National Children's Fund, Eastern and Southern Regional Office. "Combatting Antivaccination Rumours: Lessons Learned from Case Studies in East Africa." 2001.

United Nations Department of Public Information. "Global Summary of the HIV/AIDS Epidemic." DPI/2199. Background information sheet distributed at the African Summit on HIV/AIDS, Tuberculosis and Other Related Infectious Diseases, Abuja, Nigeria, April 2001. http://www.un.org/ga/aids/pdf/stats.pdf.

United States Department of Health and Human Services. "Guidance for Clinical Trial Sponsors: Establishment and Operation of Clinical Trial Data Monitoring Committees." March 2006. http://www.fda.gov/downloads/RegulatoryInformation/Guidances/ucm127073.pdf.

Varmus, Harold, and David Satcher. "Ethical Complexities of Conducting Research in Developing Countries." *New England Journal of Medicine* 337, no. 14 (1996): 1003–5.

Vaughan, Megan. *Curing Their Ills: Colonial Power and African Illness.* Stanford: Stanford University Press, 1991.

Wallace, David. *Unit 731: Japan's Secret Biological Warfare Unit in World War II.* New York: The Free Press, 1989.

Wamae, C. N., C. Mwandawiro, E. Wambayi, S. Njenga, and F. Kiliku. "Lymphatic Filariasis in Kenya since 1910, and the Prospects for Its Elimination: A Review." *East African Medical Journal* 78, no. 11 (2001): 595–603.

Wamae, C. N., W. Nderitu, and F. M. Kiliku. "*Brugia Patei* in a Domestic Cat from Lamu Island, Kenya." *Filaria Links* 2 (1997): 7.

Wawer, Maria J., and David Serwadda. "Randomized Trial of Male Circumcision: STD, HIV and Behavioral Effects in Men, Women and the Community Protocol." *Circ Gates Protocol Stage 2* (March 27, 2007): 1–56. http://files.figshare.com/458570/Text_S1.doc

Wawer, Maria J., Frederick Makumbi, Godfrey Kigozi, David Serwadda, Stephen Watya, Fred Nalugoda, Dennis Buwembo, et al. "Circumcision in HIV-Infected Men and Its Effect on HIV Transmission to Female Partners in Rakai, Uganda: A Randomised Controlled Trial." *Lancet* 374, no. 9685 (2009).

Washington, Harriet. *Medical Apartheid: The Dark History of Medical Experimentation on Black Americans from Colonial Times to the Present.* New York: Anchor, 2006.

Webb, James L. A., Jr. "The First Large-Scale Use of Synthetic Insecticides for Malaria Control in Tropical Africa: Lessons from Liberia, 1945–1962." *Journal of the History of Medicine and Allied Sciences* 66, no. 3 (2011): 347–76.

———. *Humanity's Burden: A Global History of Malaria.* Cambridge: Cambridge University Press, 2009.

———. *The Long Struggle against Malaria in Tropical Africa.* Cambridge: Cambridge University Press, 2014.

Webel, Mari. "Ziba Politics and the German Sleeping Sickness Camp at Kigarama, Tanzania, 1907–1914." *International Journal of African Historical Studies* 47, no. 3 (2104): 399–424.

Weindling, Paul. *Nazi Medicine and the Nuremberg Trials: From Medical War Crimes to Informed Consent.* New York: Palgrave Macmillan, 2004.

———. "Rooting New Communities in Tanzania's Sleeping Sickness Concentrations." April 16, 2013. Accessed February 9, 2014. http://ssrn.com/abstract=2251606.

Weiss, Brad. "Electric Vampires: Haya Rumours of the Commodified Body." In *Bodies and Persons: Comparative Perspectives from Africa and Melanesia,* edited by M. Lambek and A. Strathern, 172–94. Cambridge: Cambridge University Press, 1998.

———. *The Making and Unmaking of the Haya Lived World: Consumption, Commoditization and Everyday Practice.* Durham: Duke University Press, 1996.

Weiss, H. A., M. Quigley, and R. Hayes. "Male Circumcision and Risk of HIV Infection in Sub-Saharan Africa: A Systemic Review and Meta-Analysis." *AIDS* 14, no. 15 (2000): 2361–70.

Wendland, Claire. *A Heart for the Work: Journeys through an African Medical School.* Chicago: University of Chicago Press, 2010.

———. "Moral Maps and Medical Imaginaries: Clinical Tourism at Malawi's College of Medicine." *American Anthropologist* 114, no. 1 (2012): 108–22.

"What Is the Purpose of Medical Research?" *Lancet* 381, no. 9864 (2013): 347.

White, Luise. "Blood Brotherhood Revisited: Kinship, Relationship, and the Body in East and Central Africa." *Africa* 64, no. 3 (1994): 359–72.

———. *Speaking with Vampires: Rumor and History in Colonial Africa.* Berkeley: University of California Press, 2000.

Whyte, Susan Reynolds. "Penicillin, Battery Acid and Sacrifice: Cures and Causes in Nyole Medicine." *Social Science & Medicine* 16, no. 23 (1982): 2055–64.

———. *Questioning Misfortune: The Pragmatics of Uncertainty in Eastern Uganda.* Cambridge: Cambridge University Press, 1997.

Wollensack, Amy. "Closing the Constant Garden: The Regulation and Responsibility of U.S. Pharmaceutical Companies Doing Research on Human Subjects in Developing Nations." *Washington University Global Studies Law Review* 6, no. 3 (2007): 747–71.

Women and Girls Empowerment Project. "Rakai District Uganda: Fact Sheet." 2012.

Worboys, Michael. "The Discovery of Colonial Malnutrition between the Wars." In *Imperial Medicine and Indigenous Societies,* edited by David Arnold, 208–25. New York: Manchester University Press, 1988.

World Health Organization. *The Global Elimination of Lymphatic Filariasis: The Story of Zanzibar.* Geneva: World Health Organization Press, 2002.

———. "Overview: HIV/AIDS." Accessed January 26, 2014. http://www.afro.who.int/en/clusters-a-programmes/dpc/acquired-immune-deficiency-syndrome/overview.html.

———. *Progress Report 2000–2009 and Strategic Plan 2010–2020 of the Global Programme to Eliminate Lymphatic Filariasis: Halfway towards Eliminating Lymphatic Filariasis.* Geneva: World Health Organization Press, 2010.

———. "Questions and Answers on Malaria Vaccines." October 2013. http://www.who.int/immunization/topics/malaria/vaccine_roadmap/WHO_malaria_vaccine_q_a_Oct2013.pdf.

World Medical Association. "World Medical Association Declaration of Helsinki: Ethical Principles for Medical Research Involving Human Subjects." *Journal of the American Medical Association* 310, no. 20 (2013): 2191–94.

Wylie, Diana. *Starving on a Full Stomach: Hunger and the Triumph of Cultural Racism in Modern South Africa.* Charlottesville: University of Virginia Press, 2001.

Ylvisaker, Marguerite. "Lamu in the Nineteenth Century: Land, Trade and Politics." PhD diss., Boston University, 1979.

Zimmerman, Andrew. "'What Do You Really Want in German East Africa, Herr Professor?': Counterinsurgency and the Science Effect in Colonial Tanzania." *Society for Comparative Study of Society and History* 48, no. 2 (April 2006): 419–61.

INDEX

CPSIA information can be obtained
at www.ICGtesting.com
Printed in the USA
LVHW111136021221
705063LV00008B/85